Evaluating Gas Network Capacities

MOS-SIAM Series on Optimization

This series is published jointly by the Mathematical Optimization Society and the Society for Industrial and Applied Mathematics. It includes research monographs, books on applications, textbooks at all levels, and tutorials. Besides being of high scientific quality, books in the series must advance the understanding and practice of optimization. They must also be written clearly and at an appropriate level for the intended audience.

Editor-in-Chief

Katya Scheinberg
Lehigh University

Editorial Board

Santanu S. Dey, *Georgia Institute of Technology*
Maryam Fazel, *University of Washington*
Andrea Lodi, *University of Bologna*
Arkadi Nemirovski, *Georgia Institute of Technology*
Stefan Ulbrich, *Technische Universität Darmstadt*
Luis Nunes Vicente, *University of Coimbra*
David Williamson, *Cornell University*
Stephen J. Wright, *University of Wisconsin*

Series Volumes

Koch, Thorsten, Hiller, Benjamin, Pfetsch, Marc E., and Schewe, Lars, editors, *Evaluating Gas Network Capacities*
Corberán, Ángel, and Laporte, Gilbert, *Arc Routing: Problems, Methods, and Applications*
Toth, Paolo, and Vigo, Daniele, *Vehicle Routing: Problems, Methods, and Applications, Second Edition*
Beck, Amir, *Introduction to Nonlinear Optimization: Theory, Algorithms, and Applications with MATLAB*
Attouch, Hedy, Buttazzo, Giuseppe, and Michaille, Gérard, *Variational Analysis in Sobolev and BV Spaces: Applications to PDEs and Optimization, Second Edition*
Shapiro, Alexander, Dentcheva, Darinka, and Ruszczynski, Andrzej, *Lectures on Stochastic Programming: Modeling and Theory, Second Edition*
Locatelli, Marco and Schoen, Fabio, *Global Optimization: Theory, Algorithms, and Applications*
De Loera, Jesús A., Hemmecke, Raymond, and Köppe, Matthias, *Algebraic and Geometric Ideas in the Theory of Discrete Optimization*
Blekherman, Grigoriy, Parrilo, Pablo A., and Thomas, Rekha R., editors, *Semidefinite Optimization and Convex Algebraic Geometry*
Delfour, M. C., *Introduction to Optimization and Semidifferential Calculus*
Ulbrich, Michael, *Semismooth Newton Methods for Variational Inequalities and Constrained Optimization Problems in Function Spaces*
Biegler, Lorenz T., *Nonlinear Programming: Concepts, Algorithms, and Applications to Chemical Processes*
Shapiro, Alexander, Dentcheva, Darinka, and Ruszczynski, Andrzej, *Lectures on Stochastic Programming: Modeling and Theory*
Conn, Andrew R., Scheinberg, Katya, and Vicente, Luis N., *Introduction to Derivative-Free Optimization*
Ferris, Michael C., Mangasarian, Olvi L., and Wright, Stephen J., *Linear Programming with MATLAB*
Attouch, Hedy, Buttazzo, Giuseppe, and Michaille, Gérard, *Variational Analysis in Sobolev and BV Spaces: Applications to PDEs and Optimization*
Wallace, Stein W. and Ziemba, William T., editors, *Applications of Stochastic Programming*
Grötschel, Martin, editor, *The Sharpest Cut: The Impact of Manfred Padberg and His Work*
Renegar, James, *A Mathematical View of Interior-Point Methods in Convex Optimization*
Ben-Tal, Aharon and Nemirovski, Arkadi, *Lectures on Modern Convex Optimization: Analysis, Algorithms, and Engineering Applications*
Conn, Andrew R., Gould, Nicholas I. M., and Toint, Phillippe L., *Trust-Region Methods*

Evaluating Gas Network Capacities

Edited by

Thorsten Koch
Technische Universität Berlin
Konrad-Zuse-Zentrum für
Informationstechnik Berlin
Berlin, Germany

Benjamin Hiller
Konrad-Zuse-Zentrum für
Informationstechnik Berlin
Berlin, Germany

Marc E. Pfetsch
Technische Universität Darmstadt
Darmstadt, Germany

Lars Schewe
Universität Erlangen-Nürnberg
Erlangen, Germany

Society for Industrial and Applied Mathematics
Philadelphia

Mathematical Optimization Society
Philadelphia

Copyright © 2015 by the Society for Industrial and Applied Mathematics and the Mathematical Optimization Society

10 9 8 7 6 5 4 3 2 1

All rights reserved. Printed in the United States of America. No part of this book may be reproduced, stored, or transmitted in any manner without the written permission of the publisher. For information, write to the Society for Industrial and Applied Mathematics, 3600 Market Street, 6th Floor, Philadelphia, PA 19104-2688 USA

Trademarked names may be used in this book without the inclusion of a trademark symbol. These names are used in an editorial context only; no infringement of trademark is intended.

CONOPT and CONOPT4 are trademarks of ARKI Consulting & Development A/S.
FLOWMASTER is a trademark of Lincoln Industrial Corporation.
GAMS is a trademark of Gams Development Corp.
GANESI is a trademark of PSI Energy Oil and Gas.
GUROBI is a trademark of Gurobi Optimization, Inc.
IBM ILOG CPLEX is developed and supported by IBM, Inc. IBM ILOG CPLEX is a registered trademark of IBM, Inc. www.ibm.com.

KNITRO is a registered trademark of Ziena Optimization LLC.
Maple is a trademark of Waterloo Maple, Inc.
PIPESIM is a trademark of Schlumberger Software.
SNOPT and QPOPT are trademarks of Stanford University and UC San Diego.
SPS is a trademark of GL Noble Denton.
STANET is a trademark of Ingenieurbüro Fischer-Uhrig.
WINTRAN is a trademark of Gregg Engineering, Inc.

Figure 1.1 reprinted with permission from BGR, Hannover, Germany.
Figure 1.2 reprinted with permission from BP.
Figure 1.3 reprinted with permission from ENTSOG.
Figures 1.4–5, 2.1–2, 2.4, 2.6a, and 2.9 reprinted with permission from Open Grid Europe GmbH, Essen, Germany.
Figures 2.3, 2.5, and 10.1–5 reprinted with kind permission from Springer Science+Business Media.
Figure 2.6b reprinted with permission from Siemens AG.
Figures 7.1, 7.4, and 7.5 reprinted with permission from Taylor and Francis.

Publisher	David Marshall
Acquisitions Editors	Sara Murphy and Elizabeth Greenspan
Developmental Editor	Gina Rinelli
Managing Editor	Kelly Thomas
Production Editor	Lisa Briggeman
Copy Editor	Matthew Bernard
Production Manager	Donna Witzleben
Production Coordinator	Cally Shrader
Compositor	Techsetters, Inc.
Graphic Designer	Lois Sellers

Pictured on the back cover are the following members of Project ForNe (asterisks indicate contributors to this book):
Front row, left to right: *Isabel Wegner-Specht, Andrea Zelmer, *Claudia Stangl, *Veronika Kühl, *Uwe Gotzes, Marie Steckhahn, *Lars Schewe, Jácint Szabó, *Marc C. Steinbach. *Back row, left to right:* Martin Grötschel, *Alexander Martin, *Martin Schmidt, *Hernan Leövey, Debora Mahlke, *Antonio Morsi, Frank Nowosatko, *Armin Fügenschuh, *Jesco Humpola, *Rene Henrion, Djamal Oucherif, *Imke Joormann, *Marc E. Pfetsch, *Bernhard Willert, *Christine Hayn, *Klaus Spreckelsen, *Benjamin Hiller, *Werner Römisch, Ansgar Steinkamp, *Björn Geißler, *Thomas Lehmann, Jörg Bödeker, *Radoslava Mirkov, Gerald Gamrath, *Nina Heinecke, *Jonas Schweiger, *Robert Schwarz, *Andris Möller, *Rüdiger Schultz, *Ralf Gollmer, *Thorsten Koch.

Library of Congress Cataloging-in-Publication Data

Evaluating gas network capacities / edited by Thorsten Koch, Technische Universität Berlin, Berlin, Germany [and 3 others].
       pages cm. – (MOS-SIAM series on optimization)
    Includes bibliographical references and index.
    Summary: "This book deals with a simple sounding question whether a certain amount of gas can be transported by a given pipeline network. While well studied for a single pipeline, this question gets extremely difficult if we consider a meshed nation wide gas transportation network, taking into account all the technical details and discrete decisions, as well as regulations, contracts, and varying demand. This book describes several mathematical models to answer these questions, discusses their merits and disadvantages, explains the necessary technical and regulatory background, and shows how to solve this question using sophisticated mathematical optimization algorithms."– Provided by publisher.

    ISBN 978-1-611973-68-6

1. Gas pipelines–Mathematical models. 2. Gasses–Transport properties. 3. System analysis. 4. Mathematical optimization. I. Koch, Thorsten, 1967- editor.

TN879.5.E93 2015

665.7'44–dc23   2014035425

 is a registered trademark.    is a registered trademark.

List of Contributors

Dagmar Bargmann
Open Grid Europe GmbH,
Essen, Germany

Dr. Mirko Ebbers
Open Grid Europe GmbH,
Essen, Germany

Prof. Dr. Armin Fügenschuh
Zuse Institute Berlin (ZIB),
Berlin, Germany
Current affiliation:
Helmut-Schmidt-Universität
der Bundeswehr, Hamburg,
Germany

Dr. Björn Geißler
Friedrich-Alexander-
Universität
Erlangen-Nürnberg, Erlangen,
Germany

Dr. Ralf Gollmer
Universität Duisburg-Essen,
Duisburg, Germany

Dr. Uwe Gotzes
Open Grid Europe GmbH,
Essen, Germany

Christine Hayn
Friedrich-Alexander-
Universität
Erlangen-Nürnberg, Erlangen,
Germany

Nina Heinecke
Open Grid Europe GmbH,
Essen, Germany

Dr. Holger Heitsch
Humboldt-Universität zu
Berlin, Berlin, Germany

PD Dr. René Henrion
Weierstraß Institut für
Angewandte Analysis und
Stochastik (WIAS), Berlin,
Germany

Dr. Benjamin Hiller
Zuse Institute Berlin (ZIB),
Berlin, Germany

Jesco Humpola
Zuse Institute Berlin (ZIB),
Berlin, Germany

Imke Joormann
Technische Universität
Darmstadt, Darmstadt,
Germany

Prof. Dr. Thorsten Koch
Zuse Institute Berlin (ZIB),
Berlin, Germany

Veronika Kühl
Open Grid Europe GmbH,
Essen, Germany

Dr. Thomas Lehmann
Zuse Institute Berlin (ZIB),
Berlin, Germany
Current affiliation: Siemens
Corporate Technology,
Erlangen, Germany

Ralf Lenz
Zuse Institute Berlin (ZIB),
Berlin, Germany

Hernan Leövey
Humboldt-Universität zu
Berlin, Berlin, Germany

Prof. Dr. Alexander Martin
Friedrich-Alexander-
Universität
Erlangen-Nürnberg, Erlangen,
Germany

Dr. Radoslava Mirkov
Humboldt-Universität zu
Berlin, Berlin, Germany
Current affiliation:
TriSolutions GmbH,
Hamburg, Germany

Andris Möller
Weierstraß Institut für
Angewandte Analysis und
Stochastik (WIAS), Berlin,
Germany

Dr. Antonio Morsi
Friedrich-Alexander-
Universität
Erlangen-Nürnberg, Erlangen,
Germany

Antje Pelzer
Open Grid Europe GmbH,
Essen, Germany

Prof. Dr. Marc E. Pfetsch
Technische Universität
Darmstadt, Darmstadt,
Germany

Prof. Dr. Werner Römisch
Humboldt-Universität zu
Berlin, Berlin, Germany

Jessica Rövekamp
Open Grid Europe GmbH,
Essen, Germany

Dr. Lars Schewe
Friedrich-Alexander-
Universität
Erlangen-Nürnberg, Erlangen,
Germany

Prof. Dr. Martin Schmidt
Leibniz Universität Hannover,
Hannover, Germany
Current affiliation: Friedrich-
Alexander-Universität
Erlangen-Nürnberg, Erlangen,
Germany

Prof. Dr. Rüdiger Schultz
Universität Duisburg-Essen,
Duisburg, Germany

Robert Schwarz
Zuse Institute Berlin (ZIB),
Berlin, Germany

Jonas Schweiger
Zuse Institute Berlin (ZIB),
Berlin, Germany

Klaus Spreckelsen
Open Grid Europe GmbH,
Essen, Germany

Dr. Claudia Stangl
Universität Duisburg-Essen,
Duisburg, Germany

Prof. Dr. Marc C. Steinbach
Leibniz Universität Hannover,
Hannover, Germany

Isabel Wegner-Specht
Humboldt-Universität zu
Berlin, Berlin, Germany

Bernhard M. Willert
Leibniz Universität Hannover,
Hannover, Germany

Contents

Foreword xi

Preface xiii

I   Fundamentals 1

1   Introduction 3
T. Koch, M. E. Pfetsch, J. Rövekamp
    1.1   Mathematical characteristics and challenges 3
    1.2   Stationarity 4
    1.3   Basic tasks 5
    1.4   State of the art and new methodology 6
    1.5   Fundamentals of gas transmission 6
    1.6   Transport networks 11

2   Physical and technical fundamentals of gas networks 17
A. Fügenschuh, B. Geißler, R. Gollmer, A. Morsi, M. E. Pfetsch,
J. Rövekamp, M. Schmidt, K. Spreckelsen, M. C. Steinbach
    2.1   Gas transport 18
    2.2   Gas properties 19
    2.3   Gas network elements 23
    2.4   Gas network structures 37
    2.5   Gas network representation 42

3   Regulatory rules for gas markets in Germany and other European countries 45
U. Gotzes, N. Heinecke, B. Hiller, J. Rövekamp, T. Koch
    3.1   Overview of gas market regulation in Europe and Germany 46
    3.2   Current rules for using gas transmission networks 48
    3.3   Current rules for determining capacities 56
    3.4   Challenges for gas transmission system operators 62
    3.5   Summary and outlook 64

4   State of the art in evaluating gas network capacities 65
D. Bargmann, M. Ebbers, N. Heinecke, T. Koch, V. Kühl, A. Pelzer,
M. E. Pfetsch, J. Rövekamp, K. Spreckelsen
    4.1   Background for capacity evaluation and simulation 67
    4.2   Generation of scenarios 68

	4.3	Network control options in simulation	78
	4.4	Simulation	81
	4.5	Interpretation of calculation results	83
	4.6	Conclusions	84

II Validation of nominations — 85

5 Mathematical optimization for evaluating gas network capacities — 87
L. Schewe, T. Koch, A. Martin, M. E. Pfetsch

5.1	The building blocks of our hierarchy	88
5.2	Abstract problem statement	94
5.3	Additional modeling considerations	96
5.4	Pre- and postprocessing	98
5.5	Overview of the literature	100
5.6	Overview of our approaches	101

6 The MILP-relaxation approach — 103
B. Geißler, A. Martin, A. Morsi, L. Schewe

6.1	An MINLP model for the validation of nominations	103
6.2	An MILP relaxation of the MINLP model	114

7 The specialized MINLP approach — 123
J. Humpola, A. Fügenschuh, B. Hiller, T. Koch, T. Lehmann, R. Lenz, R. Schwarz, J. Schweiger

7.1	Passive pipe networks	124
7.2	From passive pipe networks to gas networks with active devices	130
7.3	Element modeling	134
7.4	Conclusion	143

8 The reduced NLP heuristic — 145
R. Gollmer, R. Schultz, C. Stangl

8.1	Reduction of variables	146
8.2	Constraints for active elements	150
8.3	Objective function	154
8.4	Summary of the model	155
8.5	Heuristics to fix binary decisions	156
8.6	Conclusion	162

9 An MPEC based heuristic — 163
M. Schmidt, M. C. Steinbach, B. M. Willert

9.1	Model	165
9.2	MPEC regularization	175
9.3	Solution technique: A two-stage approach	176

10 The precise NLP model — 181
M. Schmidt, M. C. Steinbach, B. M. Willert

10.1	Component models	182
10.2	Objective functions	207
10.3	Relaxations	208
10.4	A concrete validation model	209

11 What does "feasible" mean? — 211
I. Joormann, M. Schmidt, M. C. Steinbach, B. M. Willert

- 11.1 Feasible network operation ... 211
- 11.2 Availability and accuracy of model data ... 213
- 11.3 How "feasible" are solutions of our models? ... 214
- 11.4 NLP validation vs. network simulation ... 216
- 11.5 The interpretation of ValNLP solutions ... 221
- 11.6 Analyzing infeasibility in a first stage model ... 224

12 Computational results for validation of nominations — 233
B. Hiller, J. Humpola, T. Lehmann, R. Lenz, A. Morsi, M. E. Pfetsch, L. Schewe, M. Schmidt, R. Schwarz, J. Schweiger, C. Stangl, B. M. Willert

- 12.1 Introduction ... 233
- 12.2 Results for the MILP-relaxation approach ... 236
- 12.3 Results for the specialized MINLP approach ... 242
- 12.4 Results for the reduced NLP heuristic ... 249
- 12.5 Results for the MPEC based heuristic ... 252
- 12.6 Results for the validation NLP ... 260
- 12.7 Comparison of the decision approaches and combined solver ... 264

III Verification of booked capacities — 271

13 Empirical observations and statistical analysis of gas demand data — 273
H. Heitsch, R. Henrion, H. Leövey, R. Mirkov, A. Möller, W. Römisch, I. Wegner-Specht

- 13.1 Descriptive data analysis and hypothesis testing ... 274
- 13.2 Reference temperature and temperature intervals ... 279
- 13.3 Univariate distribution fitting ... 280
- 13.4 Multivariate distribution fitting ... 283
- 13.5 Forecasting gas flow demand for low temperatures ... 286

14 Methods for verifying booked capacities — 291
B. Hiller, C. Hayn, H. Heitsch, R. Henrion, H. Leövey, A. Möller, W. Römisch

- 14.1 Motivation and outline of the approach ... 292
- 14.2 Sampling statistical load scenarios for verifying booked capacities ... 295
- 14.3 Generating quantiles for verifying booked capacities ... 301
- 14.4 Modeling capacity contracts ... 303
- 14.5 An adversarial heuristic for generating booking-compliant nominations ... 305
- 14.6 Methods to verify booked capacities ... 308
- 14.7 Computational results for verifications of booked capacities ... 310
- 14.8 Conclusions ... 314

15 Perspectives — 317
C. Hayn, J. Humpola, T. Koch, L. Schewe, J. Schweiger, K. Spreckelsen

- 15.1 Physical models and transient effects ... 317
- 15.2 Modeling flow situations ... 318
- 15.3 Determining maximal capacities ... 319
- 15.4 Extending the network ... 320

		15.5	Making it work in practice	322
		15.6	Outlook	323

A Background on gas market regulation — 325
J. Rövekamp

    A.1   Legislative power, authorities, and organizations 325
    A.2   Chronology of European and German gas market regulation 327
    A.3   Ongoing and future activities 330

Acronyms — 331

Glossary — 333

Regulation and gas business literature — 339

Bibliography — 345

Index — 361

Foreword

If you have tried it for yourself ...

When setting up the rules for the gas grid access, the Federal Network Agency had to deal with the following argument again and again: "With the new rules of the game you will only alter the financial relations, not the gas flows. Gas will still come from Russia, Norway, and The Netherlands, and consumers will not change their behavior anyway." On these grounds we were asked to completely abolish any management of capacities in the gas grid altogether and to just collect the costs of the network operation from the end-consumers—as is the practice with electricity.

We had our doubts in this respect. More precisely, we were much more optimistic. We assumed that competition would well be able to bring about lasting changes in supply relationships: The operational mode of storage facilities, but also import flows, would undergo dramatic changes during our transition from the old world of monopolies to the new, heterogeneous world of competition.

We proved to be right. Gas flows are constantly changing and capacity bookings are still an important factor. Instead of abolishing the management of capacities in the gas grid we further developed and differentiated them. The aim was and still is to design capacity products such that they hamper trade as little as possible but at the same time account for network requirements in a safe way.

In order to understand these network requirements, we have made a small and incomplete attempt to calculate for ourselves how much fixed capacity a network operator can offer its customers. The result was simple: Zero. Due to the high number of entries and exits, every meshed network has so many conceivable use cases that it is easy to substantiate that at least one use case is not representable.

The result was not satisfactory and we have started to look for assumptions and methods that improve the situation. It became clear pretty soon that we had to be careful with any kind of simplification, because it is the extreme individual case that must remain controllable or be excluded. Otherwise it is not possible to offer fixed capacities.

We soon got stuck in the jungle of possibilities and combinations. And our appreciation for the technicians at the gas network operators grew even more. In the control and marketing decisions they make every day, these people take up responsibility to operate the gas grid in ever-growing market areas according to the needs and requirements of transport customers as well as adjoining and downstream networks. They do this on the basis of their experience of what the grid is capable of and how it reacts to certain conditions. There is no standardized methodology for this.

This book changes little for the technician on site. However, it means vast changes for theoretical considerations of the capabilities of a gas grid. What we have tried once on a very small scale has been done from A to Z by a really huge team of mathematicians: This involves research on the border of the capability of today's methods, and is therefore interesting and challenging for mathematicians. We do believe this and we are enthusiastic about their method: Millions of possible supply relationships are combined with millions of system settings and checked for permissibility. The fact that these calculations do not take millions of years is really great art.

The result is reassuring: There are indeed fixed capacities that the network operators can offer their customers.

Peter Stratmann
Federal Network Agency

Preface

Structure of this book and how to read it

This book is divided into three parts:

Part I Fundamentals,
Part II Validation of nominations,
Part III Verification of booked capacities.

Part I gives fundamental information about the planning problems arising in gas transport and is structured as follows: Chapter 1 gives an introduction to the main topics of this book and describes the overall setting. Chapter 2 will define the notation and technical basics of *gas networks* as used throughout the book. Chapter 3 explains why we have to deal with the evaluation of gas network capacities in the first place (other than for technical reasons) by giving all the legal background needed. (Appendix A provides further details.) Chapter 4 describes the state of the art of gas transport planning using simulation.

Part II deals with approaches for the validation of nominations problem. In Chapter 5 we move from simulation to optimization in order to deal with the validation of nominations (NoVa) problem. Chapters 6–9 describe different approaches for obtaining discrete decisions, which are validated by a high precision nonlinear program (NLP), presented in Chapter 10. Chapter 11 reviews the methodology and puts it into perspective, discussing the interpretation of the results. Chapter 12 gives computational results using all the methods described in Chapters 6–10.

Part III discusses approaches for the verification of bookings. Chapters 13 and 14 deal with the generation of realistic nominations, taking into account past, future, and legal aspects; this will lead to an approach to deal with the verification of bookings. We will also present some computational results for the verification of bookings. Finally, Chapter 15 will provide an outlook on what further, more involved questions might be answered by building upon the results presented in this book.

As a rule of thumb, the following combinations of chapters can be read independently:

▷ Chapter 2 gives an independent introduction to gas transport modeling.
▷ Chapters 3 and 4 give an introduction into the current legal conditions and state of the art. Appendix A provides further background into the legal setup.
▷ Chapter 5 provides the basic setting for Chapters 6–10. Each of the five latter chapters is independent from the others.
▷ Chapter 11 requires some knowledge about Chapters 5 and 6–10; in particular requires Chapters 6 and 10.
▷ The computations presented in Chapter 12 require knowledge about the individual approaches, i.e., Chapters 6–10.

▷ Chapters 13 and 14 describe the approach for the verification of bookings and present some computational results.

Let us finally list some additional information on the structure of this book:

▷ A glossary for terminology of gas transport starts on page 331.
▷ An index is given starting on page 361.
▷ There are two tables inside the front and back covers of this book, respectively, that contain a list of physical and technical quantities and constants that are used throughout this book.
▷ Moreover, this book contains two lists of literature. The first starting on page 339 contains references for legal or gas business related information; the corresponding citations are marked in brackets, e.g., [GasNZV 2005]. The second starting on page 345 contains mathematical literature; its citations use parentheses, e.g., Domschke et al. (2011).

Goals of this book

The main goal of this book is to provide an introduction to the field of gas transport planning. We highlight the many interesting mathematical questions that arise in this context. Moreover, we describe the above mentioned new approach for evaluating the capacity of a gas network in detail.

In particular, this book provides the following information:

▷ Chapter 2 provides a compact description of the main mathematical concepts and formulas needed to describe gas transport.
▷ We describe the legal framework for gas transport (Chapter 3) and the state of the art (Chapter 4).
▷ Four approaches to find (discrete) decisions for the active elements in order to transport a given load situation are developed (Chapters 6–10). The outcome of these methods is used in order to find a solution of a more accurate NLP model (Chapter 10).
▷ Chapters 13 and 14 provide an approach to automatically generate stressing load situations.
▷ Finally, we present computational results of the proposed methods in Chapters 12 and 14.

We thus see this book just as a first step towards the development of mathematical concepts and results that arise in this area with the goal of providing practically useful methods.

Acknowledgments

This book is one of the results of the research project

Untersuchung der technischen Kapazität von Gasnetzen
(Investigation of the technical capacity of gas networks)

supported by the German Federal Ministry for Economic Affairs and Energy (Bundesministerium für Wirtschaft und Energie – BMWi).

The goal of the project was to improve the mathematical methods used for mid- to long-term capacity planning of gas transportation networks. It was conducted between 2009 and 2012 and has been triggered by another project, called *Forschungskooperation Netzoptimierung* (ForNe) (Research cooperation network optimization) that was initiated by Germany's largest gas transport system operator Open Grid Europe GmbH (OGE) due to the challenges imposed by new regulations on gas transportation. ForNe has been executed by OGE and Friedrich-Alexander-Universität Erlangen-Nürnberg, Humboldt-Universität zu Berlin, Technische Universität Darmstadt, Leibniz Universität Hannover, Universität Duisburg-Essen, Weierstraß Institut für Angewandte Analysis und Stochastik (WIAS), and Zuse Institute Berlin (ZIB). In particular, the authors of this book were involved. As of this writing ForNe is still running, working on bringing the results obtained into daily practice.

We would like to express our gratitude to

- the Bundesnetzagentur (BNetzA) for their support of the project, in particular Peter Stratmann, Stefan Wolff, and Markus Backes;
- Projektträger Jülich for their support, in particular Rainer Schneider, Christoph Jessen, and Johannes Tambornino;
- OGE for making available the data used for the computational experiments;
- Thomas Liebling for his support from the MOS-SIAM series on optimization side.
- the people at SIAM, in particular Sara Murphy, Elizabeth Greenspan, and Lisa Briggeman, who supported us and waited faithfully for the manuscript;
- Henrik Pilz and Jakob Witzig for their help with preparing the figures;
- Siemens (Andrea Klottig) for allowing us to use the picture in Figure 2.6.

Part I

Fundamentals

Chapter 1
Introduction

Thorsten Koch, Marc E. Pfetsch, Jessica Rövekamp

This book deals with a simple sounding question:

Can a certain amount of gas be transported by a given network?

For a single pipeline with gas entering at one end and leaving at the other end, this question is well studied and can be answered in a comparatively easy way. But how about for an 11 000 km long meshed nationwide gas transportation network? In this case, both precisely specifying the question and giving a definitive answer become difficult.

What is so difficult about gas networks? Imagine a rubber raft on the ocean with 10 inlets and 100 holes. Now 10 people on the raft are blowing air into the raft, while air leaves through the holes. If they blow too hard, the holes will widen; if they blow too weakly, the boat will sink. Your job is to orchestrate the people blowing to make sure the raft stays afloat.

A more general version of the above question is determining the maximal amount of gas that a given network can transport, i.e., we ask for the *capacity* of the network. This is a question that gas transport companies and political administrations are regularly faced with. As we will describe later, the capacity is one of the main concepts in legislative regulations.

1.1 • Mathematical characteristics and challenges

There are two important things to keep in mind when discussing the capacity of gas transport networks:

▷ Everything is connected. If you change the pressure for some inflow or outflow, it can affect large parts of the network.

▷ There is no such thing as an isolated capacity of a pipeline within the network.

These two issues arise because of the interdependency between gas flow and pressure. The latter issue distinguishes gas networks from many telecommunication settings in which there is a certain bandwidth for each link. As a consequence, classical network flow theory (see, e.g., Ford and Fulkerson (1956), Korte and Vygen (2007)) does not suffice to describe the behavior of gas networks. In particular, the max-flow/min-cut theorem (or variations of it) is not applicable to determine the capacity.

What makes the determination of the capacity even more complicated is the fact that gas networks contain so-called active elements. These are devices that can be controlled by the gas network dispatchers (operators). The most basic active elements are valves, which can open or close a pipeline. But this implies that the topology of the network is variable, and even configurable. Moreover, gas networks contain compressors that allow increasing the pressure in order to compensate for the pressure loss occurring when transporting gas through pipelines. The operation of compressors depends on several parameters and is described by so-called characteristic diagrams.

Mathematically, the problem of determining whether a certain amount of gas can be transported through a network has the following characteristics:

▷ The physical behavior of gas is described by systems of ordinary/partial differential equations (ODEs/PDEs)—the so-called Euler equations.
▷ Nonlinear equations and inequalities describe other components like compressors.
▷ Integral/discrete variables arise from the control of active elements.
▷ The problem is large- to huge-scale: ultimately, one would like to handle cases like the whole European gas network.
▷ The problem also has stochastic components, as we will describe below.

In conclusion, this yields an extremely challenging problem that can only be solved for instances of limited size and with limited accuracy by today's methods. This book discusses new approaches to deal with instances of practical size and high accuracy.

1.2 ▪ Stationarity

In addition to the above characteristics, the operation of gas network transportation is actually dynamic and transient. In fact, gas travels at about 20 km/h, and consequently changes in the amount of gas flow require time to reach the far ends of the network. On the other hand, gas is compressible, and the network itself can store quite a large amount of gas. Thus, asking for capacities without specifying the current state of the network is not a well-defined question.

In this book we concentrate on mid- and long-term capacity planning. Thus, capacities have to be determined for an unknown time in the future. Consequently, the state of the network at that time is also unknown. If one wants to play it safe, the usual approach in such cases would be to assume a worst-case state, e.g., an empty network. This would result in an unreasonably low capacity of the network. Furthermore, it is unclear how long the capacity in question should be provided and whether there are any requirements on the state of the network afterwards. On the other hand, we might assume that enough gas is present in the initial state and then deliver this stored gas, leaving the network empty afterwards. A compromise between these two extreme approaches would be to require that the state at the beginning is a reasonable average state and that the state at the end is not worse than at the beginning. This can be achieved by a steady state or stationary state.

In this book we will consider only the stationary case. As a consequence, no initial state is needed and the solution is sustainable. Moreover, no duration of the capacity request is needed. One could view this as having an average initial state and requesting a constant capacity for an unlimited time.

While not true in a strict mathematical sense, usually the stationary case is more restrictive than the transient case, i.e., if we find a stationary solution, it will be possible to obtain a feasible transient network control, provided the initial state is suitable.

Of course, we lose the freedom of transient control by restriction to a stationary solution. In particular, we cannot model storage in the network, the so-called linepack, and compressor stations cannot be operated dynamically. On the other hand, one has to acknowledge that the resulting large-scale stationary discrete-continuous nonsmooth control problems are hard to solve, and the transient case is even harder to solve. We will investigate in Chapter 11 more rigorously the meaning of a feasible solution of a stationary problem.

1.3 • Basic tasks

We will now describe the two main tasks for gas network transportation that we deal with in this book. For more background on the mathematical modeling and (legal) definitions we refer the reader to Chapters 2 and 3, respectively.

Since around 2005, European legislation has forced gas trading and gas transport to be separate operations, typically performed by different companies. Network operators are solely responsible for the transportation of gas. On the other hand, gas traders only need to specify where they want to inject gas, at so-called entry points, or extract gas, at so-called exit points. The exchange with the corresponding other supplier or consumer is done via the so-called virtual trading point.

For every entry and exit of the network, network operators sell independent capacity rights to transfer gas into or out of the network up to a certain maximal amount. The acquisition of these rights is called booking, and the use of such a right is called nomination. For any given time frame, some of the holders of these rights have to get together and form a *balancing group* of traders providing and consuming gas from the network such that each nominates an amount of gas yielding a balanced amount in total.

Therefore, we call a vector defining the amount of gas entering and leaving on each entry and exit, respectively, a nomination.[1] We require a nomination to always be balanced, i.e., the sum of flows at the entries equals the sum of the exits.

The resulting load situation (nomination) has to be transported by the network operator during this time frame. Note that when the rights to participate are sold to the traders (possibly years before the concrete execution of the right), the network operator has no knowledge on what particular balancing groups might team up later. Therefore, the network operator has to decide which capacity rights to sell, way in advance of the actual nomination. This motivates the basic question of this book:

What is the capacity of the network and how can it be computed?

In fact, a clear mathematical definition of the term *capacity* is not easy. In this book, we will generally take the practical viewpoint that the *capacity* of a gas network is determined by the set of all "reasonable" nominations for which there exists a feasible control of the network. The restriction to consider only "reasonable" nominations prevents an overly pessimistic estimation of the capacity. Of course one needs to specify what "reasonable" refers to. Generally, statistical data for the past can be used for this task.

Thus, in order to estimate the capacity, in the second part of this book, we will present an approach to answer the fundamental question "Can a particular nomination be transported?" We call the corresponding task the *validation of nomination* (NoVa) problem.

In the final part of the book, we will show how to answer the question "Can all reasonable nominations be transported?" by constructing a number of particular nominations. These constructed nominations then give an estimation of the capacity of the network.

---

[1] Beware that this is a little different from what "to nominate" means in the daily gas business.

The corresponding task is called *verification of booked capacities* in this book. The main property of this setting is the usage of statistical information to select only likely nominations.

1.4 • State of the art and new methodology

As we will describe in Chapter 4, the current state of the art in gas network transportation planning is "expert-driven simulation": Experienced planners determine nominations that are "stressing" for the network. Then they try to determine controls for the active elements and test whether the nomination can be transported using simulation software.

Since the considered nominations and the controls are hand-crafted, only a small number of nominations can be checked. Thus, in order to reduce the number of tested nominations, one assumes that every convex combination of feasible nominations is feasible. Unfortunately, it can easily be shown that the feasible area is not convex and nonlinear (i.e., not polyhedral). Furthermore, when using this approach it is extremely important to find the right balance between *extremal*, i.e., allowed by the contracts but for some reason highly unlikely or even impossible in practice, and *reasonable*. If the nominations are too extreme, they might become infeasible; if they are too reasonable, something worse might occur in practice.

Thus, the aim is to replace this process by software that automatically determines stressing nominations and corresponding controls of active elements (or determines that the nominations are infeasible).

By introducing this automated method to test the feasibility of a nomination, the number of tested nominations becomes much less restrictive, especially as these tests can be computed in parallel. For the first time, this allows not only testing extremal nominations, but also nominations "inside" of the potential feasible area. Using empirical observations, these "inside" nominations can be generated using statistical methods. As mentioned before, the border between extreme and reasonable nominations is not well defined. By using many nominations, probability arguments can be applied to explore this "soft" border.

In the end, we arrive at a totally different way to answer the initial question. It is fully automated and it provides quantitative answers instead of just "yes" or "no." We hope that this change in paradigm will only be the beginning of a new way of planning gas transport networks.

1.5 • Fundamentals of gas transmission

After having described the main tasks of this book, we give a brief introduction to the importance of gas consumption and the basics of gas transport. The presented numbers mainly refer to Europe and Germany, since we use available data for a large German network. For more details about gas supply, transport, and storage, especially in the technical sector, we refer the reader to Cerbe (2008), Finnemore and Franzini (2002), Homann and Hüning (1997), Hüning and Eberhard (1990), Katz (1959), Lurie (2008), Menon (2005), and Osiadacz (1987).

1.5.1 • Sources of supply

Of all the various sources of gas supply, most of the gas currently consumed worldwide comes from *conventional deposits*. In this context, conventional deposit means that the

1.5. Fundamentals of gas transmission

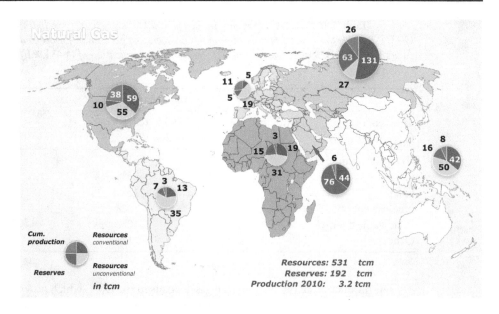

Figure 1.1. *Potential of gas (819 × 10¹² m³) [BGR2011, page 24]. Reserves are currently proven and economically recoverable gas sources, while resources are nonproven or not yet economically recoverable.*

gas source is accessible and exploitable with traditional and profitable mining techniques and will flow out without taking further measures. However, gas production from *unconventional* sources has been increasing significantly since early 2000. From these sources, unconventional gas does not flow freely and needs sophisticated drilling techniques and further measures like so-called *fracking*, where it is necessary to blast rock layers several thousand meters deep in order to access gas pores and wash them out using chemicals. Unconventional gas includes coal-bed methane, shale gas in dense rock formations, tight gas in thick sandstone or limestone horizons, aquifer gas, and gas hydrate [UBA2011, page 2]. Both conventional and unconventional gas sources have their roots in anaerobic decay of organic substances under pressure deep under the earth's surface.

The size of gas deposits varies from region to region around the world. Russia has the largest deposits. Large deposits can also be found in the USA and the Middle East, while Europe (excluding Russia) has relatively small amounts of gas deposits. In total, the worldwide potential adds up to 819×10^{12} m³ of gas, 98×10^{12} m³ of which have already been recovered. The remainder is divided into roughly 192×10^{12} m³ of currently proven and economically recoverable reserves and about 529×10^{12} m³ of nonproven or not yet economically producible resources. Of these, 311×10^{12} m³ are conventional and 218×10^{12} m³ are unconventional resources, excluding aquifer gas and gas hydrates (see Figure 1.1). These potential gas resources of aquifer gas and gas hydrates, which might not be recoverable, are estimated to have a potential of over 2000×10^{12} m³, i.e., many times more than conventional gas deposits, depending on improved mining methods and rising gas and energy prices [UBA2011, page 1ff], [BGR2010, page 12].

Note that the volumes of natural gas given in this section give only a rough estimation of its energy value, since the calorific value varies among the sources. Taking a calorific value of 11.63 kW h/m³ would give an approximation of the energy value; see also Table 1.1.

Table 1.1. *Primary energy mix in Europe and Germany in 2010 including fuels [Eurostat GIEC]. Roughly one quarter of the consumption is covered by gas both in Europe and Germany. The values have been converted to gas m³ from toe, assuming a calorific value of 11.63 kW h/m³, i.e., 1 toe = 1000 m³.*

	Europe	Germany
Total energy consumption	$\approx 1760 \times 10^9 \, m^3$	$\approx 336 \times 10^9 \, m^3$
Gas consumption	$\approx 442 \times 10^9 \, m^3$	$\approx 73 \times 10^9 \, m^3$
Crude and mineral oil	35%	34%
Gas	25%	22%
Coal	16%	23%
Nuclear energy	13%	11%
Renewable energy	10%	9%
Others	1%	1%

The production of unconventional gas deposits in the USA, which now belongs to the world's largest gas producers and can almost cover its own needs, is the most advanced [BGR2011, foreword and page 22]. There are also large unconventional gas deposits in Europe, above all in Poland. When taking the USA as an example, it becomes clear that the mining of these deposits poses a major challenge to technology and environmental protection. This is the reason why many European states issue licenses for the production of unconventional gas but, to date, have not permitted their actual mining or have revoked licenses. Nevertheless, 30% of the cumulative exploitation of $3.3 \times 10^{12} \, m^3$ was already covered by unconventional gas in 2011.

The gas sources described above are supplemented by smaller quantities from other sources like the production of *biogas*. Biogas, which can be generated from biomass and the biodegradable part of household and industrial wastes as well as landfill or sewage gas, accounts for a major proportion of these renewables [EEG2012, §3]. Biogas can be fed into the gas *transmission system* and mixed with the carrier stream provided that all official technical requirements are satisfied, e.g., regarding the removal of undesired components. Generally, biogas is fed into distribution and regional systems close to the end consumer with low pressure.

It is the declared aim of the German government to sustain a minimum of 35% of the electricity supply (or 18% of total energy consumption) with renewable energies by 2020 and to continuously increase this share further [EEG2012, §1 (2)]. Hence the production of biogas, but potentially also of hydrogen and methane from surplus electricity, will become a more important source of gas throughout the world in the near future.

1.5.2 ▪ From the source to the consumer

Gas is an important source of primary energy. Table 1.1 shows that gas covered almost one quarter of European and German energy consumption in 2010.

As can be seen from Table 1.2, the largest proportion (36%) of this gas comes from indigenous production in Europe. The most important countries exporting gas to Europe are Russia, Norway, Algeria, and some African states. With an indigenous production of 16%, Germany mainly purchases gas from Russia, Norway, and The Netherlands.

Natural gas mainly consists of methane, but also contains ethane, propane, butane, and other higher order hydrocarbon compounds. However, after exploitation, natural gas also contains inert gases such as nitrogen and carbon dioxide as well as other sub-

1.5. Fundamentals of gas transmission

Table 1.2. *European and German gas procurement portfolio in 2009 [Exxon2009; Eurogas2010]. While Germany is supplied mainly by pipeline transport, Europe imports gas also via tanker vessels from Africa, South America, and the Gulf states.*

Europe		Germany	
EU 27 (indigenous production)	36%	Germany (indigenous production)	16%
Russia	23%	Russia	36%
Norway	20%	Norway	26%
Algeria	10%	Netherlands	18%
Others (Nigeria, Lybia, etc.)	11%	Denmark and others	4%

Table 1.3. *Exemplary gas composition of different sources.*

Composition (Vol.-%)		H-gas	L-gas	LNG	Biogas
CH_4	Methane	93	82	83.2	65
N_2	Nitrogen	1.1	14	0.9	1
CO_2	Carbon dioxide	1	0.7	—	34
C_2H_6	Ethane	3	2.7	11.8	—
C_3H_8	Propane	1.3	0.4	3.5	—
C_4H_{10}	Butane	0.6	0.2	0.6	—
Standard density	(kg/m^3)	0.78	0.83	0.85	1.15
Molar mass	(kg/kmol)	17.49	18.49	19.04	25.67
Calorific value	(kW h/m^3)	11.5	9.8	12.7	7.2
	(MJ/m^3)	41.4	35.3	45.8	25.9

stances such as sulphur and water. The composition of gas differs depending on its origin, which impacts the energy content and gas properties. Therefore different gas qualities are subsumed under two categories: *H-gas* (high calorific gas) and *L-gas* (low calorific gas); see e.g., Mischner and Schewe (2009), page 212 and Hölzel (1988), pages 1–10. Gas can also be transported in liquified form as so-called *liquified natural gas (LNG)*. Its composition also depends on its origin and processing. A comparison of typical gas compositions depending on the type of the source can be found in Table 1.3.

Excluding LNG, these different gases are generally transported by pipeline to German end-consumers or exported through Germany to neighboring countries. Domestic end-consumers are gas-fired power stations, industrial customers, and private households, each of which is geared towards different gas quality bands. It is therefore permitted to mix the different gases up to the technical limits of network and end-consumer appliances. As mentioned above, a large part of the gas arriving in Germany is, however, not consumed in Germany but generally moved on to southern and western regions of Europe. Here, too, certain regulations apply to the composition and quality of the exported gas. Gas pipeline systems are therefore planned such that they connect the production and consumption areas by the transport of certain gas quality bands from the supply source to defined sales areas.

Producing and transporting gas is most efficient if performed on a constant and continuous basis. The end-consumers' gas consumption, however, fluctuates considerably and depends, among other things, on different factors such as the time of day, temperature, season, and business cycles. Moreover, gas prices which are subject to supply and

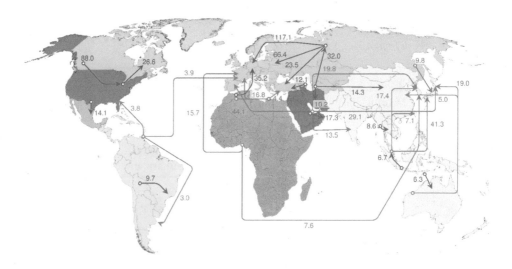

Figure 1.2. *Major trade movements worldwide (in 10^{12} m^3) [BP Review 2012, page 29]. LNG transport covers 32% of total gas transport worldwide.*

demand lead to market situations where it is economically advantageous to purchase from one source and sometimes from another. Consequently, there are significant imbalances between supply sources and consumption with impacts on gas transmission. These can be offset, on the one hand, by varying the way of operating the network and thus providing a certain level of gas buffer within the transmission network, the so-called *linepack*. On the other hand, gas storage facilities are also required to temporarily balance out major fluctuations in volume by injecting and withdrawing gas. Gas storage facilities can compensate for gas shortages even in emergencies when, for example, transmission routes cannot be used for a certain time, because of technical failures or when gas imports fail to arrive. Nevertheless, both transport and storage businesses are unbundled by regulation, and these compensatory mechanisms are mostly controlled by *transport customers*, not by asset operators; see Chapter 3 for more information.

1.5.3 ▪ Transport and storage

Gas is mainly transported through pipeline systems and by tanker vessels; see Figure 1.2 for an illustration of the global trade movements of natural gas. Pipeline-based transmission is the alternative most frequently used in Central Europe, since sources can generally be connected to the consumption regions by pipelines over land or relatively short distances over sea. Gas storage facilities allow storing energy and offer the possibility to balance fluctuations between supply and consumption of gas.

1.5.3.1 ▪ Pipeline-based gas transmission

Pipeline-based transmission uses pressure differences for transportation. As a compressible fluid, gas flows from the higher to the lower pressure (except possibly for pipes that bridge large height differences). The decreasing diameters and pressure stages of gas pipelines towards distribution subnetworks ("telescope-shape") generally dictate the direction of gas transport. Nevertheless, the transmission system is typically intermeshed to offer different transport route options.

In order to compensate for friction losses during transmission, gas is compressed at compressor stations. Moreover, other active network elements such as valves and control valves are provided to control different route options and to separate the different pressure stages from one another. Control valves have individual regulators that allow reducing gas pressure or limiting the maximal passing gas volume. Valves permit the use and closure of different transmission routes. Chapter 2 contains detailed explanations of the different network elements.

Depending on the type and design of the pipeline system, compressor stations are located at intervals of about 90–150 km along main transmission pipelines Hauenherm (1977) page 184 and up to 400 km along international or intercontinental long-distance transmission pipelines Fasold (2010) page 531. Figure 1.3 provides an overview of the main European gas pipeline system including the new pipeline *Nord Stream* and the planned large-scale projects *South Stream* and *Nabucco*.

As can be seen, gas pipelines are placed both onshore and offshore. Offshore gas pipelines can have nominal pressures of more than 220 bar; see Fasold and Wahle (1998). On land, large high-pressure transmission pipelines generally have diameters between 0.4 m and 1.4 m (see Mischner, Huke, and Möhlen (2010), page 224) and can withstand nominal pressures from 16 bar up to 100 bar; see Fasold (2010), page 531 and Hauenherm (1977), page 183. In Germany, they are laid underground at a depth of around 1–1.2 m; see Mischner (2009), page 593. Regional and distribution networks generally tend to have lower pressure stages. End-consumer networks are rated for gauge pressures of 25.5 mbar only up to 4 bar (Streicher and Feßmann (1977), page 208).

In Germany, there are a total of some 750 companies which transmit gas. However, only 17 *transmission system operators (TSOs)* are responsible for long-range gas transportation. The majority of gas transport companies are downstream network operators at regional and local levels like municipal utilities: so-called *distribution system operators (DSOs)* [BNetzA Monitoring 2009]. By reason of the multitude of network operators and the high investment costs, networks are overlapping and main pipelines are shared between several owners.

1.5.3.2 ▪ Storage

Compared to electricity, gas has the advantage that even large volumes of energy can be stored. There are several options with different characteristics for storing gas. Smaller gas volumes are stored in surface gas storage facilities or for shorter time periods in gas pipeline systems to balance fluctuations in consumption. Large gas volumes are stored in underground gas storage facilities. There are two important types: cavern and aquifer storage facilities. Whereas cavern storages in gas-impermeable salt deposits are leached drop-shaped cavities, aquifer storage facilities are located in porous rock structures. Sometimes these are fully depleted oil or gas fields. These types of storage facilities differ greatly, e.g., in their injection and withdrawal rates and their gas losses.

Since gas storage facilities are owned by transport customers rather than TSOs, they will mostly be ignored for all the planning problems that we will address in this book.

1.6 ▪ Transport networks

Gas transport networks are intermeshed pipelines which are actuated and safeguarded by active elements. The single elements appearing in such a network will be described in detail in Section 2.3, and groups of elements appear in Section 2.4. In this section, we want to give an impression of how these elements and groups are included in an entire

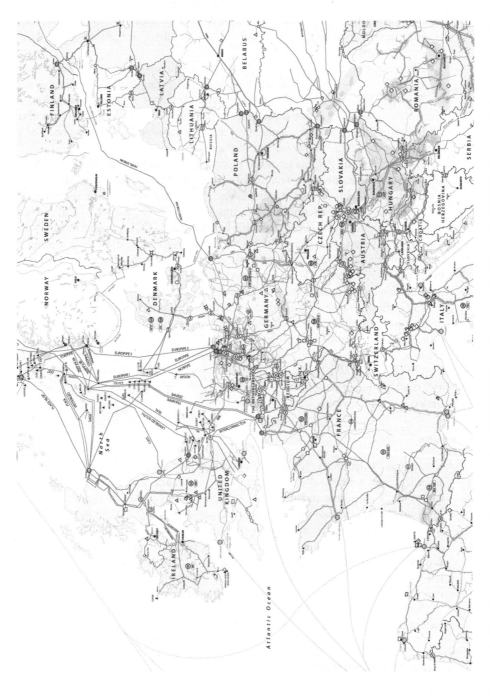

Figure 1.3. *Major European gas pipelines (excerpt from ENTSO-G Transmission Capacity Map 2012 [ENTSOGMap2012]). Due to its central geographical position, Germany serves as an important gas hub for Europe. (Source: www.entsog.eu, 2012.)*

high-pressure network. We use the German system and the network of Open Grid Europe GmbH (OGE)—Germany's largest gas transport company—as an example.

Note that we do not consider storages or other assets like liquefied natural gas (LNG) terminals or exploitation sites as network components, but model them as entry and exit points.

The existing German high-pressure pipeline system is operated by 17 market-area-spanning transmission system operators (TSOs) (for further detailed information on these TSOs see Rövekamp (2015)). The information published jointly in the Netzentwicklungsplan (NEP) 2013 [FNBGas] provides an indication of the dimensions of the German high-pressure gas network system. Table 1.4 gives an overview of all German TSOs and their network systems. Some pipelines are owned and operated jointly by several TSOs by fractional or joint venture ownership, such as the MEGAL, the TENP, the NETRA, the NEL, and the OPAL. Note that this results in a certain double-counting in the table. In total, the German high-pressure pipeline system consists of around 35 000 km of pipeline with roughly 3300 exit points.

As already seen in Section 1.5, the German high-pressure gas network serves as an important gas hub for Europe, mainly transporting the Russian and Norwegian gas to southern and western European regions. The network structure which is necessary to supply domestic end-consumers is depicted in Figure 1.4. While L-gas (black pipeline system) is normally not exported, but consumed fully in Germany, the H-gas network (red pipeline system) is built for domestic consumption and transit and has a high share of gas imports and exports.

The largest TSO in Germany is Open Grid Europe GmbH (OGE), whose network is subject to our studies. Including joint ventures and fractional ownership, the dimensions of the network considered in the remainder are shown in Table 1.4 and Figure 1.5. It can be seen that OGE including complete or fractional ownership operates more than one third of the pipeline system, exit and interconnection points, compressor stations, compressor machines, and total installed power.

For the purposes of simulating and optimizing the network, it is divided into three parts, i.e., the OGE H-gas north (purple), the OGE H-gas south (red), and the OGE L-gas (green) systems, which are dealt with separately.

Table 1.4. *Dimensions of the German high-pressure gas network (see [FNBGas]) for 2012. Please note that data from fractional or joint venture ownership might be counted twice, which does not allow proper accumulation. Column 5 gives the total installed power of compressor stations in MW. Column 6 lists the number of cross-border interconnection points. In column 8, we present the simultaneous domestic peak load in MWh for 2012, while column 9 gives the total domestic exit load in TWh for 2012.*

TSO	Pipeline network km	Compressor facilities stations #	machines #	power MW	Connection points #	Exit points #	Exit peak load (2012) MWh	Total exit load (2012) TWh
Bayernets	1313	1	1	8	3	151	16173	71
Fluxys (incl. TENP)	1010	4	15	181	3	23	14579	83
Gascade	2300	9	25	420	5	83	62000	165
Gasunie (incl. NETRA)	3260	9	28	158	6	185	42620	236
GRTgaz D (incl. MEGAL)	1167	6	23	286	3	17	57563	375
GTG Nord	321	0	0	0	1	51	10629	35
jordgas (incl. NETRA)	408	2	7	57	1	0	9751	0
NEL/Fluxys NEL/GOAL	440	0	0	0	1	n.a.	n.a.	n.a.
Nowega	710	0	0	0	0	112	6015	22
Ontras	7242	2	5	38	2	518	41920	159
OPAL/LBGT	470	1	3	96	2	1	34400	n.a.
OGE (incl. TENP, MEGAL, NETRA)	12000	27	97	1000	17	1000	131456	725
Terranets	1965	2	7	33	5	203	19963	73
Thyssengas	4211	6	15	120	5	1063	22200	67

1.6. Transport networks

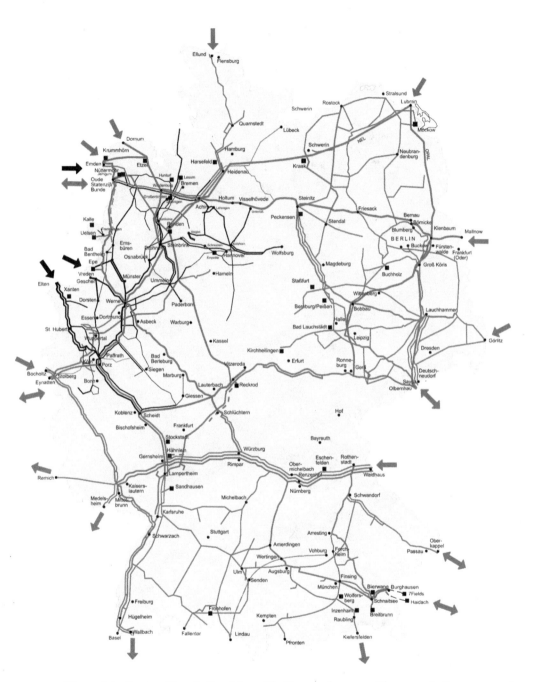

Figure 1.4. *German H-gas (red) and L-gas (black) network systems. The arrows indicate entry and exit nodes. Gas storages are represented by black squares. (Source: OGE.)*

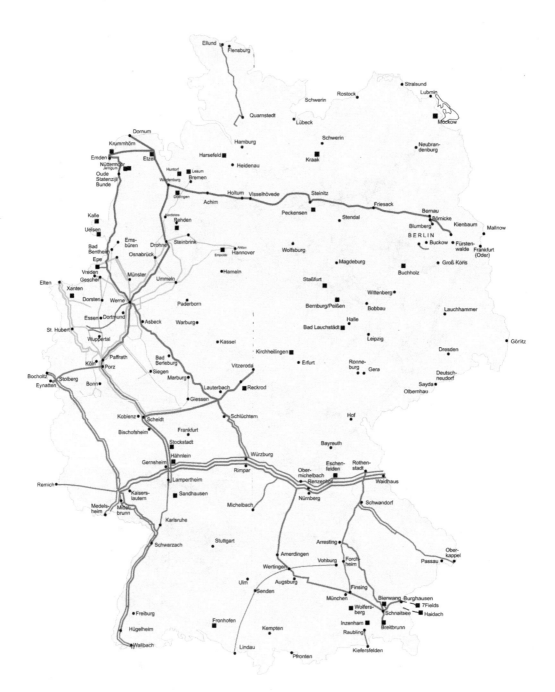

Figure 1.5. *Network systems of OGE. The network model is split into three parts: H-gas north (purple), H-gas south (red), and L-gas (green). (Source: OGE.)*

Chapter 2

Physical and technical fundamentals of gas networks

Armin Fügenschuh, Björn Geißler, Ralf Gollmer,
Antonio Morsi, Marc E. Pfetsch, Jessica Rövekamp,
Martin Schmidt, Klaus Spreckelsen, Marc C. Steinbach

Abstract *This chapter describes the fundamentals of gas transport. This includes an introduction to the basic terminology and basic physical laws with respect to natural gas. Then the basic elements needed to represent gas networks are discussed: pipes, resistors, valves, control valves, and compressor machines and drives. These elements can be grouped into larger entities like compressor groups and subnetwork operation modes.*

Natural gas is distributed through large and complex pipeline networks; for instance, the total length of the European network is more than 100 000 km. The main part of the network consists of interconnected gas pipelines. The gas flow is controlled or affected by additional components like compressors, (control) valves, and resistors. The goal of this chapter is to describe these components as well as their physical and technical properties. See also the tables inside the front and back covers of this book for a list of the physical/technical quantities and constants introduced in this chapter, respectively.

The contents of this chapter are taken from many sources in the literature and are well known. For general literature concerning the description of gas networks, we refer the reader to the books by Cerbe (2008), Lurie (2008), Osiadacz (1987), and Zucker and Biblarz (2002), as well as the article by Fasold and Wahle (1992). Concerning the modeling of compressor machines, see Odom and Muster (2009) for a recent survey.

Let us add a general comment on the literature in this area: The formulas on gas transport and their derivations that are used in both theory and practice are sometimes not derived in scientific literature, but appear in documentation of simulation software or working papers, e.g., of the Pipeline Simulation Interest Group (PSIG). Moreover, some literature, in particular, more technical articles, is not available in English. Whenever possible, we (also) cite an English article. There are, however, cases for which we do not know any English source and have to refer to German literature.

Note that certain parts that we describe in the following are *virtual* components of the network (e.g., resistors) and do not directly represent physical entities. As such they are based on modeling *decisions*—in contrast to plain descriptions of physical laws. Moreover, all given formulas represent only an abstraction of reality, since certain practical effects are not included. This is clearly motivated by the trade-off between an accurate description of reality and the possibility of solving the resulting models. Furthermore, note that our presentation is influenced by the goal to treat the stationary case.

This chapter is structured as follows: Section 2.1 gives a short introduction into the basic concepts of gas transport from a technical point of view. In Section 2.2, we present in detail the properties of different kinds of natural gas before we give a description of each type of network element: pipes (Section 2.3.1), resistors (Section 2.3.2), valves (Section 2.3.3), control valves (Section 2.3.4), and compressor machines and drives (Section 2.3.5). These different types of network elements can be combined into larger entities: control valve stations (Section 2.4.1), compressor groups (Section 2.4.2), and subnetworks involving subnetwork operation modes (Section 2.4.3) such as compressor stations. Finally, Section 2.5 introduces the graph representation for gas transport networks.

Before we begin, let us remark that we generally use SI (base) units in this book, while different units are used throughout the gas literature (e.g., the old *barg*, which measures pressure in bar, zero-referenced to the atmospheric pressure).

2.1 ▪ Gas transport

Gas is fed into the network at certain points, so-called entries, and may be withdrawn from the network at other points, so-called exits. The terms *entry* and *exit* abstract from reality, in the sense that an entry may be, e.g., a natural gas field as well as an interconnection point from another gas network. Also the term exit may refer to an interconnection point as well as to a certain customer such as a factory or a gas power plant.

The network operator is responsible for the successful transport of gas, while its customers are allowed to nominate any amount of gas per time unit at the entries and exits subject to contractual and legal constraints. To this end, a nomination is issued in terms of thermal power, supplied at a certain set of entries and withdrawn at a certain group of exits. In addition, the customers are legally bound to issue only *balanced nominations*, meaning that the total power nominated at entries must be equal to the total power nominated at exits. Moreover, contracts normally require a minimal gas pressure at exits. In contrast to gas consumers and natural gas fields, certain points in the network are *hybrid points*, i.e., they can be used bidirectionally and therefore must be treated as exits at one point of time and must be considered entries at another point of time. Examples are network interconnection points and gas storages. In situations where a gas storage is filled, it must be considered an exit, while it takes the role of an entry when gas is withdrawn from the storage. Similarly interconnection points change from entries to exits and vice versa depending on the flow directions.

In order to facilitate a successful gas transport, gas flow, pressure, and composition have to be controlled using the components of the network. For example, pipes in regional subnetworks are designed for rather low pressure levels, while large transport pipelines can be operated at high pressure. So-called control valves allow reducing pressure from high to low pressure pipelines at transition points. Moreover, when gas is transported through a pipeline network, it experiences a pressure drop. In order to transport gas over large distances, pressure has to be increased at so-called compressors. Furthermore, a network contains a number of valves, which might be open or closed, and which allow routing the gas through the network.

In the following, valves, control valves, and compressor groups are called *controllable* or *active* network elements. All other elements are called *passive*. Besides pipes, a gas network usually contains a number of further passive elements, e.g., filtering or measuring stations.

The gas network is usually remotely controlled by human dispatchers. From a central operating room they continuously monitor the status of the network and make decisions according to the current and forecast flow situation. They control the settings of active elements under regular operational conditions. In case of sudden emergencies, such as an overheating of gas or a loss of pressure due to a pipeline burst or leakage, an automated pipeline shut-down in the affected parts of the network limits the impact of such rare events. The task of network operators is to maintain a stable flow of gas through the network between the entries and exits such that this flow does not violate technical restrictions such as maximum pressure levels of pipelines, or minimum entry pressure levels at compressor groups, for example. Furthermore, pressure levels and the amount of flow have to be within a specified interval at the entries and exits due to contractual requirements. Operating networks within these limits should be achieved using a minimal amount of energy that is consumed, mainly by compressors.

Dispatching is performed dynamically over time, i.e., transient gas flows are considered. For long-term planning, which is the focus of this book, it is also interesting to consider steady state or stationary states. See Section 5.3.1 for a discussion of this.

2.2 ▪ Gas properties

One main component to describe the behavior of gas is the so-called *equation of state* connecting *mass, pressure p, volume V*, and *temperature T*. These quantities are sufficient to determine the inner state of a simple thermodynamical system for gas (i.e., systems with a single homogeneous gas mixture). The latter three are known as thermal state quantities. A state quantity is a property of a system that depends only on the current state of the system and not on the process leading to that state. There are two types of state quantities: inner and outer states. The investigation of the outer state (e.g., position, velocity) is the purpose of mechanics, as described in the following sections for each network element. Thermodynamics deals with the inner state, i.e., the equation of state.

For an *ideal gas*, the so-called *equation of state for ideal gases* or, in short, *ideal gas law* (see, e.g., Menon (2005)) applies:

$$pV = \tilde{n}RT = Nk_\text{B}T,$$

where V denotes the gas volume in m³, $\tilde{n}$ is the *amount of substance* of the gas in mol, and $R = 8.3144621$ J/mol/K $= N_\text{A} k_\text{B}$ is the *universal gas constant*. Here, the *Avogadro constant* is denoted by $N_\text{A} = 6.02214129 \times 10^{23}$/mol and $k_\text{B} = 1.3806488 \times 10^{-23}$ J/K denotes the *Boltzmann constant*. The *number of particles* is given by $N = \tilde{n} N_\text{A}$. The assumptions for an ideal gas are that molecules in the gas are point-like and do not interact with each other.

A *real gas*, however, shows a different relation between pressure, temperature, and volume than ideal gases, especially because of interaction between the gas particles. This resulting deviation of a real gas from an ideal gas is described by the so-called *compressibility factor*, which is defined as

$$z := \frac{pV}{\tilde{n}RT} = \frac{pV}{Nk_\text{B}T}.$$

In order to derive an equation of state for a real gas from the ideal gas law, one can then apply the so-called *virial expansion* (see Plischke and Bergersen (2006) or Hill (1960)), which is the following power series in N/V:

$$z = 1 + \sum_{i=1}^{\infty} B_{i+1}(T)\left(\frac{N}{V}\right)^{i}. \tag{2.1}$$

The coefficient B_i is called the ith *virial coefficient*. One of the main benefits of the virial expansion is that $z = z(p, V, N, T)$ is written as a function that depends only on N/V and T, i.e., $z = z(N/V, T)$. Equivalently, the compressibility factor can be formulated as a power series in p, and hence in a form $z = z(p, T)$ that depends only on p and T. Such a dependency,

$$z = 1 + \sum_{i=1}^{\infty} \tilde{B}_{i+1}(T) p^{i}, \tag{2.2}$$

is more adequate for our purposes and is assumed throughout this book. Note that the compressibility factor z is equal to 1 for an ideal gas. Moreover, the virial expansion can be used to derive arbitrarily many approximations of the compressibility factor, i.e., of state equations, by truncating higher order terms in (2.1) or (2.2). With the abstract introduction of the compressibility factor z as above, the relationship

$$pV = \tilde{n}RTz = Nk_B T z, \tag{2.3}$$

in terms of the thermal state quantities p, V, and T, is known as the *thermodynamical standard equation of state for real gases* Starling and Savidge (1992).

In practice, the computation of z must be performed approximately. The most accurate and most general approximation of the compressibility factor known is given by the GERG-2008 equation of state Kunz and Wagner (2012). Other high accuracy approximations, at least for the case of natural gas transport through pipeline networks, are constituted by the AGA-8 DC-92 equation Starling and Savidge (1992) and the GERG-2004 equation Kunz et al. (2007). These three equations can be derived from the virial expansion (2.1) (or (2.2)) and thus from the laws of statistical thermodynamics. These formulas require detailed knowledge about the composition of gas. However, there are also a number of empirical formulas for the compressibility factor that depend only on the *reduced pressure* $p_r = p/p_c$ and the *reduced temperature* $T_r = T/T_c$ of a gas mixture. Here, p_c denotes the *pseudocritical pressure* and T_c is the *pseudocritical temperature* (measured in Pa and K, respectively). Below or at the so-called *critical temperature*, a pure gas may be liquefied under pressure. Above the critical temperature, this is impossible independent of the pressure. Based on this temperature, the *critical pressure* is defined as the minimum pressure which would suffice to liquefy a pure gas at its critical temperature. For a gas mixture, critical temperature and critical pressure are called pseudocritical. When approaching the *pseudocritical point* (p_c, T_c) in a pressure-temperature phase diagram, differences between the two aggregate phases vanish, i.e., properties of the gaseous and liquid phases (e.g., density) are indistinguishable and can no longer be regarded as being different phases. Instead, above this point the substance is in a homogeneous supercritical phase.

Two common formulas that are used to compute the compressibility factor z and are purely based on the reduced pressure and temperature of a gas mixture are the formula of Papay (see Papay (1968) and Saleh (2002), Chap. 2),

$$z(p,T) = 1 - 3.52 p_r e^{-2.26 T_r} + 0.247 p_r^2 e^{-1.878 T_r}, \tag{2.4}$$

and an equation from the American Gas Association (AGA) (see Králik et al. (1988)):

$$z(p,T) = 1 + 0.257 p_r - 0.533 \frac{p_r}{T_r}. \tag{2.5}$$

2.2. Gas properties

The Papay equation is known to be accurate up to 150 bar, whereas the AGA equation is known to be accurate only up to 70 bar (see LIWACOM (2004)).

Finally, besides the virial equation of state (2.1) and the thermodynamical standard equation of state (2.3), several other empirical equations with different ranges of validity exist. Among those, the cubic equations of state of Waals (1873), Benedict, Webb, and Rubin (1940), Redlich and Kwong (1949), Soave (1972), Peng and Robinson (1976), Stryjek and Vera (1986), and Stryjek and Vera (1986) are commonly used in engineering (see, e.g., Rao (2003)). In the remainder of this book we restrict ourselves to the application of the thermodynamical standard equation of state for real gases (2.3) if not stated otherwise. Hence, all laws and formulas involving thermodynamics are based on this fundamental relationship and might differ if an alternative equation of state is desired. Nonetheless, most equations of state can be rewritten in a form compatible with Eq. (2.3) and Eq. (2.1).

Another important quantity in thermodynamics is the *specific isobaric heat capacity* c_p. It measures the amount of energy required to raise the temperature of one kilogram of gas by one Kelvin at constant pressure. For our purposes, the introduction of the related *molar isobaric heat capacity* $\tilde{c}_p = c_p m$, where m denotes the *molar mass*, is convenient. For real gases, the specific or molar isobaric heat capacity is typically split up into an ideal gas term and a real gas correction:

$$c_p(p,T) = \frac{\tilde{c}_p(p,T)}{m} = \frac{1}{m}\left(\tilde{c}_p^0(T) + \Delta \tilde{c}_p(p,T)\right).$$

Here, $\tilde{c}_p^0$ is the molar heat capacity of ideal gas, $\Delta \tilde{c}_p$ is a correction for real gas, and m is the molar mass. The ideal gas term

$$\tilde{c}_p^0(T) = \tilde{A} + \tilde{B}\,T + \tilde{C}\,T^2,$$

which is independent of p, is modeled using a quadratic data fit. Higher order polynomial approximations are likewise possible, but quadratic fits are typically sufficient in the context of gas transport networks. The coefficients $\tilde{A}, \tilde{B}, \tilde{C}$ within the polynomial are called *heat capacity coefficients*. The real gas correction within the above formula is given by (see Doering, Schedwill, and Dehli (2012))

$$\Delta \tilde{c}_p(p,T) = -R \int_0^p \frac{1}{\tilde{p}}\left(2T \frac{\partial z}{\partial T} + T^2 \frac{\partial^2 z}{\partial T^2}\right) d\tilde{p}.$$

Here, R is again the universal gas constant, and $z = z(p,T)$ denotes the compressibility factor as introduced above.

A thermodynamical effect to mention at this point is the interdependence between changes in pressure p and changes in gas temperature T, which is known as the so-called *Joule–Thomson effect* (see, e.g., Oliveira (2013)). The Joule–Thomson effect is due to the interaction of gas molecules. When molecules are attracting each other and pressure is reduced, the distance of the molecules grows and mechanical work has to be performed to compensate for the attraction. The required energy comes from the kinetic energy of the molecules, which leads to a decrease of gas temperature. When the gas temperature exceeds the so-called *inversion temperature*, the molecules now repel each other, and an expansion of the gas leads to an increase of its temperature. However, in real-world gas transport networks these effects can usually be ignored and we can assume that the temperature increases as pressure is increased and vice versa. The change in the gas temperature caused by the Joule–Thomson effect can be obtained as an ODE solution involving the

Joule–Thomson coefficient μ_{JT}:

$$T_{\text{out}} - T_{\text{in}} = \int_{p_{\text{in}}}^{p_{\text{out}}} \mu_{\text{JT}}(p, T)\, dp, \tag{2.6}$$

with

$$\mu_{\text{JT}}(p, T) = \frac{T^2}{p}\frac{R}{\tilde{c}_p}\frac{\partial z}{\partial T}. \tag{2.7}$$

The above formula can be derived by a straightforward application of the thermodynamical standard equation for real gases (2.3) to the definition of the Joule–Thomson coefficient (see, e.g., Oliveira (2013)).

Besides the physical properties introduced so far (especially pressure and temperature), the flow at every point in space and time of the gas in a network is a matter of particular interest. For operational purposes, gas flow is usually specified in terms of *mass flow q* (measured in kg/s). In contrast, network users, i.e., transport customers, usually nominate an amount of *thermal power* (measured in W). Thermal power P is related to mass flow via the equation

$$P = q H_c,$$

where H_c is the *calorific value*, which depends on the chemical composition of the gas. Thus, the properties of the gas mixtures supplied to the network are of significant importance. Usually, *low-calorific gas* (L-gas) with a calorific value of about 36 MJ/kg and *high-calorific gas* (H-gas) with a calorific value of about 41 MJ/kg are distinguished (see Table 1.3). Besides the calorific value, a few further important *gas quality parameters*, which have already been introduced, depend on the actual chemical composition of a gas mixture. These are the isobaric molar heat capacity coefficients $\tilde{A}, \tilde{B}, \tilde{C}$ (given in J mol^{-1}K$^{-\alpha}$, where $\alpha = 1, 2, 3$, respectively, correspond to $\tilde{A}, \tilde{B}, \tilde{C}$), the molar mass m (measured in kg/mol), as well as the pseudocritical pressure p_c and pseudocritical temperature T_c (measured in Pa and K, respectively).

There are junctions in the networks at which multiple pipes or other network elements join. At these junctions, we have *conservation of mass*:

$$\sum_{i \in I} q_i = \sum_{k \in K} q_k, \tag{2.8}$$

where I is the set of ingoing connections and K is the set of outgoing connections at the junction.

If heterogeneous gas mixtures are supplied to a nontrivial network, *gas mixtures* of different qualities meet at junctions: assume that gases with quality parameter vectors $X_i = (m_i, H_{c,i}, p_{c,i}, T_{c,i}, \tilde{A}_i, \tilde{B}_i, \tilde{C}_i)$ and mass flows q_i, $i \in I$, flow into a *junction j*. Then the mixed gas in the junction has parameters given by the vector

$$X_j = \frac{\sum_{i \in I} \frac{q_i}{m_i} X_i}{\sum_{i \in I} \frac{q_i}{m_i}}. \tag{2.9}$$

The gas temperature at junction j is determined by

$$T_j = \frac{\sum_{i \in I} c_{p,i} q_i T_i}{\sum_{i \in I} c_{p,i} q_i}. \tag{2.10}$$

Equation (2.10) can be derived from the *conservation of energy*; see Schmidt, Steinbach, and Willert (2014).

Figure 2.1. *A pipe with a diameter of* 1.2 m. *(Source: Open Grid Europe GmbH (OGE).)*

2.3 ▪ Gas network elements

In this section we describe the basic elements that commonly appear in gas networks, including virtual elements that we use as modeling devices. Every type of element has its own subsection in which we first give a phenomenological description of the element and then present a general mathematical formulation of the corresponding technical and physical effects. Since later chapters (Chapters 6–10) discuss concrete model formulations of the elements presented in this section, it serves as a foundation for the rest of the book.

2.3.1 ▪ Pipes

Natural gas transport networks consist mainly of pipelines (or pipes, for short); see Figure 2.1 for an illustration. Although the role of pipelines in gas networks is obvious, the impact of their design parameters on the gas flow needs closer consideration and will therefore be discussed in detail in the following. We first present a general pipe model and then give an approximate model for the stationary case.

2.3.1.1 ▪ The general case

From the technical point of view, the material and thickness of the pipe walls determine a limit for the pressure that the pipe is guaranteed to withstand. This limit is specified as *nominal pressure* of the pipe. Gas dynamics are influenced by the *length L* and the *diameter D* of a pipe. The longer the pipe is, the larger the pressure difference between its endpoints will be for a fixed amount of flow, and the larger the diameter is, the lower is the occurring pressure drop. In domestic gas transport networks, pipe diameters usually range between 10 cm and 1.4 m. A single gas network can contain large transport pipelines of more than 100 km of length and rather short pipes of lengths of less than 10 m.

The *inclination* of a pipe can have a significant impact on the gas flow. For example, in hilly areas, the endpoints of the pipe might be located at different levels causing a change of pressure due to gravity.

In general, the shape of the cross section of a pipe is also of importance. However, since virtually all gas pipelines in the considered transport networks are cylindrical, we restrict ourselves to this case and refer to Menon (2005) for a detailed description of the treatment of different pipe geometries. We describe selected properties of the pipe walls in the following.

Pipes are normally constructed of carbon steel with varying surface characteristics, depending on the manufacturing process. The pressure drop occurring in a pipe mainly arises because of friction, which is due to the roughness of the material of the inner pipe wall. The roughness of a pipe is a measure for the vertical deviation of the inner surface of the pipe wall from its ideal form and is thus measured in m. Pipes of different roughness also vary in their frictional resistance. Besides material roughness, the frictional resistance of a pipe is also influenced by other factors like, e.g., the presence of weld seams, the curvature of the pipe, corrosion processes, and the deposition of dirt and dust. All these factors are herein summarized in an "equivalent" surface roughness, the so-called *integral roughness* of the pipe. An estimate for the integral roughness of a real pipe can be obtained either experimentally or via measurements of optical or feeler instruments (see Kamnev (1966)).

Using the concept of the integral pipe roughness constitutes a simplification, in which effects, e.g., due to surface irregularities, are summarized. A detailed modeling of all such effects can be done for the investigation of gas dynamics in a single pipe using techniques from computational fluid dynamics. This leads to a system of three-dimensional PDEs with an additional choice of treating resistance due to the pipe walls; see, e.g., Anderson, Jr. et al. (2009), Batchelor (2000), and Landau and Lifshitz (1987) for an introduction and more details. In this book, however, we concentrate on simulating and optimizing entire gas networks, where many pipes and other elements are interconnected. Thus, a more macroscopic view is adequate in our context.

As mentioned earlier, we will restrict ourselves to the case of *cylindric* pipes and to the modeling of *one-dimensional* flow in the pipe direction x. Under these assumptions the mass flow q is related to gas density ρ and velocity v via

$$q = A\rho v, \tag{2.11}$$

where $A = D^2 \pi/4$ denotes the constant *cross-sectional area* of the pipe.

The gas dynamics within a single pipe is described by the following set of nonlinear, hyperbolic PDEs (see Feistauer (1993); Lurie (2008)), often referred to as *Euler equations*:

$$\frac{\partial \rho}{\partial t} + \frac{1}{A}\frac{\partial q}{\partial x} = 0, \tag{2.12}$$

$$\frac{1}{A}\frac{\partial q}{\partial t} + \frac{\partial p}{\partial x} + \frac{1}{A}\frac{\partial (qv)}{\partial x} + g\rho s + \lambda(q)\frac{|v|v}{2D}\rho = 0, \tag{2.13}$$

$$A\rho c_p \left(\frac{\partial T}{\partial t} + v\frac{\partial T}{\partial x}\right) - A\left(1 + \frac{T}{z}\frac{\partial z}{\partial T}\right)\frac{\partial p}{\partial t}$$
$$- Av\frac{T}{z}\frac{\partial z}{\partial T}\frac{\partial p}{\partial x} + A\rho v g s + \pi D c_{\text{HT}}(T - T_{\text{soil}}) = 0. \tag{2.14}$$

The *continuity equation* (2.12) and the *momentum equation* (2.13) describe the *conservation of mass* and the *conservation of momentum*, respectively, while *energy conservation* is expressed by (2.14). Here, $v = v(x,t) \in \mathbb{R}$ is the velocity in the direction of the pipe, g denotes the *gravitational acceleration* (with standard value $9.80665\,\text{m/s}^2$), and $s \in [-1,1]$ denotes the (constant) *slope* of the pipe, i.e., the tangent of its inclination angle. The *soil temperature* is denoted by T_{soil}, and c_{HT} denotes the *heat transfer coefficient*, given in $\text{J}/(\text{m}^2\,\text{K}\,\text{s})$, which expresses how heat is exchanged between the gas in the interior of the pipe and the surrounding of the pipe. This coefficient depends on the material and the thickness of the pipe wall. Thus, these have consequences on the fluid mechanical behavior of the gas.

2.3. Gas network elements

Frictional forces are expressed via the so-called *friction factor* $\lambda = \lambda(q)$. For the computation of λ several phenomenological formulas are known from the literature. These formulas differ for *turbulent* and *laminar* flow, and hence depend on the *Reynolds number* (see Lurie (2008) and Saleh (2002)),

$$Re(q) = \frac{D}{A\eta}|q|, \tag{2.15}$$

where η denotes the *dynamic viscosity* of the gas. The flow is *turbulent* if $Re(q) \geq Re_{\text{crit}} \approx 2320$ and *laminar* otherwise.

In the laminar case, the formula of Hagen-Poisseuille (see Finnemore and Franzini (2002)) should be used to compute the friction factor:

$$\lambda(q) = \frac{64}{Re(q)}. \tag{2.16}$$

For the turbulent case, the implicit equation of *Prandtl and Colebrook* (also known as the Colebrook–White equation) constitutes the most accurate approximation of reality (Saleh (2002), Chap. 9):

$$\frac{1}{\sqrt{\lambda}} = -2\log_{10}\left(\frac{2.51}{Re(q)\sqrt{\lambda}} + \frac{k}{3.71D}\right). \tag{2.17}$$

A number of explicit approximations of (2.17) are known. We mention the equation of Hofer (see Hofer (1973); Mischner (2012)),

$$\lambda(q) = \left(-2\log_{10}\left(\frac{4.518}{Re(q)}\log_{10}\left(\frac{Re(q)}{7}\right) + \frac{k}{3.71D}\right)\right)^{-2}, \tag{2.18}$$

and the formula of Nikuradse (see Nikuradse (1933); Nikuradse (1950); Mischner (2012)):

$$\lambda = \left(2\log_{10}\left(\frac{D}{k}\right) + 1.138\right)^{-2}, \tag{2.19}$$

which can be derived from (2.18) for $Re \to \infty$.

In order to complete the system (2.11)–(2.14) of five unknowns and four equations, a suitable equation of state has to be added. As mentioned previously, we use the thermodynamical standard equation for real gases

$$p = \rho R_s T z \tag{2.20}$$

for this purpose. This is an equivalent reformulation of Eq. (2.3), which is obtained using the definition of *density*, $\rho = \tilde{n}\, m/V$, and the specific gas constant, $R_s = R/m$.

2.3.1.2 • Approximations in the stationary and isothermal case

In this section we focus on the *stationary case*, i.e., the gas is in a steady state and all time derivatives in (2.12)–(2.14) are equal to zero. In this case, the continuity equation (2.12) simply states that the mass flow along the pipe is constant, i.e., $\partial_x q := \partial q/\partial x = 0$. Moreover, we only consider the *isothermal case* here. Thus, the gas temperature is considered to be constant, and the energy equation (2.14) can be neglected. The remaining momentum equation (2.13) states how the pressure change along the pipe depends on the amount of mass flow and the technical parameters of the pipe. In this book, we often use a further simplification of that equation: a quadratic approximation (see Bales (2005); Lurie (2008), which can be derived as follows.

We note first that the ram pressure term $\partial_x(qv)/A$ contributes less than one percent to the sum of all terms under normal operating conditions; see Wilkinson et al. (1964). Hence we assume that $\partial_x(qv)/A$ can be neglected. Moreover, we assume that the gas temperature T and the compressibility factor z can be approximated by suitable constants along the entire pipe, say, by mean values T_m and z_m. Finally, assume that only pipes with constant slope s are considered. Then the stationary momentum equation (2.13) can be rewritten as

$$\frac{\partial p}{\partial x} + g \rho s + \lambda(q) \frac{|v|v}{2D} \rho = 0. \tag{2.21}$$

Lemma 2.1. *For $s \neq 0$, the solution $p(x)$ to (2.21) with initial value $p(0) = p_{\text{in}}$ is given by*

$$p(x)^2 = \left(p_{\text{in}}^2 - \tilde{\Lambda} |q| q \frac{e^{\tilde{S}x} - 1}{\tilde{S}} \right) e^{-\tilde{S}x} \tag{2.22}$$

with

$$\tilde{S} := \frac{2gs}{R_s z_m T_m}, \quad \tilde{\Lambda} := \lambda(q) \frac{R_s z_m T_m}{A^2 D}.$$

Proof. In (2.21), we replace the gas velocity v by the mass flow q using (2.11) and the gas density ρ by the pressure p using the equation of state (2.20). This yields

$$\frac{\partial p}{\partial x} + g \frac{p}{R_s z_m T_m} s + \lambda(q) \frac{|q|q}{2A^2 D} \frac{R_s z_m T_m}{p} = 0,$$

where we use the assumption that the gas temperature and compressibility factor are constants T_m and z_m, respectively. Multiplication by $2p$ leads to

$$\frac{\partial}{\partial x} p^2 + \tilde{S} p^2 = -\tilde{\Lambda} |q| q.$$

If we now substitute $y = p^2$, we end up with the first-order linear ordinary differential equation (ODE)

$$\frac{\partial}{\partial x} y + \tilde{S} y = -\tilde{\Lambda} |q| q, \quad y(0) = p_{\text{in}}^2. \tag{2.23}$$

This ODE can be solved analytically by "variation of constants," and we arrive at

$$y(x) = p(x)^2 = \left(-\tilde{\Lambda} |q| q \frac{1}{\tilde{S}} e^{\tilde{S}x} + p_{\text{in}}^2 + \tilde{\Lambda} |q| q \frac{1}{\tilde{S}} \right) e^{-\tilde{S}x},$$

where the last two terms in parentheses represent the integration constant obtained from the initial value $y(0) = p_{\text{in}}^2$. This concludes the proof. □

By evaluating the solution of Eq. (2.22) at $x = L$ (with $p(L) = p_{\text{out}}$) and fixing the notation $\Lambda := \tilde{\Lambda} L$ and $S := \tilde{S} L$, we finally obtain a well-known relationship of inlet and outlet pressures and the mass flow through the pipe (see, e.g., Lurie (2008)),

$$p_{\text{out}}^2 = \left(p_{\text{in}}^2 - \Lambda |q| q \frac{e^S - 1}{S} \right) e^{-S} \tag{2.24}$$

2.3. Gas network elements

with

$$\Lambda = \lambda(q)\frac{R_s z_m T_m L}{A^2 D} = \left(\frac{4}{\pi}\right)^2 \lambda(q)\frac{R_s z_m T_m L}{D^5}, \quad S = \frac{2gsL}{R_s z_m T_m}. \quad (2.25)$$

Note that the pressure $p(x)$ according to Eq. (2.22), and hence $p(L) = p_{\text{out}}$ according to Eq. (2.24), are not defined for horizontal pipes, i.e., if the slope s is zero. The solution for this case is obtained by solving the (trivial) ODE (2.23) with $\tilde{S} = 0$, or by taking the limit for $s \to 0$ (equivalently $\tilde{S} \to 0$) in (2.22) using l'Hôpital's rule.

Lemma 2.2. *For $s = 0$, the solution $p(x)$ to Eq. (2.21) with initial value $p(0) = p_{\text{in}}$ is given by*

$$p(x)^2 = p_{\text{in}}^2 - x \tilde{\Lambda} |q| q, \quad (2.26)$$

with $\tilde{\Lambda}$ as defined in Lemma 2.1.

Evaluating Eq. (2.26) at $x = L$ now yields the pressure loss formula for horizontal pipes:

$$p_{\text{out}}^2 = p_{\text{in}}^2 - \Lambda |q| q. \quad (2.27)$$

It remains to choose appropriate approximations for the constant values z_m and T_m as required by Lemma 2.1. A good estimate of the mean values would be most suitable. Since $z_m = z(p_m, T_m)$ is typically defined by Eq. (2.4) or Eq. (2.5), z_m can be obtained from an adequate mean value of p_m. In fact, there exists an elegant closed-form expression for p_m which is often used throughout the literature (see, e.g., Saleh (2002)). It depends only on p_{in} and p_{out} and is superior to a simple arithmetic mean.

Lemma 2.3. *Let $p(x)$ be given as in Lemma 2.2, and let*

$$p_m := \frac{1}{L}\int_0^L p(x)\,dx$$

be the mean pressure along the pipe. Then

$$p_m = \frac{2}{3}\left(p_{\text{in}} + p_{\text{out}} - \frac{p_{\text{in}} p_{\text{out}}}{p_{\text{in}} + p_{\text{out}}}\right). \quad (2.28)$$

Proof. Initially, we seek a closed form expression for $p(x)$ independent of q and any mean values. To obtain such a formula we eliminate the flow term by multiplying Eq. (2.26) by L and subtracting Eq. (2.27) multiplied by x. After solving for $p(x)$, we can rewrite the mean pressure as

$$p_m = \frac{1}{L}\int_0^L p(x)\,dx = \frac{1}{L}\int_0^L \sqrt{p_{\text{in}}^2 - \frac{x}{L}(p_{\text{in}}^2 - p_{\text{out}}^2)}\,dx.$$

Evaluating the integral then yields the desired formula (2.28). □

Without involving the energy equation (or approximative solutions of it), a simple arithmetic mean for the average temperature

$$T_m := \frac{1}{2}(T_{\text{in}} + T_{\text{out}}) \quad (2.29)$$

is the most adequate choice.

Finally, we would like to mention that Eq. (2.24) (or (2.27)) may be equivalently formulated in terms of the *volumetric flow rate under normal conditions* Q_0, using the relation

$$q = A\rho v = \rho_0 Q_0.$$

Here, ρ_0 denotes the normal density, which is the density under *normal conditions*. Both can be obtained from the equation of state for real gases (2.20), and hence ρ_0 is simply a constant, at least if mixing effects are ignored.

2.3.1.3 • Further effects

We remark that, as it is the case with virtually any mathematical model for complex physical phenomena, not all practically relevant effects can be taken into account. For example, unexpected pressure losses can be caused by leakages, which mostly occur due to corrosion. To check for leaks, to clean the pipes, and for measuring purposes, gas transport companies make use of so-called *"pipeline pigs."* These are sets of tools that are installed within the pipelines. For further details on this topic, see, e.g., Kennedy (1993).

2.3.2 • Resistors

In addition to the pressure loss resulting from friction of the flow through the pipes, there are certain gas properties and network components that also induce a pressure loss, which has to be accounted for. Causes for such pressure losses are, e.g., flow diversion and turbulence in shaped pieces, measurement devices, curvature of the piping within compressor stations and pressure regulators, filter systems, reduced radii, and partially closed valves. These effects are often quite complicated, and no accurate models are available for most of them. Resistors are a surrogate modeling tool used for representing these forms of pressure loss.

Resistors influence pressure in the same way for both directions of the flow. Thus, p_in and p_out in the following formulas refer to the pressure at the inlet and the outlet node of the resistor, which depend of the flow direction. Similarly, v_in refers to the gas velocity at the inlet.

There are two forms of resistors being used. In the first form a resistor causes a nonlinear pressure loss according to a type of Darcy–Weisbach formula (Finnemore and Franzini (2002); Lurie (2008)) with parameters ζ and D. The pressure loss depends on the drag factor ζ, the density of the entering gas ρ_in, and its velocity v_in:

$$p_\text{in} - p_\text{out} = \frac{1}{2}\zeta \rho_\text{in} v_\text{in}^2, \tag{2.30}$$

where the velocity is computed from the mass flow q and the resistor's fictitious cross-sectional area, $A = D^2 \pi/4$, as follows:

$$v_\text{in} = \frac{q}{A\rho_\text{in}}. \tag{2.31}$$

The corresponding parameters have to be fitted to measurements of the actual pressure loss. This Darcy–Weisbach form can be used if there are sufficient measurement data for the affected region. Otherwise, a simpler form of resistor model is used, based on an estimate of the pressure loss.

In this second form, resistors incur a fixed pressure loss ξ in the flow direction:

$$p_\text{in} - p_\text{out} = \text{sgn}(q)\xi, \tag{2.32}$$

2.3. Gas network elements

Figure 2.2. *A ball valve. (Source: OGE.)*

where

$$\operatorname{sgn}(x) = \begin{cases} 1 & \text{if } x > 0, \\ 0 & \text{if } x = 0, \\ -1 & \text{if } x < 0. \end{cases}$$

Due to the Joule–Thomson effect (see Section 2.2), the temperature of the gas decreases when passing through a resistor.

2.3.3 • Valves

Valves are active network elements that can be controlled by the network operators. From our point of view, they can be closed or open. In practice, valves can also be *partially closed* in order to control the gas velocity. In this case, we model the partly closed valve as a resistor (see Section 2.3.2).

Valves are used to route the gas flow for parts of the transport network or to block gas flow for maintenance in subnetworks. In addition, they are frequently used in compressor or control valve stations for inner station piping (see Section 2.4.1 and Section 2.4.3). Depending on the concrete valve type, the switching between these states is technically realized in different ways. Most of the valves in the transport networks under consideration are *gate* or *ball valves*. The opening and closing of gate valves is simply realized by raising or lowering the gate wall. If the gate wall is raised, the gas flows through the element, while a closed valve blocks the complete gas flow. Ball valves control the gas flow with a ball that has a centered cylindrical hole; see Figure 2.2 for an illustration. If the hole of the ball is in line with the ends of the valve, the gas flows through the element. If the hole is rotated by 90°, the valve blocks the gas flow. In practice, only balls with restricted diameters are assembled. Because the used valves have to fit the ambient pipes, this maximum diameter is an upper bound for pipe diameters, too.

For the following, consider a valve with mass flow q and respective inlet and outlet pressures p_{in} and p_{out}. Open valves lead to identical values of the gas state quantities pressure p, temperature T, and density ρ:

$$p_{in} = p_{out}, \qquad (2.33)$$
$$\rho_{in} = \rho_{out}, \qquad (2.34)$$
$$T_{in} = T_{out}. \qquad (2.35)$$

Because valves are elements with negligible length, there is only insignificant friction that we neglect in (2.33)–(2.35).

Closed valves prevent gas flow, yielding decoupled gas states at both sides of the valve:

$$q = 0, \quad (p_{in}, T_{in}), (p_{out}, T_{out}) \text{ decoupled.} \qquad (2.36)$$

2.3.4 ▪ Control valves

Larger transport pipes are usually operated at higher pressure than pipes in the distribution parts of the network, which have a smaller diameter and smaller nominal pressure. In order to interconnect network parts operated at higher pressure with those operated at lower pressure and also to add a means of control of the flow, *control valves*, also known as *pressure regulators*, are used. There are two types of control valves: with remote access (automated) and without remote access (nonautomated).

Control valves can be *closed*, so that there is no flow and the inlet and outlet pressures are decoupled. If the control valve is *active*, the pressure at its outlet can be reduced to a given controllable value. In this case, control valves have a fixed working direction.

The pressure reduction is accomplished by a variable valve, being capable of restricting the flow. Its degree of opening, and thus the amount of flow through it, is usually controlled by a diaphragm actuator in combination with a compression spring. The outlet pressure can be controlled by changing the compression of the spring via a screw, thus adjusting the force that the spring exerts to the diaphragm actuator. For automated control valves, the change of the force of the spring can be remotely controlled by the dispatcher by an engine that is attached to the handle of the spring. The active/closed state and the controlled output pressure for control valves with remote access thus can be freely adjusted to the load requirements of the network. In network operation planning, the outlet pressure can be considered a control variable in this case.

For nonautomated control valves (a remote access device is not present) only a manual change is possible, requiring a person to be sent to the control valve. While the downstream pressure of control valves with remote access can be controlled directly, control valves without remote access are designed to reduce the downstream pressure to a preset pressure. In this case, the preset pressure of the control valve has to be considered fixed for short-term planning purposes. This is accomplished by the device under the conditions that the upstream pressure is at least as high as the preset value and that the downstream pressure is not greater than this value. If the downstream pressure rises above this threshold, the control valve without remote access closes automatically. If the downstream pressure is less than or equal to the threshold and the upstream pressure drops below this threshold value, the control valve without remote access opens fully and is *in bypass*. In bypass, the pressure of the gas is not affected and a flow in the opposite direction is possible. It is possible to have a bypass state for automated control valves as well; this is usually modeled on the level of control valve stations; see Section 2.4.1.

For a more detailed description of the construction of control valves and the corresponding legal requirements see Cerbe (2008).

For both types of control valves, the gas temperature drops as a consequence of the Joule–Thomson effect (see Section 2.2) due to the expansion of the gas when passing through a control valve. If the pressure change and thus the temperature drop is substantial, gas hydrates might fall out or the instrument could even freeze. For this reason, a control valve in most cases is combined with a gas *preheater*; see Section 2.4.1. Preheaters use the gas, e.g., by catalytic chemical reactions, to increase the temperature of the gas before entering the variable valve. If such a device is present, the outlet temperature of the control valve is determined by the operation of the preheater. Otherwise, the temperature change follows Eq. (2.6).

Control valves may incur a restriction on the flow rate if they are combined with measurement devices. In order to express the pressure loss due to filtering and innerstation piping, up- and downstream resistors (see Section 2.3.2) may be used to model these effects; see Section 2.4.1.

2.3.4.1 • Control valves with remote access

The different states of control valves with remote access have the following consequences. Let q be the flow through a control valve and p_{in} and p_{out}, respectively, the inlet and outlet pressures. If the control valve is *closed*, flow is blocked and pressures are decoupled:

$$q = 0, \quad p_{in}, p_{out} \text{ decoupled}.$$

If the control valve is *active*, it is capable of reducing the pressure by an amount within the range $[\underline{\Delta}, \overline{\Delta}]$, leading to the following model:

$$0 \leq \underline{\Delta} \leq p_{in} - p_{out} \leq \overline{\Delta}, \quad q \geq 0. \tag{2.37}$$

2.3.4.2 • Control valves without remote access

For a control valve without remote access and a preset downstream pressure p_a^{set}, the following relations hold:

$$
\begin{aligned}
p_{out} > p_a^{set} &\implies q = 0, \quad p_{in} \text{ arbitrary}, &&(\textit{closed}) \\
p_{out} > p_{in} &\implies q = 0, &&(\textit{closed}) \\
p_{in} > p_a^{set} \text{ and } p_{out} \leq p_a^{set} &\implies q \geq 0, \quad p_{out} = p_a^{set}, &&(\textit{active}) \\
p_{out} \leq p_{in} \leq p_a^{set} &\implies q \text{ arbitrary}, \quad p_{out} = p_{in}. &&(\textit{bypass})
\end{aligned}
$$

2.3.5 • Compressor machines and drives

Compressor machines are among the most important and complex elements in gas transport networks. They are used to increase the pressure of the incoming gas to a higher outflow pressure. Thus, compressor machines satisfy the need to overcome pressure loss caused by friction in pipes (see Section 2.3.1) and to transport gas over long distances. We sometimes also simply use the term compressor for a compressor machine. Every compressor machine has an associated drive (see Figure 2.3). It is possible that more than one compressor is powered by the same drive. In present-day gas transport networks, one mainly finds *turbo compressors* and *piston compressors* in combination with one of four drive types:

▷ gas turbines,
▷ gas driven motors,

Figure 2.3. *A compressor machine. (Source: Schmidt, Steinbach, and Willert (2014).)*

▷ electric motors, and
▷ steam turbines.

These will be described in Section 2.3.5.4.

2.3.5.1 ▪ Compressor machines

Compressor machines admit certain feasible combinations of *throughput* (measured in volumetric flow Q as defined below by Eq. (2.38)) and *specific change in adiabatic enthalpy* H_{ad} (as derived below by Eq. (2.42)). The set of all possible combinations of throughput and specific change in adiabatic enthalpy is called the *feasible operating range* of the machine. The volumetric flow depends on the mass flow q through the machine and the inflow gas density ρ_{in}:

$$q = Q\rho_{\text{in}}. \tag{2.38}$$

In order to derive the specific change in adiabatic enthalpy, we have to discuss the physical process of compression by a compressor machine. To this end, we need some fundamental thermodynamical quantities. Consider a certain quantity of gas that undergoes some thermodynamical process. The *enthalpy* $\tilde{H}$ of such a physical system is a measure of its total energy, composed of *internal energy* U and the *work* $W = pV$. In differential form, enthalpy is defined by

$$d\tilde{H} = dU + d(pV) = dU + V\,dp + p\,dV;$$

see, e.g., Tahir-Kheli (2012). By the first law of thermodynamics (see again Tahir-Kheli (2012)) the change in internal energy during the process, ignoring chemical reactions, is

$$dU = d\tilde{Q} - p\,dV,$$

where $\tilde{Q}$ denotes the *heat exchange*, i.e., energy transferred between the system and its surroundings. Thus, the change in enthalpy can be written as

$$d\tilde{H} = d\tilde{Q} + V\,dp.$$

If the compression process is done thermally isolated from the surrounding, no external heat exchange can occur. Such a process with $d\tilde{Q} = 0$ is called *adiabatic*. Under the assumptions of an adiabatic compression process, the change in enthalpy reduces to

$$d\tilde{H}_{\text{ad}} = V\,dp,$$

where the subscript "ad" indicates an adiabatic compression process. Consequently, the change in enthalpy for an adiabatic compression process with inlet pressure p_{in} and outlet pressure p_{out} is

$$\tilde{H}_{ad} = \int_{p_{in}}^{p_{out}} V \, dp. \tag{2.39}$$

According to Eq. (2.39), the change in adiabatic enthalpy that is required to increase the inlet pressure p_{in} of the gas to an outlet pressure p_{out} is equivalent to the work done on the system for this change in pressure.

In the following, let V_{in} and V_{out} denote the inlet and outlet volumes of the quantity of gas under consideration. Since the compression is adiabatic and reversible according to the second law of thermodynamics, it is *isentropic*. For an isentropic compression process one can derive the relationship

$$p_{in} V_{in}^{\varkappa} = p_{out} V_{out}^{\varkappa}, \tag{2.40}$$

using again the first and second laws of thermodynamics and the assumption of an ideal gas Tahir-Kheli (2012). The constant $\varkappa \neq 0$ is called *isentropic exponent*; a suitable choice for real gases will be discussed below. Solving Eq. (2.40) for V_{out} yields

$$V_{out} = \left(\frac{p_{in}}{p_{out}}\right)^{\frac{1}{\varkappa}} V_{in}. \tag{2.41}$$

Since this equation is valid for arbitrary p_{out} and V_{out}, we can apply Eq. (2.41) with $p_{out} = p$ and $V_{out} = V$ to Eq. (2.39):

$$\begin{aligned}
\tilde{H}_{ad} &= \int_{p_{in}}^{p_{out}} \left(\frac{p_{in}}{p}\right)^{\frac{1}{\varkappa}} V_{in} \, dp \\
&= p_{in}^{\frac{1}{\varkappa}} V_{in} \int_{p_{in}}^{p_{out}} \frac{1}{p^{\frac{1}{\varkappa}}} \, dp \\
&= p_{in}^{\frac{1}{\varkappa}} V_{in} \frac{\varkappa}{\varkappa - 1} \left[p^{\frac{\varkappa-1}{\varkappa}}\right]_{p_{in}}^{p_{out}} \\
&= p_{in} V_{in} \frac{\varkappa}{\varkappa - 1} \left[\left(\frac{p_{out}}{p_{in}}\right)^{\frac{\varkappa-1}{\varkappa}} - 1\right].
\end{aligned}$$

Now we can apply the thermodynamical standard equation for real gases (2.3) to eliminate $p_{in} V_{in}$ and obtain

$$\tilde{H}_{ad} = \tilde{n} R T_{in} z_{in} \frac{\varkappa}{\varkappa - 1} \left[\left(\frac{p_{out}}{p_{in}}\right)^{\frac{\varkappa-1}{\varkappa}} - 1\right].$$

Here, the compressibility factor is denoted by z_{in}, and T_{in} is the gas temperature at the inlet node of the compressor.

The (adiabatic) enthalpy $\tilde{H}$ ($\tilde{H}_{ad}$) as derived above is proportional to the size of the thermodynamical system; more specifically, to the amount of substance $\tilde{n}$. It is convenient to introduce a *specific enthalpy* $H := \tilde{H}/M$, which is independent of the amount of substance $\tilde{n}$ and independent of the mass M. Finally, with the relation $M = \tilde{n}m$, one

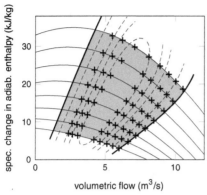

(a) Characteristic diagram of a turbo compressor. Specific change in adiabatic enthalpy H_{ad} vs. volumetric flow rate Q: Dashed lines represent isolines for adiabatic efficiency η_{ad}, thin solid lines represent isolines for compressor speed n. The left thick solid line represents the surgeline, the right thick solid line represents the chokeline. All curves are the result of least-squares fits with respect to the measurements "+".

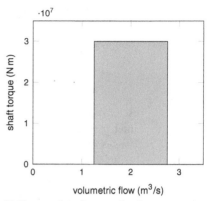

(b) Characteristic diagram of a piston compressor.

Figure 2.4. *Characteristic diagrams of compressors (feasible operating ranges are marked gray). (Source: OGE.)*

obtains the fundamental formula for the specific change in adiabatic enthalpy of a compression process:

$$H_{\text{ad}} := \frac{\tilde{H}_{\text{ad}}}{M} = R_s\, T_{\text{in}}\, z_{\text{in}}\, \frac{\varkappa}{\varkappa - 1} \left[\left(\frac{p_{\text{out}}}{p_{\text{in}}} \right)^{\frac{\varkappa - 1}{\varkappa}} - 1 \right], \qquad (2.42)$$

where $R_s = R/m$ is the specific gas constant as introduced previously. The derived unit of specific change in adiabatic enthalpy H_{ad} is J/kg. Note that the *adiabatic head* (measured in m), frequently also abbreviated as H_{ad} in the literature, is H_{ad}/g.

The isentropic exponent $\varkappa$ used above in fact depends on gas pressure and temperature. Several approximations of $\varkappa$ exist that differ in complexity and accuracy; see Section 10.1.10.2 and Schmidt, Steinbach, and Willert (2014) for a more detailed discussion. Except for Chapter 10, we will approximate the isentropic exponent by the constant 1.296 in this book.

To realize different combinations of throughput and specific change of adiabatic enthalpy within the feasible operating range, compressors can be operated at different speeds n influencing both specific change in adiabatic enthalpy and throughput. It is common in gas engineering to visualize feasible operating ranges in so-called *characteristic diagrams* (see, e.g., Odom and Muster (2009); Percell and Ryan (1987). Depending on the machine type (e.g., turbo compressor or piston compressor), the characteristic diagram may also depend on the *adiabatic efficiency* η_{ad} of the machine, i.e., the quotient of conducted and emitted power, as we explain next; see Figure 2.4 for examples.

The *power P* required for compression depends on the amount of compressed gas (mass flow q), the realized specific change in adiabatic enthalpy H_{ad}, and the *adiabatic efficiency* $\eta_{\text{ad}} \in [0,1]$ of the compression process:

$$P = \frac{q H_{\text{ad}}}{\eta_{\text{ad}}}. \qquad (2.43)$$

Here, $q H_{\text{ad}}$ is the theoretical power required by the adiabatic compression process, and η_{ad} is an estimate used to describe the deviation from the actual power P (e.g., according to mechanical losses) for a real machine.

Together with the specific *energy consumption rate* b of the corresponding drive (see below), the power required for compression determines the amount of energy that is consumed by the machine. Depending on the corresponding drive, the consumed power is either electricity or gas from the network used as fuel. Electric energy is delivered by specific electric motors, whereas gas is transformed to mechanical energy by gas turbines or gas driven motors (see Section 2.3.5.4 for a detailed description of drives). The amount of electric energy is directly given by the specific energy consumption rate b. The amount of fuel gas consumption is given by

$$q_{\text{fuel}} = \frac{b\, m}{H_{\text{u}}}, \qquad (2.44)$$

where H_{u} is the *lower calorific value* of the gas (see Cerbe (2008)). As before, m denotes the molar mass of the gas.

The compression of the gas leads to an increase of the gas temperature. To prevent overheating, most of the compressor groups (see Section 2.4.2) contain a *gas cooler* that decreases the outflow temperature if a threshold is exceeded. This threshold mostly depends on the heat resistance of the internal coating of the pipes. Standard thresholds are about 35 °C to 50 °C. Because the gas temperature increase depends on the chosen operation mode of the group, the heat resistance may exclude some modes from the principally possible ones.

At turbo and piston compressors, two different technical processes increase the gas pressure. Turbo compressors add energy to the gas by a rotating edge runner. Here, the conducted energy depends on the rotational speed of the runner. In contrast to that, the gas gets compressed by a crankshaft in piston compressors. Turbo compressors are capable of compressing larger amounts of gas, while piston compressors achieve higher compression ratios for smaller amounts of gas. Since piston compressors can be operated for very long time periods, they are typically used as storage compressors. Typical values of the pressure ratio realized by turbo compressors vary between 1.35 to 1.5, whereas piston compressors can achieve values up to 4. Typical maximum power values may vary between 5 MW and 25 MW.

Finally, we remark that it is possible to operate a compressor beyond the left boundary of its characteristic diagram. This is called *pump prevention*, since it is realized by the reinsertion of compressed gas in order to keep a certain level of flow through the compressor. For this exceptional mode of operation, gas coolers are essential to prevent overheating.

2.3.5.2 • Turbo compressors

An exemplary characteristic diagram of a turbo compressor can be seen in Figure 2.4(a). We follow the standard technique to model all curves in the characteristic diagram of a turbo compressor by quadratic polynomials (see Odom and Muster (2009)),

$$\psi(x; a) = a_0 + a_1 x + a_2 x^2, \qquad (2.45)$$

or biquadratic polynomials,

$$\chi(x,y;A) = \begin{pmatrix} 1 \\ x \\ x^2 \end{pmatrix}^T \begin{pmatrix} a_{00} & a_{01} & a_{02} \\ a_{10} & a_{11} & a_{12} \\ a_{20} & a_{21} & a_{22} \end{pmatrix} \begin{pmatrix} 1 \\ y \\ y^2 \end{pmatrix}. \tag{2.46}$$

The coefficients $a \in \mathbb{R}^3$ and $A \in \mathbb{R}^{3\times3}$ are obtained from least-squares–based data fits for given technical measurements of the compressor machine. In the following, a superscript term on a or A specifies the coefficient vector or matrix of the quantity to which it corresponds. The *isolines* of compressor speed $n \in [\underline{n}, \overline{n}]$ are given by the implicit equation

$$H_{\text{ad}} = \chi(Q, n; A^{\text{speed}}). \tag{2.47}$$

They determine the lower boundary (with minimum compressor speed $\underline{n}$) and upper boundary (with maximum compressor speed $\overline{n}$) in Figure 2.4(a). The isolines of adiabatic efficiency,

$$\eta_{\text{ad}} = \chi(Q, n; A^{\text{eff}}), \tag{2.48}$$

determine the power required to realize a given working point of the compressor (see (2.43)). The left and right boundaries of the characteristic diagram are given by the *surgeline* and *chokeline*, respectively,

$$\psi(Q; a^{\text{surge}}) \geq H_{\text{ad}}, \quad \psi(Q; a^{\text{choke}}) \leq H_{\text{ad}}. \tag{2.49}$$

2.3.5.3 ▪ Piston compressors

Piston compressors are characterized by box-shaped feasible operating ranges in the coordinates volumetric flow rate Q and *shaft torque M*; see Figure 2.4(b) for an illustration. The shaft torque is defined as

$$M := \frac{V_{\text{o}} H_{\text{ad}}}{2\pi \eta_{\text{ad}}} \rho_{\text{in}}. \tag{2.50}$$

Here, V_{o} is the operating volume of the piston compressor. The volumetric flow rate is proportional to the speed of the machine,

$$Q = V_{\text{o}} n, \quad n \in [\underline{n}, \overline{n}].$$

Depending on the technical specification of the piston compressor, the *maximum torque* is (implicitly) given by one of the following three restrictions:

$$\frac{p_{\text{out}}}{p_{\text{in}}} \leq \overline{\varepsilon}, \quad p_{\text{out}} - p_{\text{in}} \leq \overline{\Delta p}, \quad \text{or} \quad M \leq \overline{M}.$$

Here, $\overline{\varepsilon}$ and $\overline{\Delta p}$ are the maximal pressure ratio and the maximal pressure increase, respectively. Note that p_{in} and p_{out} are connected to M via H_{ad} (see (2.50)).

2.3.5.4 ▪ Drives

Drives deliver the energy required for the compression realized by compressor machines. We distinguish drives that produce energy from electricity or from gas. If the energy is produced from gas, the required amount of gas depends on the gas state and its calorific value (see (2.44)). The gas is taken from the inlet of the corresponding compressor and, thus, has the pressure of the incoming flow.

2.4. Gas network structures

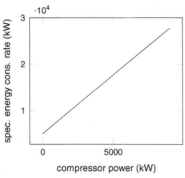
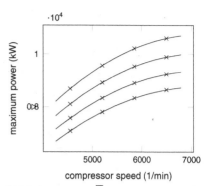

(a) Specific energy consumption rate b vs. the compressor power P.

(b) Maximum power $\overline{P}$ vs. compressor speed n for different ambient temperatures T_{amb}.

Figure 2.5. *Characteristic diagrams of a gas turbine. (Source: Schmidt, Steinbach, and Willert (2014).)*

As for compressors, we follow the common approach in industry and model all physical and technical relationships with quadratic or biquadratic functions (see LIWACOM (2004)). Every drive has a specific *energy consumption rate*

$$b(P) = \psi(P; a^{\text{energy}}) \tag{2.51}$$

and a *maximum power* $\overline{P}$ that can be delivered for the compression process. Depending on the concrete drive type, the latter may be a function of compressor speed and the ambient temperature T_{amb} at the compressor,

$$\overline{P}(n, T_{\text{amb}}) = \chi(n, T_{\text{amb}}; A^{\text{max-power}}), \tag{2.52}$$

or independent of the ambient temperature, i.e.,

$$\overline{P}(n) = \psi(n; a^{\text{max-power}}). \tag{2.53}$$

This value $\overline{P}$ restricts the power available for compression,

$$P \leq \overline{P}.$$

As for characteristic diagrams of turbo compressors, the coefficients in (2.51), (2.52), and (2.53) are obtained from least-squares data fits.

Figure 2.5 shows typical plots of the specific energy consumption rate function b and the maximum power function $\overline{P}$ for a gas turbine. Here, the maximum power function depends on the ambient temperature T_{amb}, too, leading to different maximum power functions for different values of T_{amb}. Figure 2.6 shows an example for a gas turbine as used in drives.

2.4 ▪ Gas network structures

In real-world gas transport networks, the basic network elements described in the last section are combined in complex ways to build a flexible network structure that may support various transport situations. It is thus appropriate to introduce further modeling elements that allow us to express the high-level structure of groups of basic network elements. These high-level modeling elements are discussed in this section. This includes

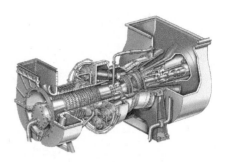

(a) Example of a gas turbine. (Source: OGE.)

(b) Schematic plot of a gas turbine. (Source: Siemens AG.) © Siemens AG

Figure 2.6. *Gas turbines.*

Table 2.1. *Technical symbols of gas network elements.*

Network element	Symbol
Resistor	
Valve	
Control valve	
Compressor (group)	

control valve stations (Section 2.4.1), compressor groups (Section 2.4.2), and compressor stations (Section 2.4.3).

Other common substructures are *loops*, which are parallel pipes. They are usually built with connecting valves at regular distances, enabling different ways of operating them. They may be used as two parallel pipes at the same conditions (i.e., pressure levels) by opening all the valves, or independently at different conditions by closing all the valves. By closing a subset of the valves, it is even possible to use parts of a loop in both directions. In contrast to compressor stations, we do not have a high-level model for loops. Instead, we model them directly by using the corresponding network structure made up of pipes and valves.

For a graphical representation of the network elements in the following, we use the symbols in Table 2.1.

2.4.1 ▪ Control valve stations

As mentioned in Section 2.3.4, control valves are usually combined with preheaters. Moreover, valves might be used to control flow and allow a bypass of the control valve. To represent elaborate piping in connection with the control valve, resistors can be added as well. This yields a subnetwork, which we call a *control valve station*; see Figure 2.7 for an example.

Note that the concrete layout may depend on the underlying mathematical optimization model and network design. For instance, Figure 2.7 does not explicitly contain a preheater, in contrast to the modeling of Chapter 10 in Figure 10.4; the latter, however, does not contain an explicit bypass. Consequently, control valve stations represent an

2.4. Gas network structures

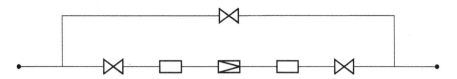

Figure 2.7. *Diagram of a control valve station.*

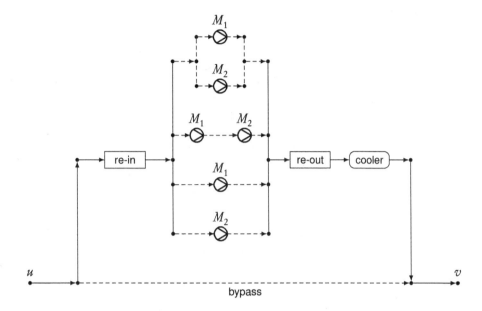

Figure 2.8. *A schematic plot of a compressor group with two machines, all possible configurations, and the corresponding bypass mode. The gas flow follows at most one of the dashed routes, i.e., either the bypass without any pressure loss, or one of the compressor configurations, in which case there is some pressure loss due to piping which is modeled by additional resistors.*

abstract modeling component that contains at least one control valve and additional network elements such as valves, preheaters, and resistors.

2.4.2 • Compressor groups and configurations

In the following two sections, we describe subnetworks that contain compressors. The literature often uses the term compressors or compressor stations. These terms are sometimes used synonymously and sometimes not. In this section, we describe what we mean by a compressor group and later describe how we build a second layer of aggregation—what we then call a compressor station.

Small groups of compressor machines and drives are often used together in such a way that the gas enters through a single pipe, is routed through some of the compressor machines, and leaves via a single pipe. To model this, we introduce so-called compressor groups; see Figure 2.8. These entities encapsulate a set of compressor machines (together with the corresponding drives) that can be operated in different predefined ways. In practice it is often the case that one machine per group is specified as a so-called *backup compressor*. This compressor is not used in planning scenarios and is only specified for safeguarding against failures. In our categorization, compressor groups are active elements,

Figure 2.9. *The compressor station in Waidhaus, Germany. (Source: OGE.)*

i.e., network operators can control the *operation mode* of the groups. Technically possible modes are *active*, *closed*, and *bypass*, similar to control valves. We first describe the closed and bypass modes and then concentrate on the more complicated active mode.

If a compressor group is closed, it behaves like a closed valve. Thus, the gas flow is zero, and inflow and outflow gas states (i.e., gas pressure, temperature, and density) are decoupled. In bypass mode, the gas flows *around* the group and is therefore not affected by any part of the group. Since the group piping that bypasses the machines is very short, no significant friction effects occur, similar to (control) valves. In what follows, we neglect these insignificant friction effects completely and assume that inflow and outflow gas states are equal (see also Section 2.3.3).

If the compressor group is active, internal parallel or serial combinations of active compressors can be chosen. As a rule of thumb, parallel arrangements are capable of compressing a larger amount of gas, whereas serial combinations yield higher compression ratios. Due to technical limitations, not every arrangement of this type is possible, but the network operators can choose a finite set of arrangements, called *configurations*. In contrast to the bypass mode, the inner group piping may lead to a significant pressure drop, which we model by additional up- and downstream resistors as for control valves (see Section 2.3.4).

2.4.3 • Compressor stations and subnetwork operation modes

Frequently, collections of compressor machines and valves are connected to more than two pipelines and may be used in various ways to route gas from some of those pipelines to other ones, increasing or even decreasing the pressure as necessary. Compressor groups are not sufficient to model such complex structures. Instead, for each such *compressor station*, an explicit subnetwork that reflects all possible routes of gas through the compressor station is required. Compressor stations are often located at intersections of several pipelines, and they are also used to route the gas between the connected pipeline systems; see Figure 2.9 for an example.

2.4. Gas network structures

Similar to compressor groups, compressor stations internally allow for multiple paths that the flow of gas can actually take. The desired path is again selected by switching a cascade of individual valves, compressor groups, and (sometimes) control valves in the right way. This is carried out by human dispatchers, who are in charge of controlling the network's active elements. Their objective is to maintain a feasible flow of gas that ideally requires a minimum amount of energy, mainly consumed by active compressors. For this, valves can be opened or closed. Compressor groups and control valves can be activated, bypassed, or closed. If they are active, the level of operation and the configuration has to be chosen.

As we have already discussed for compressor groups in Section 2.4.2, some potentially possible switching combinations are not technically possible or not practically meaningful. In fact, only a very small subset of states might be relevant for operational purposes. The relevant states of the subnetwork can be represented by either an *inner* or an *outer* description. An inner description is a finite list of all allowed states for the controllable network elements of the subnetwork, where for each state of the subnetwork and each controllable element it is specified whether this element is either open (or active) or closed (or inactive) in this state. If the element is open, then additional bounds on the flow can optionally be specified. This option is typically used to specify the flow direction. An outer description is a list of linear constraints that describe the interdependency of active elements, for example, "if element A is open, then B is open, and C is closed," or "one of the elements A, B, C must be open, but not all three together."

Both descriptions are used in practice. From a polyhedral (although not from an algorithmical) point of view they are in fact equivalent: It is always possible to determine the outer description, if the inner is given, and vice versa. In fact, the inner description specifies a finite set of points (control decisions) $s^1, \ldots, s^k \in \{0,1\}^n$, where $s^i_j = 0$ represents the fact that the controllable network element j is closed in state i, and $s^i_j = 1$ represents that it is open (or active). The convex hull of these points, $Q = \text{conv}\{s^1, \ldots, s^k\}$, is a bounded polyhedron, i.e., a polytope. The theorem of Minkowski and Weyl states that Q can alternatively be described as $\{x \in \mathbb{R}^n \mid Ax \leq b\}$ for some $A \in \mathbb{R}^{m \times n}$ and $b \in \mathbb{R}^m$, i.e., by finitely many linear inequalities given by the rows of this system; see, e.g., Nemhauser and Wolsey (1988); Ziegler (1994). This second form gives an outer description by linear constraints.

Note that the two descriptions can drastically differ in size. For example, the standard cube $[0,1]^n$ is the convex hull of 2^n points (and not fewer), but can be described by $2n$ linear inequalities. A notoriously open question is whether there exists an algorithm that converts between the two descriptions in time that is polynomial in the combined size of the input and output data; see, e.g., Avis, Bremner, and Seidel (1997).

Note that an outer description $Ax \leq b$, specified by rules on active elements as mentioned above, is usually *not* tight in the sense that in general

$$\text{conv}\{x \in \{0,1\}^n \mid Ax \leq b\} \subsetneq \{x \in \mathbb{R}^n \mid Ax \leq b\}.$$

However, since the set on the left-hand side is obviously a polytope, there exists a (tight) linear description of it. This tighter formulation often has computational advantages, when available.

The conclusion is that active elements no longer appear individually, but are combined in blocks, and the decision on one element of such a block affects several other elements belonging to the same block. In this sense, the operational decisions (modes) on the state of the elements of the subnetwork are combined, and we thus speak of *subnetwork operation modes*. Figure 2.10 shows four examples of different subnetwork operation modes;

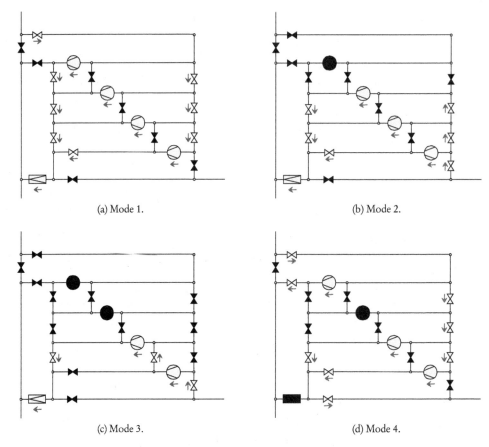

Figure 2.10. *Four different subnetwork operation modes for a large compressor station. Each of them leads to a different flow of the gas through the station—elements colored dark are "closed." Mode 1: Gas flow from north to south, quad parallel compression. Mode 2: Flow from east to south, triple parallel compression. Mode 3: Flow from east to south, double serial compression. Mode 4: Flow from north to east with double parallel and to south with single compression.*

see Table 2.1 for the symbols representing the different components. Each of the four modes leads to a different flow of the gas through the station.

We remark that subnetwork operation modes are not only used to model feasible states of compressor stations, but also other subnetworks that occur outside of physical stations. For example, groups of valves that are scattered over a large distance along parallel pipelines with joints can act as a subnetwork. This allows the operator to use these parallel pipelines as loops over a certain distance.

The operational knowledge in addition to the network's physical structure needs to be encoded in subnetwork operation modes, so that it becomes a useful representation of reality and it can be used in the models of the subsequent chapters.

2.5 • Gas network representation

The different elements, groups, and stations as discussed in the previous sections appear as elements in gas networks. In this section, we introduce the representation of a gas network in terms of a directed finite graph $G = (V, A)$ with nodes V and arcs A. This graph

2.5. Gas network representation

representation will be used throughout the rest of this book and provides a basis for the derivation of the mathematical models for the problem of the validation of nominations in Chapters 6–10. Note that G will not contain self-loops, but parallel arcs might occur.

As a graph general notation, we use $\delta^-(u) := \{a = (v,u) \in A\}$ and $\delta^+(u) := \{a = (u,v) \in A\}$ for the set of incoming and outgoing arcs of $u \in V$, respectively. The set of all incident arcs is $\delta(u) := \delta^-(u) \cup \delta^+(u)$.

The set of entries is denoted by V_+ and the set of exits is denoted by V_-. All other nodes, i.e., junctions of network elements that are neither exits nor entries, are collected in the set V_0. Thus, the node set of a directed graph of a gas network is the disjoint union of these three sets, i.e., $V = V_+ \dot\cup V_- \dot\cup V_0$; see also Section 2.1.

The set of arcs $A = A_{\text{passive}} \dot\cup A_{\text{active}}$ consists of the set A_{passive} of arcs representing passive network elements and the set A_{active} of arcs representing active network elements. Pipes (Section 2.3.1) and resistors (Section 2.3.2) are passive network elements. The set of pipes is denoted by A_{pi}, and A_{rs} denotes the set of resistors. The set of resistors is further subdivided into the set $A_{\text{nl-rs}}$ of resistors causing a nonlinear pressure drop and the set $A_{\text{lin-rs}}$ of resistors causing a fixed pressure reduction in flow direction, i.e., $A_{\text{rs}} = A_{\text{nl-rs}} \dot\cup A_{\text{lin-rs}}$. Another type of passive network element, that is, only used for modeling purposes and thus not mentioned so far, is a so-called *short cut*. Short cuts can be thought of as very short pipes not causing any pressure reduction. We denote the set of short cuts by A_{sc}. For further details on short cuts we refer the reader to Section 5.1.3. Altogether, we have $A_{\text{passive}} = A_{\text{pi}} \dot\cup A_{\text{rs}} \dot\cup A_{\text{sc}}$.

The set of active network elements is made up of valves (Section 2.3.3), control valves (Section 2.3.4), and compressors (Section 2.3.5). Arcs representing valves are elements of the set A_{va}. The set of control valves is given by A_{cv}, and by A_{cm} we denote the set of compressor machines. Therefore, we have $A_{\text{active}} = A_{\text{va}} \dot\cup A_{\text{cv}} \dot\cup A_{\text{cm}}$. Finally, the set of control valves is further subdivided into the set $A_{\text{cv}}^{\text{aut}}$ of (automated) control valves equipped with a remote access device and the set $A_{\text{cv}}^{\text{man}}$ of manually operated (nonautomated) control valves. Thus, $A_{\text{cv}} = A_{\text{cv}}^{\text{aut}} \dot\cup A_{\text{cv}}^{\text{man}}$.

The mass flow q on an arc $a \in A$ is denoted by q_a. For the pressure p at a node $u \in V$, we write p_u. Other quantities associated to a certain arc or node of the networks are indexed by respective subscripts in the same way. The sign of q_a depends on the direction of the arc; i.e., if gas flows from node u to node v on an arc $a = (u,v)$, we have $q_a > 0$, and if gas flows from node v to node u, we have $q_a < 0$.

Chapter 3

Regulatory rules for gas markets in Germany and other European countries

Uwe Gotzes, Nina Heinecke, Benjamin Hiller,
Jessica Rövekamp, Thorsten Koch

Abstract *With a share of around one quarter of the primary energy consumption, natural gas is a crucial energy source for Europe and Germany. Thus it is of high economical and political relevance. The European Union and Germany have imposed standardized regulatory rules in order to establish a liberalized European internal gas market. These rules affect both gas trading and gas transportation: On the trading side, they concern the booking and use of transportation capacity rights. On the transportation side, network planning and capacity calculation are now regulated and the respective planning processes need to be adjusted substantially. This chapter describes the current rules in Germany and Europe with an emphasis on German regulation. In particular, we point out the impact of the regulatory rules for gas transportation.*

Gas is an important component of the European and German energy mix. Both Europe and Germany covered roughly one quarter of their primary energy consumption with gas in 2010 [Eurostat GIC]. In addition to the complex physical and technical conditions of gas transportation addressed in the preceding chapter, the European and German regulatory frameworks impose additional challenges on the gas industry. As for many other branches of business, Europe aims at an open internal market in the energy sector. Thus, the organization of gas transportation has undergone many changes in recent years and will have to adapt continuously in the future.

In this chapter, Section 3.1 gives an overview of the European regulation, and in particular the German regulation. In Sections 3.2 and 3.3 we detail the current gas market organization and regulation rules that are most relevant for this book, namely those concerning the usage of a gas network and the calculation of gas network capacities. Finally, we explain in Section 3.4 the challenges faced by network operators that result from physical and legal constraints, and conclude with an outlook. An overview on the various bodies and institutions involved in the European and German regulatory process is given in Section A.1 of the Appendix. Moreover, Section A.2 provides a chronological account of the relevant different ordinances, directives, guidelines, laws, and agreements affecting the German energy sector. For further understanding of regulation and unbundling in Europe and Germany, we refer the reader to [REP 2013] and [PWC 2012].

The main goal of this chapter is to give an account of the implications of the regulatory rules on the planning and operation of gas networks. This also deals with complex mathematical questions, which will be discussed in more detail in the following chapters. It will become apparent that the legal definitions and requirements given are not always applicable in a straightforward way. Thus, sometimes the regulatory framework needs interpretation, and requirements are realizable only with difficulty. In particular, it is nontrivial to come up with mathematically precise models and interpretations of the legal requirements.

3.1 ▪ Overview of gas market regulation in Europe and Germany

The transportation of gas is not only a technical task. Due to its economic importance it is also a focus of political considerations and thus subject to regulation. Before sketching the legal details, we briefly review some terminology used in the regulation context that we will also use in the following.

The transportation of gas is usually termed "gas transmission." However, "transmission" in a stricter sense refers to typically high-pressure interregional gas networks that connect import and export points with industrial consumers, public utilities, or regional networks. Gas transport in a regional network is called "distribution." Accordingly, the companies operating transmission or distribution networks are called transmission system operators (TSOs) and distribution system operators (DSOs), respectively. Thus, we use the general term "gas transport" whenever both transmission and distribution are implied and use the more specialized terms only when necessary. Likewise, we speak of "network operators" whenever rules apply to both TSOs and DSOs. In legal documents, transport customers, who use the gas network for transporting gas are often called *shippers*.

We start our overview with an outline of how the market was organized before the liberalization—this helps one to understand the motivation for the recent regulation measures.

3.1.1 ▪ Former market organization

Prior to advancing German gas liberalization in 2005, there were only a few German gas companies, all of which were involved in all business sectors of the gas value chain (see Figure 3.1); that is, import, exploration, exploitation, transport, storage and export, and trading.

In particular, the German gas companies were *gas traders* and *network operators* at the same time (see Figure 3.2). These gas companies held gas delivery contracts with international gas producing companies, mainly from Russia, Norway, the United Kingdom, and The Netherlands. Amounts of gas could normally be ordered with flexibility between minimal and maximal ranges and were provided at the German border (import). Typical contractual elements were minimum take and minimum payment clauses, which set the minimal amount of gas to be taken or the minimal payment to be made by the importer, respectively. Additionally, some of these companies operated production sites and explored indigenous gas. Moreover, most of them owned huge storage assets to react to supply and demand imbalances.

The gas was both provided to German consumers and exported to other European countries, usually situated in the southern and western parts of Europe like France or Austria. Gas to be exported was again provided at the German border and the trading partner usually had to organize further transport. Some consumers which were con-

3.1. Overview of gas market regulation in Europe and Germany

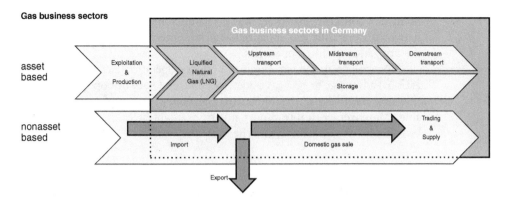

Figure 3.1. *Gas business sectors along the value chain, divided into asset-based and nonasset-based functions.*

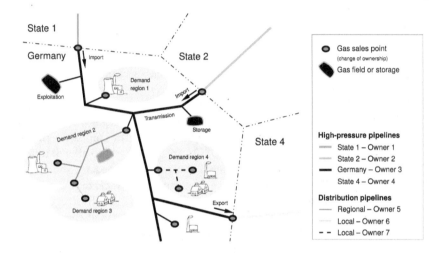

Figure 3.2. *Former gas market organization in Germany. Typically, all business activities, such as import, exploration, exploitation, transport, storage, export, and end-consumer supply, were run by a single company (depicted in black). This company transferred gas from import points, its own production sites, and storages to export points or exit points for the supply of either direct customers, power plants (supply region 1), or distribution companies (supply regions 2 and 4), which in turn supply end-customers (supply region 3).*

nected to the high-pressure network, such as gas-fired power plants and industrial plants, were provided with gas directly from the importing company. The rest of the gas was sold to connected distribution companies with low-pressure networks, which took over the ownership at the interconnection point. In the case of local distribution companies the gas was then sold to the end-consumers, for example private households. In the case of regional distribution companies, the buy-and-sell cascade was continued until the last local utility company bought the gas demanded by its consumers. The whole transport chain is depicted in Figure 3.2.

Often, utility companies and gas consumers had no alternative choice for obtaining gas, as they were usually connected to only one gas network (see Section A.2 regarding the German Energy Industry Act (EnWG) with demarcation and concession contracts). The transport path which the gas had to take from the entry point to an interconnection point or a consumer exit point was determined beforehand for planning purposes. The networks were built and enhanced regarding these paths with some flexibility between the utilization of different entries.

However, from the viewpoint of network operation and planning, it is a big organizational advantage to be a gas trader, supplier, and network operator at the same time. In this case, essentially all the information necessary to control networks efficiently is available. Moreover, critical transport situations might be easily handled or entirely avoided by suitable coordination within the company.

3.1.2 • Objectives and goals of liberalization

From the viewpoint of politics, this market organization was disadvantageous since the gas consumers had no choice of gas supplier. In particular, they depended on the given gas prizes. In order to change this, it was decided that a competitive EU-wide gas market was to be established, with rules ensuring easy handling of gas transport for gas shippers. Moreover, there should be low market entry barriers for new shippers.

To create and increase competition in the gas sector, regulation rules now require *unbundling*, i.e., separating the different business sectors along the gas value chain (see Figure 3.1) from each other. There are different forms of unbundling, reaching from the separation of business departments and software tools to the foundation of a new subsidiary. In particular, gas transport is separated from gas trading and shipping.

The market organization should be such that the companies that are now responsible for gas transport, i.e., TSOs and DSOs, are encouraged to offer as much transport capacity as possible to transport customers. This transport capacity should be easy to use for gas transport customers. Ideally, a gas transport network may be used as transparently as the Internet, without knowledge of the underlying infrastructure. Moreover, if more capacity is requested than is available (a case of so-called contractual congestion), TSOs should expand their networks such that the requested capacity becomes available.

Apart from increasing competition in the gas market, another very important goal is to ensure *security of supply*, which refers to *"safeguarding gas provision to consumers, guaranteeing the technical safety of the transmission network, ensuring a functioning internal market and being able to react with exceptional measures in case of disruption"* [Regulation (EU) No. 994/2010, Article 1]. Security of supply thus not only requires that gas be available somewhere, but also that it can be transported to the consumers. Hence TSOs should offer capacity and operate their networks such that safe and reliable transport is guaranteed.

Since unbundling purposely causes the scattering of information between TSOs, traders, suppliers, and regulatory authorities, there have to be mechanisms enabling a close cooperation to achieve these goals. In particular, all these stakeholders have to collaborate in order to ensure security of supply, since failure of any part in the gas value chain presents a risk that is harder to handle in an unbundled setting.

3.2 • Current rules for using gas transmission networks

Since the early days of liberalization, a complex set of rules has been established. In this section we discuss the ways in which transport customers may use today's gas transmission networks and the rules with which the TSOs have to comply.

3.2.1 ▪ Booking and using capacities

In order to transport gas through a gas transmission network, a transport customer has to *book capacity contracts*, i.e., buy the right to inject or withdraw gas within certain limits. Booking is usually performed by auctioning capacity contracts, which is explained in more detail in Section 3.2.3. Most capacity contracts are usually booked some time (years to weeks) before the actual transportation takes place. On the day before the transport starts, transport customers have to *nominate* the amount to which they want to use their capacity contracts with the respective TSO. This procedure enables the TSO to schedule the transport in advance. These initial nominations can be modified until two hours before the beginning of the transport due to necessary changes, e.g., in supply or demand. This process is called *renomination*. There are restrictions on how much the renomination may differ from the initial nomination. However, these apply only to transport customers having a certain market power. Effectively, renomination is usually not limited. In an extreme case, a transport customer who owns both entry and exit capacity contracts at a single point (e.g., a storage) might initially nominate on both contracts, leaving the final decision of the actual gas amount open until two hours before the transport. In addition, capacities without a nomination are auctioned as *day-ahead capacities*, which are offered one day before the actual transport takes place. Thus these unused capacities may be used to nominate further gas flows by the successful tenderer (see Section 3.2.3 for details). Figure 3.3 illustrates this daily process.

From the viewpoint of reliability, there are two types of capacity: firm and interruptible capacity. For *firm capacity*, the TSO has to guarantee that gas flows nominated on booked firm capacity contracts are always transportable if the network is in a fully functioning technical state. Thus, firm capacities have to be usable partially or completely, depending on the shipper's decision, in every flow situation of the network. In contrast, for *interruptible capacity*, the TSO is entitled to transport nominated gas flows only partially or not at all if this is required for technical feasibility. To do this, the TSO has to inform the gas transport customer about the amount of gas flow that will not be transported. However, the TSO has to take *best endeavors* to avoid an interruption.

Moreover, TSOs are obliged to determine the maximum capacity that can be provided on a firm basis, the so-called *technical capacity* (see [GasNZV 2010, §2 (13) and §9]), and to offer this capacity to the market. Firm capacities may not be interrupted, except in case of so-called *force majeure*, i.e., severe failures that cannot be anticipated and/or are beyond the TSO's control.

It may happen that transport customers require more firm capacity than can be offered by the TSO. This is called a *contractual congestion*. In general, the reason for a contractual congestion is that there is a limitation in the network infrastructure, known as *technical congestion*, that prohibits the TSO from offering more firm capacity. It may be possible to resolve short-term contractual congestions by special mechanisms as specified in so-called congestion management procedures. For instance, the auctioning of day-ahead capacity mentioned above is such a mechanism that makes booked, but unused, capacity available again.

3.2.2 ▪ Entry-exit model, market areas, and freely allocable capacities

We already mentioned that for an effective gas market it is important that gas transmission be easy to use, i.e., does not require a known transportation path and many corresponding contracts as in the old single-capacity booking model. Instead, the so-called *entry-exit model* was introduced in Germany by the German Energy Industry Act (EnWG) §20 (1b)

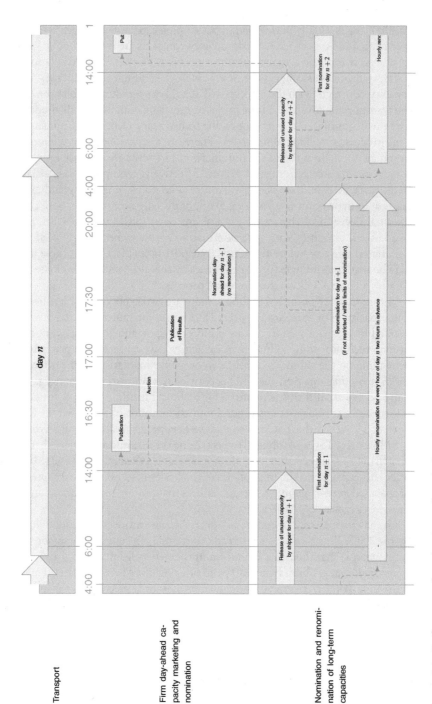

Figure 3.3. *Daily process of nomination and marketing. Gas transport customers have to notify the usage of their capacity contracts by nominations on the day before the transport starts. This enables the TSO to schedule the transport in advance. These initial nominations can be modified until two hours before the beginning of the transport. This process is called renomination.*

3.2. Current rules for using gas transmission networks

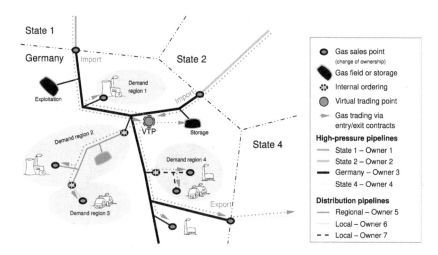

Figure 3.4. *Market area with VTP realizing the entry-exit model. The VTP is the central trading point in the market area. Capacity contracts (indicated by dotted arrows) can be booked by transport customers for transporting gas from an entry point to the VTP or for transporting gas from the VTP to an exit point, independently of the ownership of the used network infrastructure.*

in 2005, which obliged TSOs to *"design the rights on booked capacities in such a way that transport costumers are allowed to feed in gas on every entry point for withdrawal at every exit point of the network or, in case of permanent congestions, for at least a subnetwork."*[2] So TSOs have to *"offer entry and exit capacities which allow for a transaction independent of the transport path and which are independently useable and tradeable"*[3] (see also [GasNZV 2005, §4 (2)]). Transport customers may book entry and exit capacities independently, in differing quantities, and with differing durations (see [GasNZV 2010, §8 (2)]). A *market area* refers to a region of the gas network that realizes the entry-exit model, i.e., transport customers can transport gas independently of the transportation path by only holding suitable entry and/or exit contracts.

Formally, there is a *virtual trading point (VTP)* for each market area, representing all entry and exit points in that market area. Inside the market area, *"an entry contract authorizes the transport customer to use the network from the entry point to the VTP; the exit contract authorizes the transport customer to use the network from the VTP to the exit point of an end-consumer, a cross-border or market area interconnection point or a storage facility,"*[4] even if this exit point belongs to a network owned by a different network operator. Capacity contracts satisfying this requirement are called *freely allocable capacity (FAC)* contracts, since transport customers can freely choose partners from among the other customers having matching rights to virtually exchange the gas at the VTP under the sole condition that supplies and withdrawals are balanced. Figure 3.4 illustrates these concepts.

---

[2] "...die Rechte an gebuchten Kapazitäten so auszugestalten, dass sie den Transportkunden berechtigen, Gas an jedem Einspeisepunkt für die Ausspeisung an jedem Ausspeisepunkt ihres Netzes oder, bei dauerhaften Engpässen, eines Teilnetzes bereitzustellen." [EnWG 2005, §20 (1b)]

[3] "...Einspeise- und Ausspeisekapazitäten anbieten, die den Netzzugang ohne Festlegung eines transaktionsabhängigen Transportpfades ermöglichen und unabhängig voneinander nutzbar und handelbar sind." [EnWG 2005, §20 (1b)]

[4] "Der Einspeisevertrag berechtigt den Transportkunden zur Nutzung des Netzes vom Einspeisepunkt bis zum Virtuellen Handelspunkt; der Ausspeisevertrag berechtigt den Transportkunden zur Nutzung des Netzes vom Virtuellen Handelspunkt bis zum Ausspeisepunkt beim Letztverbraucher, zu einem Grenzübergangs- oder Marktgebietsübergangspunkt oder zu einer Speicheranlage." [GasNZV 2010, §3 (3)]

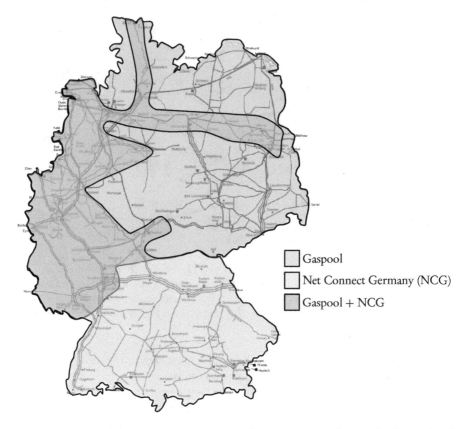

Figure 3.5. *In Germany, there are two market areas, NCG and Gaspool. Observe that there are some areas that belong to both market areas.*

Of course, the amount of gas injected into the gas network has to match the amount of gas withdrawn, at least in the long run, to operate the network in a stable way. Transport customers must therefore include their capacity contracts in so-called *balancing groups*, which are maintained by a special company, the *market area coordinator*. Within these balancing groups the usage of entry and exit contracts must be balanced over time; for each balancing group, a single party is responsible for maintaining the balance. There are hourly and daily imbalance tolerances and the party responsible for the balancing group must pay fines when they are exceeded. It is the task of the market area coordinator to supervise the balance of each balancing group and to charge fines if necessary.

The obligation of creating market areas causes the necessity of close cooperation since there are numerous network operators active in the German gas sector (see Section 1.5.3.1). TSOs must *"exploit all cooperation options with other TSOs to aim at a preferably small number of networks or subnetworks and balancing zones."*[5] The previous fragmentation of the German gas market into 19 market areas was not acceptable for the regulatory authorities, so the number of market areas has been drastically reduced. Since October 1, 2011, there have been only two gas market areas, both of them mixed H-gas and L-gas areas (see Figure 3.5 and [Gaspool; NetConnect]).

---

[5] "...alle Kooperationsmöglichkeiten mit anderen Netzbetreibern auszuschöpfen, mit dem Ziel, die Zahl der Netze oder Teilnetze sowie der Bilanzzonen möglichst gering zu halten." [EnWG 2013, §20 (1b)]

It may seem a paradox at first sight, but merging two market areas may lead to reduced capacities at some points. This may happen, for instance, if the interconnection capacity of two market areas is limited. As long as the market areas are separated, this means that the capacity at the interconnection points is limited, and so are the transports from one market area to the other. If the market areas are merged, the technical congestion at the former interconnection points remains, but is now *within* the new market area, for which there has to be free allocability. Thus, this technical congestion now limits the FAC that may be offered, which was not the case before. This is illustrated by a simple example in Figure 3.6.

The high-pressure upstream network of the two German market areas (see Figure 3.5) is operated by 17 TSOs. The majority of the network operators are DSOs on the regional or local level. Thus, during transport, the gas often has to pass several interconnection points of pipelines owned by different network operators, which are jointly responsible for the organization of the entire transport along the whole transport chain in Germany. *Internal orders* are capacity bookings between downstream and upstream network operators at interconnection points within a single market area (see Figure 3.7). All network operators are *"obliged to cooperate with one another to a degree which is necessary to enable the transport customer to book only one entry and one exit contract, even if the transport route passes several network systems connected by interconnection points."*[6] Therefore, *"downstream network operators order from their directly connected upstream network operators firm exit capacities at the interconnection points (internal order) to guarantee the permanent gas supply of end-consumers at their own network and in all downstream systems."*[7] As many upstream network operators have several internal interconnection points with a single downstream network operator, exits are often grouped into *exit zones* (which are *entry zones* for the downstream network operator). For each exit zone, a single internal order capacity is determined, which is more convenient to use for the downstream network operator (see Figure 3.7).

A more recent concept is *bundling*, which refers to the selling of only one capacity contract for both entry and exit capacity at a single cross-border or market area interconnection point. The advantage is that the customer does not have to book separately on both sides of the interconnection point, but directly gets access from one system or VTP to the other. The term *zoning* refers to the establishment of a group of points which consolidates several bundled interconnection entries and exits and requires only a single booking. Both concepts are illustrated in Figure 3.8. In Germany, the zoning and bundling of market area interconnection points is obligatory.

3.2.3 ▪ Rules for marketing capacities

German TSOs have to determine and offer both firm and interruptible capacities on a regular basis [GasNZV 2010, §11 (1)]. The capacities may be offered with different durations, including at least yearly, monthly, quarterly, and daily products. Of these capacities, *"on state or market area borders 20 percent of the yearly technical capacity of an entry point [or an exit point] is reserved for capacity products with contract durations up to two years.*

---

[6] "...verpflichtet, untereinander in dem Ausmaß verbindlich zusammenzuarbeiten, das erforderlich ist, damit der Transportkunde zur Abwicklung eines Transports auch über mehrere, durch Netzkopplungspunkte miteinander verbundene Netze nur einen Einspeise- und einen Ausspeisevertrag abschließen muss." [EnWG 2013, §20 (1b)]

[7] "Nachgelagerte Netzbetreiber bestellen bei den ihrem Netz unmittelbar vorgelagerten Fernleitungsnetzbetreibern feste Ausspeisekapazitäten an den Netzkopplungspunkten (interne Bestellung), um insbesondere die dauerhafte Versorgung von Letztverbrauchern mit Gas im eigenen und in den nachgelagerten Netzen zu gewährleisten." [GasNZV 2010, §8 (3)]

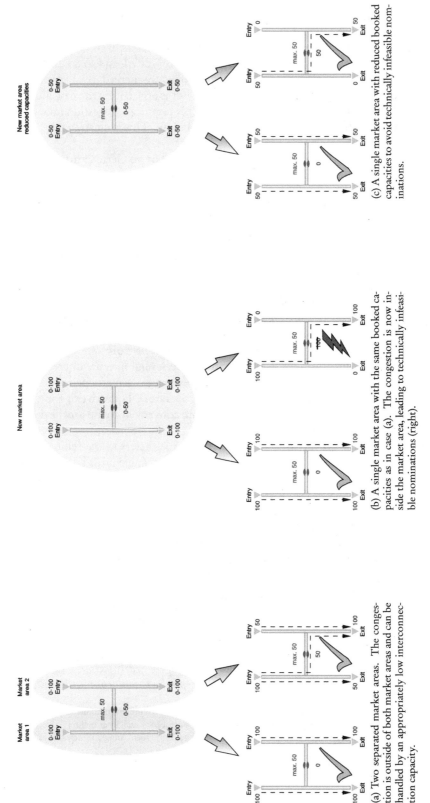

Figure 3.6. Illustration of the difficulties and limitations of firm FAC based on a simple H-shaped network with two entries and two exits, where the connecting pipeline is congested, i.e., has limited capacity. As explained in the text, shippers are allowed to use their booked capacities (intervals in upper pictures) either completely or only partially. The lower pictures show potential nominations according to the booked capacities.

3.2. Current rules for using gas transmission networks

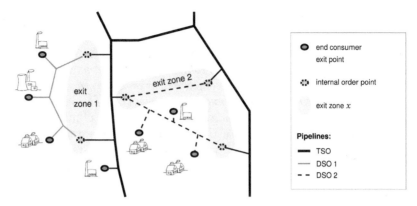

Figure 3.7. *Exit zones with internal ordering. DSOs are often connected by several interconnection points to the upstream network, which are subject to internal ordering. To simplify the procedures these interconnection points can be grouped into one exit zone.*

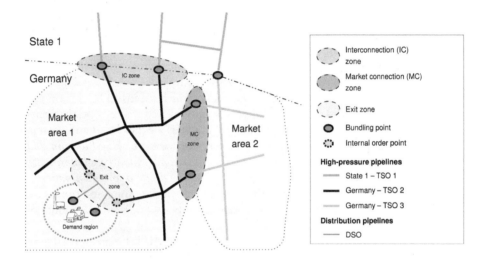

Figure 3.8. *Bundling and zoning: Bundling and zoning allows gas traders to book capacity at an interconnection zone for crossing state borders, at a market connection zone for transporting from one German VTP to another VTP, or at an exit zone to deliver a downstream supply region.*

65 percent of the technical yearly capacity of an entry point [or an exit point] might be offered with contract durations of more than four years."[8]

These standardized capacity products are sold via auctioning, since regulatory authorities regard auctioning to be a nondiscriminatory and transparent way to market capacity products. The successful bidder buys the capacity product and thus books the capacity. Germany obliged the TSOs to establish a common platform for the auctioning of primary interconnection capacities and the marketing of secondary capacity products starting from August 1, 2011 [GasNZV 2010, §12 (1)]. The platform must display all similar capacity offers by the TSOs and all demands in a transparent way for the cus-

---

[8] "An Grenzen zu anderen Staaten und Marktgebieten sind 20 Prozent der technischen Jahreskapazität eines Einspeisepunkts für Kapazitätsprodukte reserviert, die mit Vertragslaufzeiten von bis zu zwei Jahren einschließlich vergeben werden. 65 Prozent der technischen Jahreskapazität eines Einspeisepunkts dürfen mit Vertragslaufzeiten von mehr als vier Jahren vergeben werden." [GasNZV 2010, §14 (1)]

tomers [GasNZV 2010, §12 (3)]. The price of these capacities is determined by the market clearing price, which is the price where capacity offer and demand meet [GasNZV 2010, §13 (1)]. The obtained auction revenues have to be used by TSOs to increase network capacity: *"In case of permanent congestions, extra revenues have to be used for immediate measures for eliminating these congestions, or to be accrued for such measures. In case of temporary congestions, extra revenues have to be accrued for capacity-increasing measures or to be included in reduced transmission prices."*[9] Owners of interruptible capacities have the option to purchase firm capacities by auction and to convert their interruptible capacities into firm capacities in case of success [GasNZV 2010, §13 (1) and (2)].

As indicated earlier, the quantity of firm FAC offered by a TSO is estimated conservatively, since the TSO has only partial knowledge about future flow situations and the state of the gas network and thus has to resort to conservative assumptions. For short-term products, the situation is quite different since much more information is available to the TSO. This is particularly true for day-ahead capacities, which are offered one day before the actual transport takes place. For instance, it is then known to which extent long-term capacities will be used. Possible additional capacities and unused capacities should be available to the market. Thus, *"transmission system operators shall publish daily updates of availability of short-term services (day-ahead and week-ahead) based, inter alia, on nominations, prevailing contractual commitments and regular long-term forecasts of available capacities on an annual basis for up to ten years for all relevant points"* [Regulation (EC) No. 715/2009, Annex I, 3.3.3] (see also [GasNZV 2010, §40 (1.3)]).

To avoid so-called *capacity-hoarding*, which means to book firm capacities with the intention to disturb other shippers' supplying and trading options rather than to use the transport rights, different measures known as *use-it-or-lose-it* and *use-it-or-sell-it* are taken in Europe and in Germany: *"Transport customers are obliged to immediately offer entirely or partly unused firm capacities [...] on the secondary market platform or make it available for the TSO for the duration and extent of the disuse before nomination time."*[10] As long as booked firm capacities are not nominated until nomination time, TSOs are obliged to offer unused capacities on a firm day-ahead basis, taking into account possible renominations [GasNZV 2010, §16 (2)].

For long-term capacities which remain unused, TSOs are obliged to withdraw booked firm capacities with a minimum contract duration of one year if a transport customer has not used them during three months in the previous calendar year, whereof one of these months has to be between October and March (see [GasNZV 2010, §16 (3)]).

3.3 ▪ Current rules for determining capacities

The current regulation standardizes and harmonizes different areas of gas transport. In particular, there are many rules regarding how capacities are to be calculated and what measures may be taken or have to be taken in order to increase the firm capacities that can be offered. The main goal of the regulation is to offer the maximum firm FAC, i.e., the technical capacity, for the gas market and to drive network expansion towards higher technical capacity.

---

[9] "aa) Wenn dauerhafte Engpässe vorliegen, sind Mehrerlöse unverzüglich für Maßnahmen zur Beseitigung dieser Engpässe einzusetzen oder für solche Maßnahmen zurückzustellen. bb) Wenn vorübergehende Engpässe vorliegen, sind Mehrerlöse für Maßnahmen zur Kapazitätserhöhung zurückzustellen oder entgeltmindernd in den Netzentgelten zu berücksichtigen." [KARLA Gas, 5. c)]

[10] "Transportkunden sind bis zum Nominierungszeitpunkt verpflichtet, vollständig oder teilweise ungenutzte feste Kapazitäten unverzüglich als Sekundärkapazitäten auf der [...] Sekundärhandelsplattform anzubieten oder dem Fernleitungsnetzbetreiber für den Zeitraum und im Umfang der Nichtnutzung zur Verfügung zu stellen." [GasNZV 2010, §16 (1)]

3.3.1 ▪ Statistical capacity model

In accordance with the current German regulation, the calculation of capacities should be carried out *"based on state-of-the-art load flow simulations which also take into account network and market area crossing load flows. In particular, TSOs take into account historical and forecasted capacity utilization as well as historical and forecasted demand and counterflows on the basis of likely and realistic load flows."*[11] They also have to use their knowledge about existing and anticipated congestions in the network system (see [GasNZV 2010, §17 (4)]). Regulation (EU) No. 994/2010, Article 8 additionally obliges the use of statistical data as a basic principle for security of supply calculations within the member states. For network planning and capacity calculation, the regulatory commitment to statistics necessitates close connection to mathematics as well as knowledge about mathematical and especially statistical methods and solutions (see Chapter 13). Also it is legally important to know that state-of-the-art capacity calculation bears the permanent risk of an interruption of firm capacities by reason of statistical elements. One important way of minimizing the risk for unsound forecasts and assumptions is an extensive communication between the different market players which can, e.g., be observed in the United Kingdom. Here traders, TSOs, and authorities work together on realistic planning scenarios that gather relevant information.

Generally, internal orders are based on statistical analysis of past flow rates, which can be expressed by temperature-dependent functions. The *design temperature* for ordering at those internal exit points and zones is the lowest two-day average value of the air temperature which is reached or underrun 10 times in 20 years. It is specified in DIN EN 12831, supplementary Sheet 1, Table 1a (see [KoV, §13 (5)]).

To ensure security of supply for end-consumers, TSOs have to provide sufficient firm capacities for their gas demand at design temperature. In addition, statistical analysis can be used to obtain further information, e.g., temperature-dependent behavior of customers, as a basis to improve capacity calculation and long-term network planning.

3.3.2 ▪ Further capacity products

There are several types of capacity products that are envisaged by current regulation rules. We provide a summary in Table 3.1 and briefly discuss those not mentioned yet.

Often, the amount of firm FAC that can be provided is limited by certain factors and cannot be increased without auxiliary means. In these cases, TSOs may offer different capacity products that either specify additional conditions (so-called *conditional capacities*) or that are not freely allocable, i.e., they do not offer access to the VTP [GasNZV 2010, §9 (3.2) and (3.3)]. However, these *"specifications have to be limited as much as possible"*[12] as they are not considered to support the internal market as much as FAC. For conditional capacities, certain limitations are imposed, e.g., capacities that are only usable on a firm basis during certain times, temperature ranges, particular flow or demand situations, or between certain locations (see [BNetzA Capacity Management, §1.1, page 4]). Typically those conditional capacities can be used on an interruptible basis as FAC with VTP access. Especially common are *restrictively allocable capacities (RACs)*, that enable the transport customer to connect only certain entry points with certain exit points on a firm basis.

---

[11] "...auf der Grundlage von Lastflusssimulationen nach dem Stand der Technik, die auch netz- und marktgebietsüberschreitende Lastflüsse berücksichtigen. Die Fernleitungsnetzbetreiber berücksichtigen dabei insbesondere die historische und prognostizierte Auslastung der Kapazitäten sowie die historische und prognostizierte Nachfrage nach Kapazitäten sowie Gegenströmungen auf Basis der wahrscheinlichen und realistischen Lastflüsse." [GasNZV 2010, §9 (2)]

[12] "...diese Vorgaben sind so gering wie möglich zu halten." [GasNZV 2010, §9 (3.2) and (3.3)]

Table 3.1. *Characteristics of different capacities.*

Name of capacity	Short description	Combination possible?						
		Freely allocable capacity	Firm capacity	Interruptible capacity	Conditional capacity	Restrictively allocable capacity	Day-ahead capacity	Intraday capacity
Freely allocable capacity	Freely allocable capacity allows for transmission independent of the transport path and is independently useable and tradeable by time and quantity.		yes	yes	only interruptible	only interruptible	yes	yes
Firm capacity	Firm capacity always has to be usable fully or partially (on shippers' decision) in every flow situation of the network and may not be interrupted except in case of force majeure.	yes		no	only when conditions are met	only with limitations	yes	yes
Interruptible capacity	Interruptible capacity does not guarantee transmission. However, TSOs may only reduce the flows in case the network system is not capable of carrying extra gas flows.	yes	no		might be used as FAC	might be used as FAC	yes	yes
Conditional capacity	Conditional capacity is offered on a firm basis coupled, e.g., to a temperature range or flow situation and can be used as interruptible FAC with VTP access at different temperatures or flow situations.	only interruptible	only when conditions are met	might be used as FAC		possible but unusual	possible but unusual	possible but unusual
Restrictively allocable capacity	Restrictively allocable capacity can only be used on a firm basis to transport from certain entry points to certain exit points, but sometimes includes the right to be used as interruptible FAC with VTP access if possible.	only interruptible	only with limitations	might be used as FAC	possible but unusual		possible but unusual	possible but unusual
Day-ahead capacity	Day-ahead capacity enables shippers to transport for exactly one gas day and is sold the day before.	yes	yes	yes	possible but unusual	possible but unusual		possible but unusual
Intraday capacity	Intraday capacity enables shippers to transport for exactly a predefined period of 1 to 23 hours and is sold the day before the gas day the transport takes place.	yes	yes	yes	possible but unusual	possible but unusual	no	no

With conditional capacities the total capacity can be enhanced, FAC can be marketed in parallel on the same entry and exit points, and both products can even be combined by transport customers on an interruptible basis. This situation changes when particular entry and exit points are excluded from free allocation. At these points, only conditional capacities can be marketed. This solution may be chosen for isolated pipeline sections without an interconnection to other networks in the same market area, for instance.

Finally, as already mentioned, network operators have the option to offer interruptible capacities. Interruptible capacities cannot be offered on a firm basis, as there are flow situations in which the network reaches technical limits. As mentioned above, short-term *"interruptible capacities have to be offered by TSOs [...] in case of [contractual] congestions"* [Regulation (EC) No. 715/2009, Articles 14 (1b) and 16 (3a) and Annex I, 1.1 and 2.2.1] on a daily basis (see also [GasNZV 2010, §11 (1)]). Moreover, the share of the firm long-term capacity that is unused after nomination and that cannot be used by renomination is to be offered as firm day-ahead capacity. In case of congestions, the TSO determines entry and exit points and interrupts them such that the flow situation becomes technically feasible. Then interruptible capacities at these points are allocated and interrupted in the chronological order of their booking (see [GasNZV 2010, §13 (1)]). Booked interruptible capacities might be converted to firm capacities in case firm capacities become available, e.g., via auctioning (see Section 3.2.3). This also holds if only a part of the whole contract duration or a part of the total amount of contracted interruptible capacities is available on a firm basis (see [GasNZV 2010, §13 (2)]).

Regulation (EC) No. 715/2009, Article 14 (1b) requires that *"the price of interruptible capacity shall reflect the probability of interruption."* This obligation is a challenging task for capacity and price calculation. A comparatively straightforward possibility is to use information from an analysis of the interruptions in the past. However, for a reliable assessment of potential future interruptions it will be necessary to take into account dynamic factors such as hub prices, supply routes, supply agreements, capacity contracts, or demand forecasts. These more elaborate approaches are challenging from the modeling side, because some of this information (e.g., supply agreements) might not be available to the TSO at all.

3.3.3 ▪ Measures to enhance the technical capacities

From the market point of view, it is desirable to have as much firm FAC as possible to avoid contractual congestion. However, technical congestion may prohibit this. To overcome this, current regulation offers and regulates the use of several operational measures to increase the firm FAC that may be marketed by a TSO. These measures are diverse, allowing influence over the available capacity for long-term (years), mid-term (several months to one year), and short-term (hours to days) capacity products.

3.3.3.1 ▪ Long-term measures

The most obvious way to overcome technical congestion is to expand the gas network. Capacity enhancement by network expansion requires investment planning that effectively takes into account all capacity requests at the same time. Therefore, many TSOs used market surveys for upcoming capacity demand prior to a coordinated investment planning process. Generally, within this process nonbinding capacity requests could be submitted in a first phase, followed by a period where only one-sided binding requests were accepted by the TSO. If the TSO decided to invest in some network expansion, the customer directly bought new capacities. If no investment was made, the binding requests were canceled.

Today, with the advancing establishment of the internal European gas market, the coordination of infrastructure investments becomes more and more important to regulatory authorities. Therefore, the European Network of Transmission System Operators for Gas (ENTSO-G) has to adopt *"a nonbinding community-wide ten-year network development plan, including a European supply adequacy outlook, every two years"* [Regulation (EC) No. 715/2009, Article 8 (3b); see also Article 12 (1)]. This ten-year network development plan shall be *"based on existing and forecast supply and demand after having consulted all the relevant stakeholders"* [Directive 2009/73/EC, Article 22 (1)]. *"When elaborating the ten-year network development plan, the transmission system operator shall make reasonable assumptions about the evolution of the production, supply, consumption and exchanges with other countries, taking into account investment plans for regional and community-wide networks, as well as investment plans for storage and LNG regasification facilities"* [Directive 2009/73/EC, Article 22 (3)]. As *"a first step, the market is always tested in a transparent, detailed and non-discriminatory manner"* [Regulation (EU) No. 994/2010, Article 6 1(7)]. Moreover, the ten-year network development plan *"shall contain efficient measures in order to guarantee the adequacy of the system and the security of supply"* [Directive 2009/73/EC, Article 22 (1)]. In particular, security of supply measures and the enabling of counterflows have to be integrated (see [Regulation (EU) No. 994/2010, Article 5 (2) and Article 6]). The plan shall also contain a time frame for and information on all the investments already decided on or newly identified. The envisaged investments for the first three years are obligatory; investments scheduled later are nonbinding indications (see [Directive 2009/73/EC, Article 22 (2)]). The regulatory authority has to review and approve the investment plans and monitor the fulfillment of obligatory investments (see [Directive 2009/73/EC, Article 22 (4)–(8) and Article 41 (1g)]).

In Germany, TSOs have to develop a Gas Network Development Plan (GNDP) on April 1 of every year (see [GasNZV 2010, §17 (1)]). Comparable to Europe, among other things the development of supply and demand, market surveys, forecasted congestions, denied network accesses, and the possibilities for capacity enhancement due to cooperation with adjacent TSOs have to be taken into account (see [EnWG 2013, §15a (1)] and [GasNZV 2010, Article 17 (1)]). As on the European level, the regulatory authority supervises the preparation of the GNDP (see [EnWG 2013, §9 (2)]).

An expansion of the gas networks is, in general, rather expensive. Moreover, it is also inflexible, since today's technical congestions may not be the ones of tomorrow, due to changes in the gas market. A more flexible alternative to expanding the gas network is to allow the TSO to control the load flows to some extent, such that technical congestions may be bypassed. *Flow commitments* are *"contractual agreements with third parties who guarantee certain load flows which are adequate and necessary to enhance the offer of firm freely allocable entry and exit capacities."*[13] As an example, a flow commitment might be a customer's commitment to reduce flow at one point or to secure a certain gas flow on another point on the TSO's request (see Figure 3.9). TSOs purchase flow commitments from transport customers in a nondiscriminatory, transparent, and market-based way, e.g., by a tender process. Nevertheless, *"the extent of flow commitments has to be limited as much as possible."*[14] Flow commitments can be used to avoid network enhancements or postpone investments. Compared to new infrastructure, flow commitments are an instrument of high flexibility, which is especially interesting regarding nonlasting changes in flow and load situations over time.

---

[13] "…vertragliche Vereinbarungen mit Dritten, die bestimmte Lastflüsse zusichern sowie geeignet und erforderlich sind, die Ausweisbarkeit frei zuordenbarer Ein- und Ausspeisekapazitäten zu erhöhen." [GasNZV 2010, §9 (3.1)]

[14] "…der Umfang von Lastflusszusagen ist so gering wie möglich zu halten." [GasNZV 2010, §9 (3.1)]

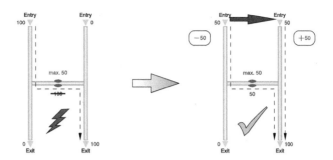

Figure 3.9. *Firm capacity enhancement by a flow commitment. The TSO has bought a flow commitment from a transport customer to switch a part of the nominated gas amount from one entry to another entry. This gas entry switch allows the marketing of more entry capacity because of the possibility of relieving the congestion if necessary.*

3.3.3.2 ▪ Mid-term and short-term measures

Using a statistical capacity model may allow, to some extent, forecasting the actual usage of a booked capacity at a point. If it is likely that the transport customers will not exhaust the available capacity, the expected remaining capacity may be marketed as well. This is the motivation for *oversubscription*, meaning to offer, up to a maximal percentage, additional firm capacities on top of the technical capacities. Of course, this implies the risk of having to interrupt transport customers because their nominations actually exceed the technical capacity. This risk is to be handled preferably by commercial means. For instance, one possibility for the TSO is to purchase options for *capacity buy-back* from customers, i.e., to be able to repurchase firm capacities in case of congestions (see [GasNZV 2010, §10 (1)]). However, before TSOs are allowed to establish oversubscription and capacity buy-back, the regulatory authority approves the scale of the technical capacity calculated by the TSOs (see [GasNZV 2010, §9 (4)]). Of extra revenues from oversubscription, TSOs might keep up to 50 percent (see [GasNZV 2010, §10 (2)]), which makes these instruments economically interesting. Nevertheless, to avoid losses which would also have to be fully covered by TSOs, the probability of the use of capacity buy-back needs to be determined. This task is very similar to determining the probability of capacity interruptions (see Section 3.3.2).

It may happen that certain load flows are difficult to handle for a TSO, since they are (unexpectedly) rather imbalanced, e.g., due to gas losses, global imbalances between gas in- and out-takes, or due to more regional differences that occur when certain parts of the network are overloaded while others are underloaded, so that the amount is balanced in total, but gas has to be distributed to less congested parts of the network. In this case, the TSO has the option to employ balancing measures, i.e., to keep the gas network in a stable and secure operational mode by increasing or reducing the gas injection into the network or by reallocating gas to other geographical areas (see [EnWG 2013, §16]). *"Balancing rules shall reflect genuine system needs taking into account the resources available to the transmission system operator. Balancing rules shall be market-based"* [Regulation (EC) No. 715/2009, Article 21 (1)], harmonized, transparent, and nondiscriminating (see [Regulation (EC) No. 715/2009, Article 21 (4)] and [EnWG 2013, §22]). For instance, balancing can be achieved by buying or selling gas as necessary to ensure the gas transport. Figure 3.10 depicts an example of regional imbalances at different times of day.

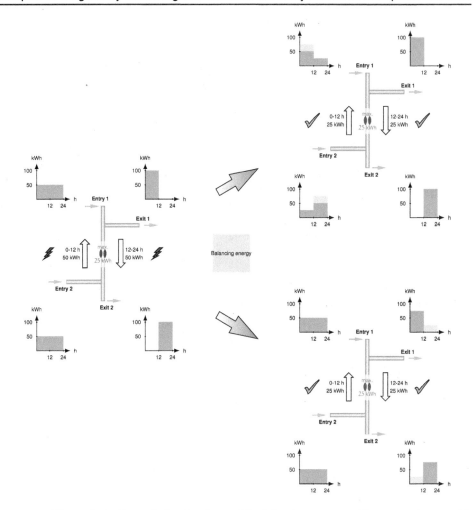

Figure 3.10. *Example showing the need for balancing measures. The entry points provide a constant supply, but the first exit is only used between 0:00–12:00 and the second exit only between 12:00–24:00. The amounts over the day are virtually in balance, but the transport cannot be realized because of congestion. To rectify this situation, the TSO has the option to either buy gas at the entry points (at entry 1 between 0:00–12:00 and at entry 2 between 12:00–24:00) or at the exit points (at exit 1 between 12:00–24:00 and at exit 2 between 0:00–12:00). Moreover, a combination of both might be possible, as long as the necessary transport does not exceed the technical limit of the congested pipeline.*

3.4 ▪ Challenges for gas transmission system operators

The European gas sector—and the German in particular—has seen tremendous changes in the last 10 years. Moreover, the changes will continue due to the "Energiewende," i.e., the German initiative to supply a large share of the energy demand using renewable energy and to stop the dependency on nuclear power as soon as possible. At least during the transformation phase, natural gas will play a major role in the German energy sector. These changes pose serious challenges for the TSOs from not only the gas transport point of view but also even more importantly the capacity planning point of view.

An important but perhaps nonobvious difficulty is the discrepancy between the simple, more conceptual use of the term "capacity" in the legislative and regulatory frameworks and the complexity of operating gas transmission networks, which has to be taken into account when determining "capacities." In fact (and perhaps unsurprisingly), even the concept of "capacity of a single point" is not well defined (apart from trivial, academic examples), since the load flow that may be handled at this point depends on the flow situation in the entire network. The mathematical reason for this is that gas flows are driven by potentials, i.e., pressures, which couple large regions of a gas transmission network. Moreover, usually there is a trade-off when declaring the capacity for a set of points, since the capacity of one point may be increased at the expense of decreasing the capacity of other points. Thus viewing entry and exit capacities at each point independently is too simplistic, which has already been acknowledged by the regulatory authorities. Similar comments apply to the concept of "congestion," since in general it is not possible to identify parts of the network that are responsible for congestion (see Chapter 11 for a detailed discussion). Mathematically, it is possible to "move" the "reason for infeasibility" in the network which is in stark contrast to classical network flow problems, where the "congestion" is localized (i.e., a graph-theoretic cut). Thus, "capacities" are not constant values, but depend on the usage of the network, the assumptions made during capacity calculation, and the capacity calculation procedure itself.

Nevertheless, TSOs need to develop procedures that deal with these issues in a reasonable and reliable way, which is, in fact, a major motivation for the research presented in this book. Since regulation authorities do not (and probably cannot) prescribe detailed procedures, every TSO can come up with its own model for capacity calculations, which poses the additional issue of obtaining consistent results with different models. This is one of the reasons for requiring a close cooperation and coordination between the TSOs. In particular, TSOs need to negotiate the ranges of entry and exit capacities on internal connection points such that the connected networks may be operated in a reliable way, as long as the gas flows are within these internal connection point capacities.

The second major challenge is the uncertainty about how the gas transmission network will be used in the future. Since the "capacity" of a point depends on the general flow situation in the network, actual network usage has a significant impact on whether or not the predicted "capacities" can be realized. As mentioned before, statistical models can (and have to) be employed to provide a sound forecast for load flows. To this end, sophisticated mathematical methods have to be developed and used with care. However, such models may only be used for points where the behavior is, in principle, predictable, i.e., depends on factors that can be modeled stochastically. This is, in general, not the case for entries, storages, and also some exit points, for instance cross-border points.

Unbundling and the introduction of big market areas together with the free allocability of capacities, bundling of capacities, and zoning have introduced a novel quality of uncertainty to the capacity planning process. First, a TSO has less reliable information about how the network is going to be used in the future, since there are now many shippers, which are distinct companies and do not involve the TSO in their sales and operations planning. Second, there may now be completely different flow situations in the network that have not been anticipated so far. Coping with these severe uncertainties requires sophisticated methods.

The situation is somewhat different for short-term, i.e., day-ahead, capacity planning. Due to the obligation to nominate the gas flows for the next day a day in advance, much more information about the flow situation is available. Just as important, the state of the network, i.e., the amount of gas stored in the pipes (the linepack), the current operation mode of the compressor stations, etc., is known as well. Thus, in principle, it is possible to predict rather precisely how much additional gas can be transported. Obviously, unused capacities can be offered to the market as day-ahead capacities according to the use-it-or-lose-it principle. However, unused capacity at one point might imply that more capacity than is available so far can be offered at other points. Therefore, a recalculation of the capacities of the entire network is required in order to maximize the capacity to offer. The daily day-ahead capacity calculations need to take into account all the information that is available and should provide a result within a relatively short time period, say one hour. To achieve this, automation of the calculation process is required.

The introduction of market areas and free allocability also impacts network planning, since FAC not only has to be offered for the existing network, but also network expansion should be guided by the FAC demands of the market. Thus, networks have to be maintained and developed in such a way that free allocability within the market area can be guaranteed. It is again crucial that TSOs cooperate with and consult each other. They are obliged to provide a common yearly GNDP in which the evolution of the German gas network is planned, taking into account the whole German gas infrastructure.

3.5 • Summary and outlook

With continuing regulation in both Europe and Germany, it is necessary not only to solve the technical and physical challenges of gas transport, but also to tackle the formidable tasks in network planning and capacity calculation outlined in Section 3.4.

It is evident that planning methods based on a few expert scenarios describing possible future gas flow situations can be complemented (see Chapter 4). To adequately address the uncertainty inherent in future gas network usage it is necessary to come up with sophisticated models that characterize realistic gas flow situations. These models need to combine stochastic forecasting with some adversarial models on how transport customers might use their capacity rights. It is thus necessary to consider (implicitly or explicitly) a wide range of potential gas flow situations, rendering futile the traditional approach of validating each potential gas flow situation manually via simulation tools (see Chapter 4).

Hence, advanced automatic methods for verifying network capacities and exposing additional available capacity are called for. These methods have to take into account, in a mathematically rigorous fashion, the underlying physical and technical constraints. Just as important, legal and contractual constraints and implications have to be considered when designing such methods. In order to be practically useful, many details have to be addressed. See Chapter 14 for a first optimization approach in this direction. The sheer size of today's gas networks, as well as the recent focus on short-term capacity planning that demands real-time capabilities, constitutes a huge challenge for network planning and thus for mathematical programming techniques.

Even if powerful methods that can cope with the *German* gas network become available within a few years, European politics has already set the stage for the next step: Currently, Europe is discussing a revised and harmonized set of rules for the internal gas market with new mechanisms and procedures in the form of a new gas target model in 2014 and beyond. The goal is to *"provide a regulatory framework that secures supplies in the long, medium and short term, which means making Europe attractive for gas imports also in the future and taking into account seasonal and short term fluctuations in gas demand"* [CEER Gas Target Model].

Chapter 4

State of the art in evaluating gas network capacities

Dagmar Bargmann, Mirko Ebbers, Nina Heinecke,
Thorsten Koch, Veronika Kühl, Antje Pelzer, Marc E. Pfetsch,
Jessica Rövekamp, Klaus Spreckelsen

Abstract *This chapter describes the current state of the art and daily practice of capacity evaluation via simulation. The presentation is based on the example of the methodology used at Open Grid Europe GmbH (OGE). The approach proceeds by generating several load conditions that have to be tested by simulation software. These selected load conditions should represent situations that are hard to realize in gas transport. In order to avoid overly restrictive load conditions, statistical and contractual information has to be integrated. Then the simulation results for these load conditions are used in order to estimate the available capacity. The simulation builds on a detailed physical model and corresponding parameter settings. Control settings for active elements have to be found by experienced planners in a relatively time-consuming process, which involves many details that are described in this chapter. This sets the stage for the approaches presented later in this book.*

The goal of this chapter is the description of the current state of the art for evaluating the capacity of a gas network as it is used in practice today.

The main idea is the following: First, certain load conditions are created that are considered to be stressing for the gas network. A load condition will be represented by a so-called scenario in this chapter, as discussed in Section 4.1. In order to generate these scenarios, a model of the amount of gas that customers supply to or discharge from the network and contract situation needs to be set up. Each generated scenario is evaluated using *simulation software*. Such software needs the specification of the network, as well as boundary and parameter settings. Moreover, the controls for active elements (compressors, valves, etc.) need to be given. Finding such controls is done in an iterative fashion, as explained in more detail below. If for each scenario an appropriate control of active elements can be found, the set of scenarios gives an estimation of the available capacity. This process rests on the assumption that these stressing scenarios approximately represent the set of feasible load conditions and that each convex combination is feasible as well. On the other hand, if for some scenario no feasible control can be found during this process, it is considered to not be realizable and thus represents a restriction on the available capacity. Consequently, the described process consists of the following steps:

1. setting up a model of customer behavior,
2. generating scenarios,
3. simulation, and
4. interpretation of the results.

Before giving a detailed description, we give a brief account of each step.

Model of customer behavior In practice, the amount of gas discharged by end-consumers depends on their current individual situation, e.g., energy used for heating is often correlated to the outside temperature. Because of this, certain stressing situations with respect to the transportation requirements for the network do not arise in practice, or arise only in hypothetical cases. For instance, it is unlikely that a high demand of gas arises in one part of the network due to cold temperatures, but the other part has low demand as in the case of high temperatures, such that gas has to be transported across the whole network. Consequently, assuming worst-case behavior of customers corresponding to the contract situation for each freely allocable capacity (FAC) exit (see Section 3.2.2) would result in unnecessarily low available capacities over the entire gas network. Thus, if statistical data on consumer behavior are available, it should be used to give a realistic evaluation of capacities, which is also suggested by the legal framework; see Chapter 3.

On the other hand, often no reliable forecast can be given by statistical data for suppliers, storage users, and certain customers. For these cases, a worst-case behavior has to be assumed, but the corresponding contract situation should be taken into account. Moreover, for long-term capacity planning a contract situation prognosis has to be used in order to deal with incomplete information because of short- or medium-term contracts. See Section 4.2.4.2 for more details.

Moreover, different customer types have to be distinguished in order to achieve a good forecast for occurring load conditions. These customer types include power stations, industrial customers, exits towards distribution system operators (DSOs), market areas, cross-border interconnection points, municipal utilities, or a storage site. Some of these act as either suppliers or consumers and they behave differently with respect to the possibility for forecasting; see Section 4.2.4.4 for more details.

Scenario generation The generated scenarios should cover all possible situations over the considered time horizon. Since simulation is a time-consuming process, only a small number of scenarios can actually be tested. Thus, one assumes that the set of feasible load conditions is convex in the sense explained below. Moreover, flow situations are aggregated such that several stressing constellations are combined into a single scenario. This aggregation is performed on the basis of identified congestions. See Section 4.2 for more details.

Simulation The basic tool for testing generated scenarios is (stationary) simulation software. Simulation needs a reasonably accurate model of the physical gas network with all active and passive elements. The network planner will change settings for all active elements in order to achieve a feasible simulation result. If no such setting is found, the scenario is assumed to be infeasible. See Section 4.4 for more information.

Interpretation of results Simulation classifies the tested scenarios into feasible and infeasible. The overall results have to be interpreted correctly by experienced planners in order to arrive at a capacity specification. The interpretation has to take into account the results of the simulations, the difference between stationary and dynamical gas network behavior, the likelihood of a physical situation that is represented by the scenarios, additional available (statistical) information, and many other aspects. See Section 4.5 for more information.

Overview of this chapter

The rest of this chapter is essentially organized according to the above process. After giving background on the organizational setting in Section 4.1, Section 4.2 deals with the generation of scenarios. Section 4.3 discusses the handling of control options in simulation and Section 4.4 discusses the process of simulation. Finally, Section 4.5 deals with the interpretation of the simulation results. We close with an outlook for the rest of the book in Section 4.6.

4.1 • Background for capacity evaluation and simulation

We start our description with an overview of the situation for capacity planning in gas transport.

As described in Chapters 1 and 3, gas pipelines are planned to connect regions of gas exploitation and consumption. Networks are designed by only considering gas from certain quality bands for transport; see Section 2.2. In addition, there are contractual agreements between network operators with respect to, e.g., supply shares, co-ownerships, joint venture relations or flow commitments; see, e.g., Section 3.3.2. These agreements, however, can only be adapted gradually to new situations and new capacity calculations by means of expansion measures, acquisitions, or renegotiations. The result is that gas transport networks and framework conditions are not fully compatible. This setting makes it difficult to set up meaningful scenarios reflecting future reality.

Due to the strict separation of trade and transport within the framework of unbundling (see Section 3.1.2), certain information is not available to players on both sides. On the trading side, the multitude of competitors does not allow single traders to take into account the supply and demand situation of the whole market. Price signals from gas exchanges, which are connected to the virtual trading points (VTPs), help us to at least get a picture of the current market situation. Long-term planning and gas purchase are more complex, since consumers have the option of a short-term supplier change. With interruptible or conditional capacity products, transport customers also have the need for risk assessment in case of interruptions and may not be able to exploit the lowest gas price of alternative supply or purchase options. These circumstances have to be taken into account for both the model of consumer behavior and the evaluation of capacity.

On the transport side, the origin of gas supply at a certain time is hardly predictable. Gas transport companies try to counteract this information deficit by using statistical data in network planning and capacity calculation (see Chapter 13 for more information). By now, statistical data analysis is also firmly rooted in the regulatory framework (see also Section 3.3.1). However, forecastability and accuracy of the customer behavior are limited due to the above mentioned flexibility of gas transports and the various influential factors involved. The used statistics are based on historical data and are suitable for making forecasts for the future behavior of transport customers mainly when the customer has only a limited scope of action, e.g., a gas-consuming end-customer. Statistical calculation models thus bear the risk of a transport failure, i.e., a load situation cannot be transported. Network operators have to include this possibility in their model considerations.

A methodology for evaluating capacities of gas networks has to answer the following two main questions. The first is how the maximal firm capacity of the network can be determined by simulation with manageable efforts and minimized risks of a transport failure. The second question is, assuming the amount of capacities in total or at certain points is insufficient for transport customers, what measures can be taken, and in which order, to increase the capacities to an adequate level with affordable costs and efforts for customers and transmission operators? Examples of these optional measures, such as the marketing of other capacity products, the usage of capacity enhancing methods, or the investment in network extensions, can be found in Section 3.3

In this chapter we describe a capacity model based on the example of the one used by OGE. For assumptions and capacity calculation methods of other German transmission system operators (TSOs) we refer, e.g., to [Gascade, *Berechnung der technischen Kapazität*], [Thyssengas], and [Terranets, *Verfahren zur Ermittlung der technischen Ein- und Ausspeisekapazitäten*]. To learn more about other European systems see, for example, [Net4Gas, *Calculations of technical capacities of the transmission system of Net4Gas*], [Energinet, *Gas in Denmark 2010 – Security of supply and development*], and [National Grid, *Gas Demand Forecasting Methodology*]. For tariffs and gas balancing fees in European countries see, for example, [KEMA 2013/Corvinus University of Budapest, *Study on Methodologies for Gas Transmission Network Tariffs and Gas Balancing Fees in Europe*].

4.2 ▪ Generation of scenarios

To determine the marketable capacity, load conditions are simulated for different scenarios in a gas transport network. Besides the load conditions, scenarios are further defined through assumptions of model parameters and so-called boundary values. Since there are infinitely many possible scenarios, a reduction of all possible scenarios to relevant and representative scenarios that have to be simulated is required. In practice, this reduction is done by means of worst-case assumptions while using statistical data.

4.2.1 ▪ Definition of scenarios

A scenario represents a particular load condition of a network that leads to a possible unique solution of the system. A *load condition* or nomination consists of given flow amounts at entries and exits that are in balance. In order to allow simulation software to find a unique solution of the system of differential algebraic equations, further values need to be specified. Consider a decomposition of the gas network at each closed valve, active or closed control valve, and active compressor machine (see Section 2.3 for more details on network elements). Then, in each component, the pressure has to be fixed in at least one point—otherwise, the solution is not unique. Moreover, different components have to be connected via an active element, i.e., the output pressure is derived from the input pressure using a *control setting*. One main example of a control setting is to preset a certain outlet pressure or flow at a compressor machine or control valve. The simulation software automatically determines the physical settings in order to meet this preset value, if this is possible.

Moreover, each network element has particular *boundary values* that specify bounds on physical values. This, for instance, includes, upper and lower pressure and flow bounds, temperature and velocity bounds, as well as ambient temperature, which might be different for each scenario. These values might arise due to physical or legal restrictions or by requirements of a consumer (i.e., lower pressure and flow bounds). A load condition can be specified by setting the lower and upper boundary values for the flow to the given value. In the following, we concentrate on those values that can be affected by contracts or customer behavior, in particular, flow and pressure bounds, as well as by ambient temperature.

Moreover, all parameters of the simulation software need to be specified, i.e., a physical model of reality has to be fixed. Parameters that have to be specified include the choice of formula for the calculation of the pressure loss and the compressibility factor or rules on how to take into account ground and air temperatures (see Chapter 2).

We call the set of load conditions, compatible boundary values that fix the pressure at entries, and parameters a *scenario*; note that the control settings of active elements are not part of a scenario. Further note that entries are usually equipped with a control valve. Thus, the fixed entry pressures effectively only specify an upper bound on the pressure, and control settings of the control valves allow reducing the input pressure.

It is also important to note that the usage of the term *scenario* differs from the *statistical load scenario* in Chapter 14, which refers to a load condition for those points that are modeled statistically; see Section 14.1. Since the term is used in both practice and stochastics, we accept the slight inconsistency in this case.

One main component of a scenario consists of entry and exit flows. In order to define these, both the contractually stipulated rights of the customers and the flows observed in the past are taken into consideration and combined with each other to determine boundary values at entries and exits. The same holds for pressure bounds.

The scenarios are then simulated and evaluated in terms of their *technical feasibility* in the existing network, i.e., whether the load condition can be realized without exceeding any of the technical limits of the network. Such scenarios are usually stationary considerations of individual temporary transport situations. Dynamic considerations are conducted only in individual cases, e.g., how long the switch of the flow direction of a long pipeline takes. We thus use only stationary calculations in the following, because we focus on long-term planning for which no dynamical information is available; see Section 5.3.1 for a discussion.

4.2.2 ▪ Assumptions for generating scenarios

The quantity of scenarios depends on the number of entry and exit points with FACs. Since all contract-compliant nominations that can be expected statistically must be technically feasible, it is impossible to consider each of the infinitely many possible scenarios. Therefore, a reduction of possible scenarios to relevant and representative scenarios is required for simulation. This reduction is done on the basis of the following assumptions.

4.2.2.1 ▪ Convexity assumption

It is assumed that the set of technically feasible nominations (seen as a vector of entry and exit flows) is convex. It is easy to see that this assumption is wrong in a strict sense. For instance, the functionality of the active elements is subject to nonconvex restrictions, e.g., compressors require a minimum flow in order to be used. Nevertheless, this simplifying assumption on the space of possible operation modes is taken in order to reduce the complexity of the model. On the positive side, dynamic effects such as the use of linepack (Section 1.5.2), intermittent operating modes (Section 11.1), or pump prevention (Section 2.3.5.1) can contribute to reducing nonconvex behavior. This makes the convexity assumption somewhat more acceptable in practice.

4.2.2.2 ▪ Constancy assumption for the subscription of customer behavior

Exit flows usually do not vary in the complete contractually permissible flow range, but only in a statistically observed range. We assume that this statistical information is valid for the future. However, this is only possible for exit flows with no contractual amendments and which can adequately be described by statistical data.

Unfortunately, it is nearly impossible to predict entry behavior, because it fluctuates extremely and is strongly influenced by price developments and the details of individual supply contracts, e.g., between gas exploitation and transport customers.

4.2.2.3 ▪ Worst-case assumption

Scenarios that are assumed to reflect particularly difficult load conditions are selected for simulation. Some of these scenarios occur in practice under extreme flow situations, while others have not been observed so far. Although such worst-case scenarios typically occur rather seldomly, they are realistic and TSOs have to deal with them. A detailed description of worst-case assumptions is the subject of the next section.

4.2.3 ▪ Generation of worst-case scenarios

When generating scenarios for capacity calculation, attempts are made to cover the most difficult transport situations in order to minimize the risk that a nomination occurs that is not technically feasible. Of course, such scenarios should still try to use the information of available statistical data. Difficult scenarios respecting statistical data can, for instance, be generated by considering particular high or low temperatures. Moreover, the worst case can be realized at points for which no statistical data are present.

In this section, we describe our approaches to generate such worst-case scenarios. When considered conjunctively, different restrictive situations for the transport network can limit the problem to one pipeline or compressor station that is intensively used and whose expansion would make all scenarios technically feasible. Such a pipeline or station is called *congestion* (see also Section 4.2.3.5).

4.2.3.1 ▪ Flow maximization

Scenarios often turn out to be difficult if maximum flow and/or maximum compression over a pipeline or station occurs. For instance, this is the case for maximizing the inflow of selected entries. Similarly, the outflow for exits can be maximized, e.g., by determining the maximum transport volume of a connecting pipe. The exit flow can be further increased by the assumption that in situations with maximum outflow, bidirectional points can also be used as exits. This increase, however, is limited by the fact that in- and outflows must be balanced.

4.2.3.2 ▪ Transport route maximization

Other difficult cases are represented by scenarios in which the length of the transport route is maximized. This is achieved, e.g., by using entries with either full or zero flow under the convexity assumption and forecasting exits with a rather high or low value for the statistical temperature-dependent outflow. By choosing high values for entries and exits further away from each other, or low values for entries and exits closer to each other, the transport route for gas can be maximized, since in principle the longer the transport route, the larger the pressure loss of gas during transport (see Chapter 2). This can lead to a shortfall in contractually stipulated minimum pressures at exits. The maximization of the transport route can be realized by scenarios with medium and warm ambient temperatures, since in these cases there is more flexibility in covering the demand of the entire network via a small number of entries.

4.2.3.3 ▪ Worst-case scenarios for exits

Load maximization for exits usually occurs when the ambient temperature is extremely low. In this case, a very large amount of gas is transported via the network, since outflows to gas power stations and household customers are extremely high. Scenarios for the coldest ambient temperature to be expected, the design temperature, are called *peak load scenarios*; see Section 3.3.1 for a definition of the design temperature. These scenarios especially test whether the connecting pipes to the end-consumers have sufficient capacity.

4.2.3.4 ▪ Worst-case scenarios for entries

Depending on the consumption situation and the composition of the relevant customer portfolios, inflows at the entries can vary for transport customers. In contrast to outflows, future behavior can only be predicted to a limited extent from past data for entries, because these data are influenced by price developments and individual details of the transport customers' supply contracts. For the maximization of entry flows, a peak load scenario with maximal exit flows does not necessarily give the worst case. Often the total network load is higher if exits close to the supplying entries have low demand and exits further away have high demand. In order to achieve a flow and transport route maximization for entries, exits are grouped and the corresponding flows are modeled by *stressing and relieving*, as explained in the following.

Many years of experience and numerous simulation calculations have shown us that certain entry combinations (for example, entries that are located close to each other) can be identified as particularly restrictive for the transport system. Other combinations are less restrictive, as they do not charge the same pipeline segments (for example, if entries from the north and the south are both supplying the network at the same time). However, since these combinations may change with varying contractual situations or investments, they must continually be adapted to new circumstances. The resulting restrictive entry combinations are tested with priority over less restrictive scenarios in the simulation. For this purpose, entries are grouped, such that the capacity of each group can be determined more easily. The distribution of capacity among the entries in the group is done subsequently with changing orders and total capacity bounds for each scenario. The classification into groups is done on the basis of experience (real network operating modes and earlier calculations), and taking into account possible load conditions and associated congestions.

Among these, groups are chosen whose inflow is to be maximized (MaxEntry). For each such group, a compensation entry group (DiffEntry) is defined to achieve a balanced scenario. Typically, scenarios are generated in which only one entry group could completely cover the network demand on its own. Moreover, different temperatures are considered in order to test high and low load settings.

Outflows are chosen as follows. Depending on their position in the network, for each scenario, exits are divided into so-called *stressing* and *relieving* exits. Exits where an increase of outflow increases the length of the transport route or the flow through a pipeline/station are called stressing exits for the scenario. The term *relieving exit* is used for some exit where an increase of the outflow causes a reduction of the length of the transport route or the load over a congestion. Figure 4.1 shows an example for flow maximization of a pipeline.

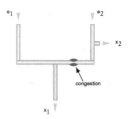

Figure 4.1. *Example illustrating the flow maximization of a pipeline. The network has two entries e_1 and e_2 as well as two exits x_1 and x_2. Between the two exits, there is a congestion (represented by the technical constriction in the pipeline), e.g., a pipeline section with large pressure losses. The lower the demand in x_2, the less outflow can be covered by entry e_2, because the congestion restricts the freely allocable transport of gas. The quantity that has to be transported through the congestion increases—if the outflow in x_1 increases—until the outflow in x_1 exceeds the transport capacity of the congestion. Thus, for the maximization of the transport route, starting with entry e_2, a high outflow is assumed for x_1 and a low outflow for x_2 within statistical limits. If the maximum capacity of e_1 is to be tested, a scenario for the maximization of the flow, starting with entry e_1 with correspondingly low load at x_1 and high load at x_2, must be considered in addition to this.*

Figure 4.2. *Systematics to determine stressing and relieving. The exits located between the entry to be maximized (MaxEntry) and the congestion (i.e., which are close to MaxEntry) are modeled as relieving exits (MinExit), whereas exits located between the compensating entry (DiffEntry) and the congestion are modeled as stressing exits (MaxExit).*

For a simple network with two entries, three exits, and only one congestion, the exemplary stressing and relieving systematics shown in Figure 4.2 is used. To analyze a specific congestion, statistically evaluated upper and lower boundaries of outflows are included in the boundary value generation.

4.2.3.5 ▪ Congestions/bottlenecks

The definition of a transport congestion or bottleneck is not trivial, since this is not a tangible object or situation, but rather an interpretation of several parameters. See Chapter 11 for a more detailed discussion of feasibility and congestions/bottlenecks.

In general, a pipeline with a substantial pressure loss causing the violation of a lower pressure limit, or a compressor station that cannot produce the desired output pressure because of a low input pressure, is interpreted as a congestion. Depending on the transport situation, different congestions can be identified. However, even when considering the same transport situation, several congestions might be located in various parts of the network.

With respect to network simulation, usually not all congestions are known a priori. Real network operating modes or congestions taken from earlier calculations provide

4.2. Generation of scenarios

some indication. If congestions are known, dedicated scenarios are generated that usually produce maximum flow at the assumed congestion.

4.2.4 ▪ Generation of boundary values

The generation of boundary values comprises three steps:

▷ evaluation of statistical data from the past;
▷ combination of statistical data with current or forecast booking situations under consideration of legal and regulatory requirements; see Section 3.2.1;
▷ generation of a balanced stationary scenario.

In the following we will provide more details for the generation of boundary values.

4.2.4.1 ▪ Effect of stationary assumption

The given statistical values must be adapted to obtain stationary scenarios. The corresponding values are measured for every hour of the day. They can be filtered by daily maxima, minima, and means. During the day, many end-consumers show a load profile with a high consumption in the morning and evening, as well as a lower consumption in between. The discrepancy between the daily maximum and daily minimum is more or less pronounced and depends on the type of end-consumer, e.g., a municipal utility, an industrial customer, a power station, or a storage site. These maxima and minima are, among others, used for the determination of stressing and relieving exits.

For stationary scenarios, we consider only situations that might occur at the same time. Thus, no flow situations are considered where one end-consumer demands maximum while another demands minimum flow. Thus, for a worst-case assumption the same filter is typically used for all end-consumers.

Statistical values are also correlated to ambient temperatures. The temperature can be given as the daily mean or the mean value for several days (see also Section 13.5). Experience has shown us that a suitable "smoothing" of measured values can be obtained by using the mean temperature over a period of four days, i.e., the mean temperatures of the measured day and the three preceding days are assigned to a measured value.

Clearly, it is not possible to consider dynamic effects such as linepack or the use of balancing energy (i.e., additional gas that can be bought by the TSO) by stationary simulations. Moreover, the required amount of fuel gas used in compressor stations (roughly 0.5% of the total flow volume for national gas transport) has to be estimated prior to the simulation in order to yield a balanced scenario.

4.2.4.2 ▪ Incorporating contracts

Besides the statistical data, the contractual situation can be considered for the construction of scenarios. More details on the available contracts and capacity products are given in Section 3.2.1.

To be able to determine the available capacities for a certain period of time, a large number of scenarios must be created and analyzed. The maximum *validity period* of a calculation is determined by the longest duration for which the contractual situation in the network, such as capacity products and flow commitments as well as the statistics and network expansion conditions, are constant. In this context, in long-term capacity planning, the shortest period is usually one day; in short-term capacity planning, however, even an hourly basis can be considered. See Figure 4.3 for an illustration of the determination of the validity period.

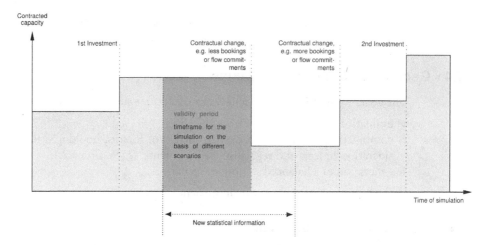

Figure 4.3. *Determination of the maximum validity period of scenarios: A calculation can only be valid for the period in which the contractual situations, the network topology, and statistics remain unchanged; the validity period is illustrated by the marked box.*

In the context of the contractual situation, not only firm capacities are considered, but also interruptible capacities, which might be relevant for capacity calculation. They offer flexibility for balancing separate in- and outflow. Especially for low temperatures, when no further firm capacities are available, interruptible capacities might be necessary to cover the demand and thus are considered in stationary scenarios. We focus, however, on the description of the calculation of firm capacities in the following.

Apart from capacity contracts, there are also so-called *interconnection agreements*, which regulate the delivery pressures between two network operators or between network operator and transport customer at a certain point. These delivery pressures might considerably limit the transport possibilities. At certain exits, gas is delivered jointly by several network operators on a proportional basis. For instance, this might be caused by split ownerships. In this case, contractually agreed supply shares which proportionally split the entire consumption of an end-consumer among the respective TSOs are applied. Vice versa, an exit zone is a combination of DSO exit points for booking a sum capacity. In this case, the distribution over individual points might follow a certain distribution key. There are regulatory efforts to also build zones between market areas and border interconnections, in particular for entries, in the future.

Another type of contract that network operators can conclude with third parties is a flow commitment (see Section 3.3.3). In contrast to all other types of contracts described here, flow commitments offer additional possibilities to market further capacities or ensure marketed capacities. If possible, flow commitments for consumers should be avoided or used only temporarily in order to bridge expansion needs. Depending on the transport problem to be solved, there are different types of flow commitments:

▷ *Minimum flow commitments* ensure that the flow at a certain point is at least a given lower value.
▷ *Storage flow commitments* link the capacity booked to the temperature at a certain point, e.g., when a load commitment guarantees to avoid outflows if the measured temperature is above 0 °C.
▷ *Reduction flow commitments* are of a similar character, enabling the reduction of entries and exits either depending or not depending on the temperature.

▷ A *distribution commitment*, which is probably the most powerful type of flow commitment, offers TSOs the possibility to increase gas to a certain level between predefined points in the network.

Apart from firm capacities, restrictively allocable capacities (RACs) (see Section 3.3.2) at certain points have some influence on capacity calculation. For instance, necessary capacities that have not yet been booked are reserved for end-consumers beyond the booking duration (see e.g., GeLiGas in Section A.2). In addition, different short booking periods which differ only slightly from each other can be "smoothed" by RACs in order to reduce the number of capacity calculations required.

As mentioned above, there are pipelines that are common property of several TSOs with different shares. For pipelines of this type, there are several limitations for the individual TSO in order to prevent the use of property shares belonging to other parties. Frequently, there are specific capacity or pressure limits or restrictions applicable for the so-called *path integral*, which is the total sum of the products of transport flow and transport distance over each pipeline section. Since TSOs cannot map their share of the pipeline system separately, they usually include the share owned by others in their calculation, e.g., in the form of a RAC contract.

In addition, there are agreements that even impose tighter restrictions on the gas composition ranges compared to the German DVGW Ordinance G260; see [DVGW-G 260]. For instance, this refers to individual interconnection agreements at border crossings or storage sites for which the EASEE gas common business practices are applicable; see [EASEE CBP]. These regulations are necessary, because an unacceptable gas composition can, for instance, trigger an increase in the bacteria population in storages. In addition, in pore storages, natural gas containing hydrogen causes certain bacteria to produce highly toxic H_2S.

All of these contracts and agreements have to be respected for sound capacity and network planning. Thus, modifications in any of these contracts automatically change the validity period. As explained above, the same holds true for changes due to network expansion measures and new statistical information.

4.2.4.3 ▪ Market area cooperation

As described in Section 3.2.2, the establishment of cross-property market areas in Germany has resulted in an urgent need for cooperation between all network operators involved. The network operators are connected to each other at physical and virtual network interconnection points. In this context, virtual network interconnection points are points that cannot be allocated to a real entry or exit, but are common property within a pipeline.

Roughly two models of market area cooperation can be distinguished. The first is the so-called network interconnection model where at least two transport networks are connected by physically existing network interconnection points. If the flows and the pressures at these points are known, one independent calculation can be made per network, as described in Section 4.2.4.4 on the basis of historical data.

The second is the so-called common capacity calculation model. In contrast to the first model, this can be used in any case, not only if physical network interconnection points exist. It is usually considered if parts of the network are shared and appear in the same market area. In this case, only virtual network interconnection points exist. Moreover, this model is useful if an isolated simulation of a separated network is not desired. The capacities calculated have to be confirmed by all partners involved in the calculation. The distribution of capacities into marketable shares per TSO is then a commercial question.

4.2.4.4 ▪ Stochastic approach for different consumer types

Historically measured exit flow data can be analyzed and used to predict the behavior of customers if the validity of the data is verified on a regular basis. To be able to make reliable forecasts, the data have to be differentiated according to consumer characteristics (see Figure 13.1). In this context, it is helpful to know which type of customer is supplied at a given point, i.e., a municipal utility, an industrial customer, a power station, or a storage site. For some of these customer types, measured flow values correlate strongly with the current temperature and the season. As mentioned above, capacity contracts at the relevant points have to be considered in order to interpret the measured data. This is explained in more detail in the next section.

Interfaces to other transmission systems, i.e., border and market region crossings, are often significantly more volatile in terms of utilization behavior, because they are used as alternatives to other sources and consumption regions. Since gas prices are subject to extreme fluctuations on the spot market, measured gas flows are also volatile. These fluctuations are particularly high at hybrid points, which can be used as both entry and exit points.

Storages are also hybrid points and traditionally serve to balance seasonal fluctuations; recently, however, they have also been used for short-term trading transactions. In many cases, a temperature-dependent tendency (injection in the summer, withdrawal in the winter) can be observed, but not reliably predicted. In particular, frequently flow commitments have been concluded for storages, so that no measured data are available without this contractual restriction to gas consumers.

The situation at network connection points is different. The usage behavior of power stations and industrial plants depends on the possibility of using alternative fuels, e.g., oil, and the type of operational management. For instance, the natural gas demand of a glass factory is relatively constant and therefore correlates directly with the production. Gas power stations are frequently employed according to a merit order list, i.e., depending on the currently offered price for electricity, and they are usually temperature independent. In contrast to this, due to their heat generation function, cogeneration power stations and heating stations are primarily temperature dependent. Even the number of heat generators is visible for some point clouds; see, for instance, Figure 13.1 for a typical example. Sometimes it is possible to derive sensible conclusions from the measured data at connection points if they are divided into workday and weekend measurements.

Domestic consumption of end-customers is usually quite inflexible and is characterized by a temperature dependency with the exception of a relatively low base load. Thus, it possible to create good forecasts for future consumption. As above, a division between workday and weekend might give a more accurate estimate. Difficulties might arise when several points are assigned to a downstream network operator for the same sales region, e.g., when power stations and industrial customers are located in a downstream area and their consumption has been measured alongside others.

For entries and exits without measured data or reliable temperature dependence, so-called worst-case assumptions on minimum and maximum flows are made, which will be explained in the following.

4.2.4.5 ▪ Combining statistics and contractual situation

For exits without statistical data, zero is assumed as a minimum flow, and the sum of contractual obligations is assumed as a maximum flow. This assumption is likely to represent the worst-case in terms of planning. Statistical considerations might help to improve these limits, but not all contracts can be statistically analyzed.

4.2. Generation of scenarios

Contracts have to be differentiated between *substitutable*,[15] i.e., statistically relevant data are available, and *nonsubstitutable*. For planning purposes, only statistical analyses of substitutable contracts should be incorporated into the contract modeling. Examples for nonsubstitutable contracts are those which have been concluded, but which will be exercised in the future, so that they are not taken into consideration in terms of measured data. Other nonsubstitutable contracts are, for instance, restrictively allocable contracts or interruptible contracts; see Section 4.2.4.2 for a discussion of such contracts. Furthermore, at certain points with collective delivery the measured values or contract values (or both) must be multiplied by a factor that reflects the individual share of the network operator.

As discussed above and illustrated in Figure 13.1 of Chapter 13, measured values have different characteristics. Whereas many exits show a relatively tight, temperature-dependent consumption behavior and therefore enable a good estimate, other values are nearly independent from the temperature and at best allow an estimation of the minimum and maximum consumption over the entire temperature range.

As explained above, these characteristics frequently are highly dependent on the type of customers located downstream or upstream. Accordingly, exits towards DSOs, power stations, industrial customers, market areas, or cross-border interconnection points are described in detail in the following.

Downstream network operators, for instance, are municipal utilities with extremely distinct temperature-dependent consumption behavior. At such points, statistical data do not have any particular importance, because customers adjust their bookings according to measurements of the previous year (see [KoV, §A.2]). For this purpose, they use a mean curve through the point cloud of plotted measured values depending on the temperature. Generally, the curve is flattened in the warm temperature range, since consumption is not increased above a certain base load, when all heatings are inactive. In the cold temperature range, the curve usually also flattens, because consumption does not increase when all heatings are already operating at their performance limit.

If all contracts at certain exits are substitutable, statistical estimates on the capacity determination can be used. For instance, the potential of a relieving exit can be used for the entire network if the minimum flow can be corrected upwards in a strongly temperature-dependent manner (see Section 4.2.3.4). Occasionally, downstream network operators have combined several points to exit zones in one booking. The distribution of capacities within these zones can be adjusted in consultation between up- and downstream network operators, e.g., depending on the season.

For power stations and industrial customers, the statistical point cloud is usually less temperature dependent. However, frequently good approximations for the minimum and maximum flow can be found. If the statistical forecast is below the maximum flow, this could be a booking that already contains quantities planned for the future. If a lower limit can be estimated for the minimum flow, this can be used in connection with its relieving potential. In this context, however, the booking situation must be taken into consideration. For instance, if a contract ends, the minimum flow can no longer be used for planning purposes. One example of such a termination can be the shut-down of a plant or the insolvency of an industrial customer.

A similar procedure may be applied at market area crossings with statistically relevant point clouds. Instead of capacity bookings at network interconnection points within market areas, there are usually market area cooperation contracts. With those contracts, adjacent network operators limit the flow range.

---

[15] Note that points with substitutable contracts are called statistical points in Chapter 14.

4.3 ▪ Network control options in simulation

The current state of the art in simulation software enables the complete modeling of all passive and active elements in the network, frequently with the option of both stationary and dynamic simulation. Depending on the software, there are various levels of detail and options. At least the use of active elements such as compressors, regulating systems, and gate valves, as well as the monitoring of technical boundary values, is standard.

Often there are several sensible choices to control active elements, each of which might lead to a feasible setting. This fact complicates the practical procedure and shows that the result is not necessarily unique.

4.3.1 ▪ State of the art in simulation software

For the modeling of complex interactions and effects in gas networks, the current state of the art is the use of simulation software, such as [AFT], [Flowmaster], [MYNTS], [PipelineStudio], [PIPESIM], [PSIGanesi], [SIMONE], [STANET], [SPS], [WinTran], or in-house developments of network operators like MCA (van der Hoeven (2004)). These programs simulate the gas physics and characteristics of the active and passive elements of the network.

Packages like these have been developed during the last decades based on earlier scientific work. For instance, PSIGANESI grew out of two dissertations and subsequent investigations (Weimann (1978); Lappus (1984); Scheibe and Weimann (1999)). Similarly, SIMONE is based on a series of mathematical investigations by the original developers (Králik et al. (1984), (1984), (1988)), and has steadily been extended and refined as a commercial product (Králik (1993); Vostrý (1993); Záworka (1993); Vostrý and Záworka (1995); Záworka (2004)). An optimization add-on based on steepest descent has also been developed (Jeníček (1993)). Thus, stationary and dynamic simulation of real gas networks can be considered a mature area of numerical mathematics.

The following presentation is motivated by SIMONE. The mathematical representation is done by physical models similar to those described in Chapter 2. Such DAE/ODE systems can be solved more or less exactly by the computing engine, depending on the respective software. The control options and boundary values of the network elements are ideally based on real-world control options and technical restrictions, e.g., the definition of the maximum flow.

In the majority of cases, both stationary and dynamic simulations can be carried out, which can be applied depending on the area of application; see Aymanns et al. (2008) for an illustration of a dynamical simulation application. For instance, for calculations on the basis of simultaneously measured values in an online calculation for calorific value tracking, a dynamic simulation is required, since the input and output volumes are not balanced and the pressure conditions in reality are influenced by dynamic effects. This behavior cannot be reproduced by stationary simulation.

Today's simulation software can reach its computing limits, due to the ever-growing size of networks to be simulated, e.g., because of the consolidation of market areas. Consequently, more research is needed in order to increase the solvable network sizes and improve the solution speed; see also Section 11.1.

4.3.2 ▪ Treatment of network elements in simulation

For the simulation of a gas network, a model-based description of the passive and active elements with their attributes and physical restrictions is needed. Interconnection points of network elements are modeled via nodes. Each network element corresponds to an arc with corresponding start and end nodes; see Chapter 2 and Chapter 5. As described in Chapter 2, pipelines are passive elements whose capacity is determined by their particular characteristics. Active elements are compressor stations, control valves, and valves that should control gas flows such that the gas network handles the given load situation as efficiently as possible. The corresponding network description and parameters are either entered into the program by means of an editor or imported as a specifically formatted text file.

The iterative adjustment of control settings to the simulated flow situation and the desired network operating mode is done manually to a large extent. In a local context, the user can use programmable functions for simplification. The software tracks whether applicable boundary values, e.g., the minimum pressure at an exit point, are respected and reports violations if necessary.

In the following, we discuss the treatment of the different network elements; see Chapter 2 for a detailed introduction.

4.3.2.1 ▪ Pipelines

Pipelines are not controllable. The adherence of technical constraints is monitored by the simulation software. Simulation also computes temperature, pressure, and flow velocity of the gas. If boundary values for these properties are violated, it might be necessary to make readjustments by means of active elements. Furthermore, effects such as the blending of different gas mixtures in the pipeline are simulated. See Section 2.2 and Section 2.3.1 for more details on gas properties and pipes, respectively.

4.3.2.2 ▪ Compressor stations

Compressor stations are the most complex active network elements; see Section 2.3.5 for an overview. In a manual simulation, compressor machines are regulated alternatively according to the following variables:

▷ output pressure,
▷ input pressure,
▷ pressure ratio,
▷ flow rate, and
▷ speed/number of rotations.

They can also be programmed in the form of certain operating modes by means of functions in the simulation software. This programming option can be extended to the active elements within a compressor station, such as valves that control the combination options of the individual compressors. In addition, internal serial and internal parallel configurations of a compressor group can be modeled by means of entering several characteristic diagrams.

Moreover, one can specify control settings that describe the behavior of active elements like compressor machines and control valves. One common setting is to preset a certain outlet pressure or flow at a compressor machine or control valve. The simulation software automatically determines the physical settings in order to meet this preset value if this is possible. Furthermore, control settings also include the settings of all active elements, e.g., valves are determined to be "open" or "closed."

4.3.2.3 ▪ Gas pressure regulating and measuring stations

Control valves allow the operator to reduce/regulate output pressure; see Section 2.3.4 for more details. Control valves are defined by the diameter relevant for the pressure loss, the input and output resistance, as well as the minimum input and output pressure. A control valve can be simulated on the basis of the following properties:

- output pressure,
- input pressure,
- pressure ratio,
- flow rate,
- "free flow" (bypass), and
- being "closed."

In addition, type-specific control valve characteristics can be used to describe the flow through the valve depending on the relation between the input and output pressure values and the opening of the control valve (in %). Thus, the control valve can be set to a certain percentage opening in the control instruction.

The temperature loss caused by the pressure reduction (Joule–Thomson effect) is calculated in the simulation. Either the existing preheating capacity or the desired output temperature can be entered. This makes it possible to integrate a check of the preheating capacity into the network calculation.

Just like compressors, control valves also allow control settings and can be controlled by functions that have been programmed by the user.

4.3.2.4 ▪ Valves

Valves have two possible states: open and closed. Just like other active elements, valve positions can be programmed depending on certain network operating modes.

4.3.2.5 ▪ Storages

In a stationary calculation, storages are considered in analogy to other entries and exits, which are analyzed in Section 4.2.4.4. In dynamic considerations, it is also possible to enter the working gas volume, as well as the injection and withdrawal rate, depending on the fill level for storages, in the simulation software.

With this information it is then possible to dynamically simulate effects such as sudden high withdrawal rates or extremely low fill levels.

4.3.2.6 ▪ Network expansion

If expansion measures have to be evaluated that aim at removing congestions, the system elements in question, e.g., parallel pipelines or compressor machines, must be entered manually into the network model. Existing simulation software cannot automatically generate expansion elements. To determine the expansion measures to be taken, several options should be simulated for different load scenarios. The results must be analyzed, compared, and evaluated. Inappropriate options are sorted out and redefined, if necessary. The resulting selection of options is then again analyzed, compared, and evaluated. This iterative process is repeated until good technical and economical solutions are found. The larger and more complex the gas network, the more expansion options must be calculated. Due to the large number of necessary manual steps, this process requires a large amount of work. Indeed, ideas for future research in this direction are discussed in Section 15.4.

4.4 • Simulation

The calculation of the capacity of gas networks by means of simulation is done by selecting scenarios defining in- and outflow to be tested. These must be balanced for a stationary calculation. However, they can fall into imbalance and will then have to be iteratively compensated for in subsequent simulations as described in Section 4.4.2 below. In addition to in- and outflows, for a simulation a set of control settings, boundary values, and parameter settings are required. The scenarios cover a validity period that depends on the factors of network expansion status, contract situation, and statistics.

In the following we describe one approach via simulation of an example using SIMONE. Different software could, for instance, treat boundary value violations and control settings for network elements or imbalances differently.

4.4.1 • Boundary values and settings

As described in Section 4.3.2, all network elements have boundary values that must not be exceeded. Typically, some of them are hard-coded for the simulation, and there is no lower or upper deviation allowed. One example of this is maximum compressor and drive capacity, which can be modeled by means of characteristic diagrams and capacity limits. Soft-coded boundary values in simulation software may be violated during simulation, e.g., the maximum gas velocity or maximum pressure in pipelines, the maximum temperature at the compressor outlet, or the minimum flow through a gas pressure measuring device and regulating station.

In addition to boundary values, active elements (compressor machines in particular) are regulated by control settings. Typically, in the case of flows or pressures, these control settings are adhered to or, at best, undercut, if it is not possible to achieve them due to other equivalent control settings or hard boundary values. In this context, control settings for pressures have priority over those for flows.

Such a hierarchy of hard boundary values, control settings, and soft boundary values serves to facilitate modeling by means of an iterative process. The solutions obtained during the process usually do not represent realistic cases, while the final solution can be used for practical planning if the process is successful. If after a simulation run some bounds on boundary values cannot be met, changing the network control by means of customized control settings might alleviate the violation or even remove it. This can be performed in an iterative fashion. If a solution without boundary value violations cannot be found in this way, this scenario is regarded as infeasible. In this case, it is assumed that the load situation in the network actually cannot be realized.

Further parameter settings that have to be chosen according to the individual scenarios are, for instance, the definition of ambient and ground temperature, as well as the use of physical calculation formulas, such as for the gas composition or the assumption of an integral heat transition coefficient.

4.4.2 • Balancing of scenarios

Stationary simulation needs balanced scenarios, i.e., the same amount of gas is injected and withdrawn simultaneously. This applies to the total balance of the network to be calculated as well as to the individual separate subnetworks, e.g., subnetworks that are supplied via a regulating valve. With knowledge of the topology of the network with respect to in- and outflows, a balanced load condition can, to a large extent, already be achieved during scenario generation. However, in Germany capacity products today are no longer sold as volume flow per time unit but as energy amount per time unit. Some

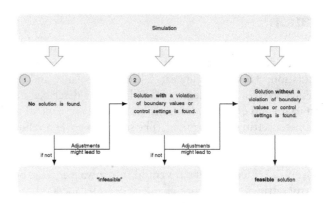

Figure 4.4. *Possible results of simulation software.*

simulation software enable the mapping of energy amounts at entries and exits and calculate the resulting volume flow by means of the gas calorific value in an interactive process, which can lead to an imbalance of the volume flows; see also Section 5.3.3 for a discussion.

In addition to the imbalances mentioned above, the internal gas consumption results in an uneven gas balance. Elements such as preheaters of control valves and, in particular, compressors require gas for their operation, unless they are operated by electricity. This so-called fuel gas is withdrawn directly from the passing gas flow. In this context, it is common practice to use around 0.5 % of the transported gas as fuel gas for national transport. Since the fuel gas consumption is calculated for an individual scenario on the basis of the compressor utilization, this can only be approximated in advance. Determining the exact demand is an iterative process that depends on the respective modeling of the active elements in the scenario. The demand varies according to the flow situation, because different machines are subject to different loads.

Certain simulation software also allow introducing a flow imbalance for scenarios that cannot be realized using the current settings of active elements. In this case, additional in- and outflows are stipulated in order to realize the scenario. The imbalance can possibly be removed by readjusting the active elements.

4.4.3 • Iterative simulation process

For a simulation, a complete set of boundary values, parameter settings, and control settings must be available. In this context, depending on the simulation software, the lack of certain data can also be compensated for by predefined values, and the simulation can still be performed. Three possible results of a simulation can arise (see Figure 4.4):

1. The simulator does not find any solution for the conditions defined.
2. The simulator finds a solution that only partly fulfills the boundary values and control settings defined.
3. The simulator finds a solution that meets all requirements.

If the simulator does not find a solution, there may be different reasons. For instance, the load situation of the network in the scenario might lead to negative pressure values at certain network elements. Another reason can be that the sum of the maximal flow into a subnetwork (usually via control valves) is smaller than the outflow at its exits. In this context, simulation software can be equipped with the option to counter imbalances occurring in the entire network or subnetworks thereof, e.g., by increasing or decreasing the inflow at entries. However, this function works only if there is an entry within such a subnetwork. For this reason, a control valve for ancillary feeding into an area is frequently controlled not on the basis of volumes, but on the basis of pressures in order to inject the residual gas volume. This works if the pressure before the control valve can be set to be large enough, so that the required residual volume can flow through the control valve into the closed area. Otherwise, the simulation of the scenario is aborted as unsolvable. On the other hand, if a solution without violation of technical bounds and settings exists, such a solution will be found by simulation. One example where a violation of settings occurs is when the gas consumption in an area is smaller than the entry volumes requested. In this case, the flow and entry values defined by the control settings are underrun (see Section 4.4.1). Another example is the exceeding of the maximum velocity of individual pipe sections.

As depicted in Figure 4.4, an iterative simulation process via adjustments of the control settings might lead from the case without any solution to a solution with a violation of some boundary values. Similarly, the latter case can possibly be transformed into the situation in which a valid solution is available. If this process fails, it is assumed that the gas network cannot realize the load situation in reality.

The running times of the simulator usually takes minutes, but may take days for very large networks. The number of different settings tested depends on the experience of the user and the complexity of the network—no general rule can be given.

4.5 ▪ Interpretation of calculation results

Among all scenarios feasible for the validity period under consideration, the minimum over the flow values calculated is determined at each individual point as well as for entry groups. This flow can definitely be represented in feasible scenarios and scenarios that are supposed to be restrictive and can be marketed correspondingly, under the convexity assumption. This approach can be supplemented by restricting the validity of the scenarios, e.g., by their temperature dependence. This approach frequently results in higher marketable capacities, because the minimum has to be formed over fewer scenarios; see Section 4.5.1 below. However, this restricts the behavior of transport customers. Moreover, in the majority of cases, the analysis of the scenarios results in an imbalanced offer of bookable capacities.

4.5.1 ▪ Minimum calculation

As described in Section 4.2, for each validity period, various restrictive scenarios are calculated. The interpretation of the results is done in different ways, depending on whether the capacity for entries or exits has to be determined.

For entries, the scenarios test at medium and/or warm temperatures how much gas can be transported into the system without pressure violations or other restrictions (see Section 4.4). For this purpose, several scenarios have to be calculated, since the inflow is strongly dependent on the temperature and consumption behavior (see Section 4.4.1).

The minimum flow of each entry regarding all considered scenarios is calculated. This can then be marketed as available capacity at this entry. The new freely allocable capacity that might have been added is derived as the technical capacity minus the capacity that has already been marketed.

Entries depending on each other in terms of flow are not tested individually, but in so-called entry groups (see Section 4.2.3.3). When interpreting the calculation results, the minimum over the entire capacity at this entry group is calculated as described above.

This technical capacity is then reasonably distributed among the individual entries, taking the minimally and maximally possible entry flows into consideration.

For exits, mainly peak load scenarios at design temperature are considered. If there are no restrictions through any of the tested scenarios, the tested value can be marketed. An increase of the technical capacity can occur, for instance, if the statistically present reference value at design temperature is higher than the technical capacity that has already been marketed at this exit. In this case, the difference is represented as FAC, so that the customers can adjust their contract value according to their consumption behavior.

If restrictions occur through one of the tested scenarios, the relation between the statistical consumption behavior and the contract value must be investigated first. If the statistical consumption behavior is higher than the contract value, the contract value is tested; the additional FAC is not handled as described above. If the contract value also results in restrictions through the scenarios, a solution must be developed in cooperation with the customer in a nondiscriminatory manner.

4.5.2 • Further capacity products

If the transport network operator wants to account for, e.g., temperature-dependent capacities, the interpretation of the simulation results as described in Section 4.5.1 must be adjusted; see Section 3.3.2 for a discussion of such capacities. In this case, at entries the minimum of the possible inflow is no longer calculated over all scenarios, but only over scenarios in a certain temperature interval. For colder temperatures, due to the higher consumption at exits, this usually leads to the fact that a higher technical capacity can be marketed at the entries. The more fine-grained these temperature-dependent entry capacities should be marketed, the more scenarios have to be calculated. For exits, the same process can be used.

4.6 • Conclusions

In this chapter we have presented the state-of-the-art approach for capacity evaluation of gas networks. The approach consists of a generation of scenarios based on statistical and contractual information. These scenarios are tested using simulation software. Experienced planners have to determine control settings of active elements and interpret the results.

As the presentation has shown, this process is quite involved and time-consuming. One main goal of this book is to propose an alternative approach that is based on the following improvements of the process described in this chapter:

▷ The most time-consuming step in the process, the iterative determination of the control settings for active elements, should be done automatically and integrated with the solution of the physical network model. This cannot be done by simulation. From Chapter 5 to Chapter 10, we introduce an optimization methodology that handles this step.

▷ The generation of scenarios should be automized, while still taking statistical information into account. Moreover, each scenario should have an attached probability that allows one to (semi)automatically evaluate the failure to find a feasible control setting. An analysis of the statistical data is presented in Chapter 13, while Chapter 14 describes a methodology to generate scenarios automatically.

In this way, the methods presented in the rest of this book improve the practical procedure in the extremely relevant area of gas transport.

Part II

Validation of nominations

Chapter 5
Mathematical optimization for evaluating gas network capacities

Lars Schewe, Thorsten Koch, Alexander Martin, Marc E. Pfetsch

Abstract *This chapter describes the way we use mathematical optimization to deal with the planning problems outlined in the preceding chapter. Our main tool is a hierarchy of different optimization models. We present different approaches that are detailed in the following chapters and discuss the corresponding modeling decisions that have to be taken.*

As discussed in Chapter 4, simulation is state of the art in gas transportation planning. In order to extend the application of simulation to a fully automatic planning process, one needs to incorporate (discrete) decisions that network operators are allowed to take for active elements. Moreover, these decisions should be optimal in some sense. Consequently, we arrive at optimization models and methods for gas transportation.

As we are interested in mid- to long-term planning, we are considering stationary gas flows. The main goal is to get (stationary) *optimization* tools that are able to match the quality of stationary solutions obtained by *simulation* tools. One classical way of achieving this goal is to set up *one* optimization model that tries to capture all relevant aspects of the problem. However, the (global) solution of such a master model for real-life networks is way beyond the capabilities of today's optimization methods, and it will probably not be possible to compute such a solution within any realistic time. Consequently, one needs to simplify and approximate certain aspects. This leads to the notorious problem of finding a good compromise between a relatively accurate modeling of the physics of the problem (as in the case of most nonlinear models) and the incorporation of the combinatorics of the problem (as in the case of many "discrete" models). Good solution methods have been developed for each of the resulting models.

Our approach is to develop a hierarchy of models that capture different aspects of the problem. The primary principle of organization is along faithfulness to the underlying physics. However, it will turn out that not all models allow a strict hierarchy in the sense that solutions from a finer model can always be "coarsened" to a solution in the coarser model. Additionally, the different network elements all need their own models. So the building blocks of our hierarchy are different models for each component. These blocks will be outlined in the following sections.

The following chapters will then show how the different components can be integrated into coherent mathematical programming models. These chapters are organized by the type of optimization model. The next four chapters (Chapter 6–9) present approaches to find discrete decisions for the active elements—we call these *decision approaches*, while the nonlinear program (NLP) approach in Chapter 10 will be used for obtaining a high accuracy solution based on the discrete decisions of the decisions approaches, which are thus *validated*.

The first two approaches solve a mixed-integer nonlinear program (MINLP) model of the underlying problem. In both cases the physics needs to be simplified. In Chapter 6, a mixed-integer linear program (MILP) model is introduced that linearizes the nonlinearities using an a priori piecewise-linear relaxation of the underlying MINLP. Chapter 7 describes a branch-and-cut approach based on spatial branching and the solution of an NLP in the leaves. A heuristic approach based on a reduced NLP is outlined in Chapter 8. Chapter 9 presents a heuristic approach using a more detailed NLP with complementarity constraints in order to represent discrete decisions. The model with the highest physical accuracy is the NLP model considered in Chapter 10. Since discrete decisions cannot be computed in this NLP model, it is mainly used to validate solutions obtained for the other models. A discussion of the feasibility of these results is given in Chapter 11. Computational results of these approaches are discussed in Chapter 12.

5.1 ▪ The building blocks of our hierarchy

The hierarchy builds upon different models for the components of the network. The models we use to describe these different components cannot necessarily be ordered strictly. Some approximations, for instance, cannot be directly compared as there are different regimes (say, turbulent flow vs. laminar flow) in which the approximations behave differently. Also NLP solvers tend to need differentiability, which sometimes makes the smoothing of functions necessary. The occurring nondifferentiable functions, however, can be adequately modeled in an MILP model; the nonlinearity of the function is the problem in that setting and introduces inexactness.

With the exception of the NLP model described in Chapter 10 we do *not* take into account the effects of temperature on the gas flow. We also assume that we are dealing with a single type of natural gas and do not need to take into account the effects of mixing different types of gas. These decisions are discussed in Section 5.3.2 and Section 5.3.3, respectively. For this reason we have not included any temperature related parameters and quantities in the following discussions. For a discussion of these we refer the reader to Chapter 10.

The following description is meant as an independent introduction into the setup of the hierarchy. Thus, some redundancies with Chapter 2 arise. For more details, we refer the reader to this chapter.

5.1.1 ▪ The abstract network

We model our gas network as a directed graph $G = (V, A)$. The state of the network is given by a *flow* q_a on each arc $a \in A$ and a *pressure* p_u on each node $u \in V$.

The purpose of the model is to express the relation between pressure and flow over each arc of the network, i.e., to relate for each arc $a = (u, v)$ the pressure p_u at node u, the pressure p_v at node v, and the amount of flow q_a.

To formally state our problem, we need additional notation. We distinguish three types of nodes: *entries*, i.e., nodes which supply gas, *exits*, nodes which demand gas, and *inner nodes*, i.e., nodes which neither supply nor demand gas. We denote the set of entries, exits, and inner nodes by V_+, V_-, and V_0, respectively. All other elements are modeled as

5.1. The building blocks of our hierarchy

Table 5.1. *Notation for arc types (excluding parameters for temperature-dependent models).*

Arc type	Set	Parameters		Symbol	Description
Pipe	A_{pi}	diameter	D_a	•———•	Sect. 2.3.1
		roughness	k_a		
		length	L_a		
		slope	s_a		
Resistor	A_{rs}			•—▭—•	Sect. 2.3.2
• flow-dependent	$A_{\text{nl-rs}}$	diameter	D_a		
		drag factor	ζ_a		
• constant	$A_{\text{lin-rs}}$	pressure reduction	ξ_a		
Valve	A_{va}			•—⋈—•	Sect. 2.3.3
Short cut	A_{sc}			•———•	
Control valve	A_{cv}			•—▷—•	Sect. 2.3.4
• automated	$A_{\text{cv}}^{\text{aut}}$	max. pressure red.	$\overline{\Delta}_a$		
		min. pressure red.	$\underline{\Delta}_a$		
• nonautomated	$A_{\text{cv}}^{\text{man}}$	preset pressure	p_a^{set}		
Compressor machines	A_{cm}	see Sect. 5.1.6		•—◯—•	Sect. 2.3.5
Compressor groups	A_{cg}	see Sect. 5.1.6		•—⊘—•	Sect. 2.4.2

arcs in the graph. These elements will be explained in the following sections. Notation for the different arc sets is given in Table 5.1; a detailed discussion is given in Chapter 2 as indicated in the table. The main distinction we make is to consider passive and active elements: Passive elements are those elements that are not directly controllable, i.e., pipes, resistors, and short cuts. The corresponding set of arcs is denoted by A_{passive}. Active elements have one or more parameters that can be controlled. The active elements we consider here are valves, control valves, and compressors. The corresponding set of arcs is denoted by A_{active}.

We note the following abuse of notation: Control valves and compressors are typically part of a more elaborate setup of elements (e.g., with additional resistors and valves), so-called groups and stations (see Sections 2.4 and 5.1.6). We use the respective symbols for compressors and control valves not only for the different elements, but also for the complete groups. More formally, the main graph contains one arc per compressor group and control valve station. Some of the models expand these arcs by introducing certain subgraphs containing arcs for each component of the station.

5.1.2 ▪ Pipes

The main parameters of a pipe $a \in A_{\text{pi}}$ are its length L_a, diameter D_a, and integral roughness k_a, which is used in the computation of the friction factor λ (see Section 2.3.1). To compute the pressure loss over a pipe we also need to take into account the height differences of its endpoints.

In modeling pipes we choose from among the following options:

Basic equations for gas flow

▷ discretized form of the Euler equations (2.12)–(2.14),

▷ algebraic form of the isothermal Euler equations (2.24),
▷ algebraic form of the isothermal Euler equation (2.24) without inclination (i.e., $s = 0$).

Friction model

▷ a smooth approximation of Hagen–Poiseuille (2.16) and Prandtl–Colebrook (2.17),
▷ Hofer's equation (2.18),
▷ Nikuradse's equation (2.19).

Compressibility factor approximation

▷ AGA8-DC92 formula (Starling and Savidge (1992)),
▷ Papay's formula (2.4),
▷ AGA formula (2.5).

Mean pressure calculation

▷ a priori estimation using lower and upper bounds (e.g., as in (10.23)),
▷ solution-dependent value (2.28).

From the point of view of exactness it is preferable to choose the discretized Euler equations using a smooth approximation of the friction factor and the AGA8-DC92 or Papay's formula for the real gas factor. These factors are pressure dependent, which introduces another set of equations which needs to be solved. This leads, however, to a far too complicated model for the approaches incorporating discrete decisions. These approaches use one of the two variants of the algebraic form of the isothermal Euler equations. To precompute the above mentioned factors, one uses a pressure estimate using the pressure bounds of the start and endpoint of the pipe. To further simplify the computations, one chooses the friction model of Nikuradse, since this model assumes that the friction factor is independent of the amount of flow. The mean pressure is estimated a priori. Both of these choices simplify pressure loss computations.

The nonlinear optimization model of Chapter 10 is the most flexible model when modeling pipes. It can also achieve the highest physical accuracy. It is able to use both the discretized Euler equations as well as the algebraic model. By default it always uses a smooth approximation of the friction factor. Both in approximating the compressibility factor and, if required, the calculation of the mean pressure, the model is able to use all the methods described above.

The MILP model of Chapter 6 uses the algebraic equation (2.24) as the basic model. It uses Nikuradse's equation to determine the friction factor and the AGA or Papay equation as approximation of the compressibility factor. The mean pressure is approximated a priori. The so-called sMINLP model of Chapter 7 makes the same choices as the MILP model with the slight exception of using only the AGA equation for the approximation of the compressibility factor. The reduced NLP model of Chapter 8 neglects the inclination of the pipes; apart from this, it makes the same choices as the sMINLP. The mathematical program with equilibrium constraints (MPEC) model of Chapter 9 is in principle able to use the same approximations as the NLP model. However, we typically choose the simplest model.

5.1.3 ▪ Short cuts

In many places of the network very short pipes are used to connect different components, e.g., in compressor stations. It is not sensible to model all of these as ordinary pipes.

We thus introduce a network element called *short cut*. Gas flowing through a short cut experiences no pressure drop and there are no restrictions on the amount of gas flow. Elements of this type are also introduced artificially to model specific network situations. A typical use is to model physical nodes in the network that can be used as entries *and* exits. This situation is represented by an entry and an exit node that are connected by a short cut. These elements are easily incorporated in all of the considered models.

5.1.4 ▪ Resistors

Resistors represent components of the network that are not modeled except for their additional resistance (see also Section 2.3.2). They are used, for instance, to model the pressure loss induced by complex piping inside compressor stations. In simulations, resistors are usually modeled using one of the two following resistor types. For flow-dependent resistors, the main parameters are a fictitious diameter D_a and the drag factor ζ_a. The equation used to model the pressure loss along the resistor is a modified Darcy–Weisbach equation (2.30). For resistors with constant pressure drop as described by Eq. (2.32) the only interesting parameter is, of course, the pressure loss ξ_a.

Most of our models use Eq. (2.30) or the constant pressure drop to model the behavior of the resistors, depending on the type of resistor. The exception is the sMINLP model, which, in both cases, uses a quadratic approximation to model resistors as a special sort of pipe. This approximation is needed since the sMINLP model uses variables only for the squares of pressures and not for the pressure itself.

The effect of these elements on gas temperature is modeled in the finer levels of the NLP model. We give an overview in Section 5.3.2.

5.1.5 ▪ Valves and subnetwork operation modes

A valve is a simple active element that can be *open* or *closed* (see Section 2.3.3). In our models we consider only those valves that can be operated remotely; all other valves are either removed (if they are closed) or treated as short cuts (if they are open). In many cases, however, the valves should not be operated independently. Instead, the dispatcher has a number of typical aggregate settings that affect groups of valves, compressors, and control valves in a bigger subnetwork. A simple example is the diagram of a compressor group in Figure 5.2: The two valves at the inlet and the outlet of the group should not be operated independently.

Subnetwork operation modes aggregate the different possibilities of switching valves and other active elements (see Section 2.4.3). They restrict the set of possible decisions. The MILP and sMINLP models both incorporate these by constructing an extended formulation of the resulting polyhedron described in Chapter 6. The reduced NLP incorporates subnetwork operation modes in its general decision making framework. The MPEC model is currently not able to deal with subnetwork operation modes. It only analyzes them to tighten bounds. The additional number of discrete decisions would make it impractical.

5.1.6 ▪ Compressors and compressor stations

As described in Chapter 2, compressor stations are an important part of gas networks. Moreover, compressors are the main contributors for the operating cost of the network, which makes correct modeling of their behavior even more important. Each compressor station is modeled by a subnetwork containing one or more compressor groups. These

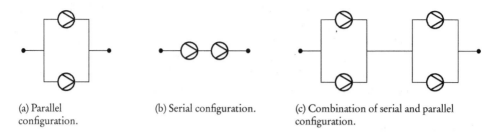

(a) Parallel configuration.

(b) Serial configuration.

(c) Combination of serial and parallel configuration.

Figure 5.1. *Exemplary compressor group configurations.*

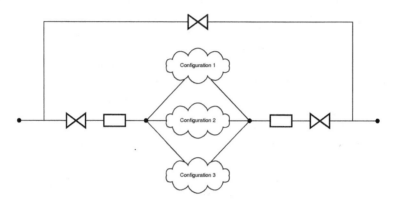

Figure 5.2. *Diagram of a compressor group.*

consist of a number of compressors, which can be used in different, predefined configurations, i.e., in series or parallel or adequate combinations (for examples, see Figure 5.1).

Two main quantities are of interest to all our models: the pressure increase Δp of the given compressor group and the *throughput*, i.e., the amount of flow through the compressor group. In general, there is a trade-off between these two quantities: Not every combination of pressure increase and throughput can be achieved. In general, serial operation of compressors is used to yield higher pressure increase and parallel operation to yield higher throughput. Additionally, there are situations where it is necessary to close the group and, thus, block the flow or bypass the whole group. When we are also interested in optimizing the operating cost of the network, another aspect is most important: the amount of energy needed to operate the compressor. Here an additional modeling trade-off is due to the fact that at different operating points the efficiency of the compressor can be significantly different.

We can choose one of the following main methods to model a compressor group:

▷ use a separate model for each compressor machine and for its drive,
▷ model each compressor machine separately as an idealized compressor,
▷ use one model for each compressor group as one entity.

To simplify the various possibilities for the operation of a compressor group we use the following general setup (see Figure 5.2): We recall that the main operation modes of a compressor group are *bypass*, *closed*, and *active*. At the inlet and the outlet of the group we have introduced two valves to close or open the group; an additional valve outside the group models the *bypass*. These three valves together allow the decision for one of the modes. In *active* mode, the group can operate in one of many different configurations

5.1. The building blocks of our hierarchy

Table 5.2. *Common and type-specific compressor quantities.*

Common quantities	
Specific change in adiabatic enthalpy	H_{ad}
Power	P
Specific energy consumption	b
Adiabatic efficiency	η_{ad}
Speed	n
Isentropic exponent	$\varkappa$
Absolute pressure increase	Δp
Relative pressure increase	ε
Turbo compressor quantities	
Minimum speed	$\underline{n}$
Maximum speed	$\overline{n}$
Surgeline	quadratic polynomial
Chokeline	quadratic polynomial
Piston compressor quantities	
Minimum speed	$\underline{n}$
Maximum speed	$\overline{n}$
Shaft torque	M
Maximum shaft torque	$\overline{M}$
Operating volume	V_o

(see Figure 5.1). To model these we do not try to replicate the original internal piping of the group, but instead put each of these configurations on a path of its own. The decision between these configurations is then taken by valves that decide on the used path. For validation using the model described in Chapter 10 it is necessary to provide the model with a configuration and a fixed setting of the valves. The models that do not use a detailed machine model need to reconstruct such settings. For this we use a heuristic procedure integrated into the validation that is explained in Section 5.4.

As mentioned in Section 2.3.5.1, the different compressors can be categorized along two independent dimensions: machine type and drive type. The *machine type* specifies how the gas is compressed by the compressor. We consider two types in this book: turbo compressors and piston compressors. The *drive type* specifies how the machine is powered. Here three main types occur: electric motors, gas turbines, and steam turbines. As the drives are considered only in the NLP model (see Section 10.1.10.3), we do not discuss them in this chapter. The parameters for all compressor models are summarized in Table 5.2.

As described in Section 2.3.5.1, the main constraints for turbo compressors relate two different quantities: the so-called specific change in adiabatic enthalpy H_{ad} and the volumetric flow Q. The feasible region can then be described in the H_{ad}-Q plane as the so-called characteristic diagram. The restrictions are given via the speed isolines for the minimum and maximum speed ($\underline{n}$ and $\overline{n}$, respectively), the surgeline, and the chokeline of the turbo compressor.

For piston compressors the constraints relate the shaft torque M and the volumetric flow Q. The constraints are again given by the minimum and maximum speed n and a specification of the maximum torque $\overline{M}$, which is either given directly or by specifying the maximum absolute ($\overline{\Delta p}$) or relative ($\overline{\varepsilon}$) pressure increase.

The MILP model of Chapter 6 uses a simplified model for each compressor. The sMINLP of Chapter 7 and the reduced NLP of Chapter 8 model compressor groups as one entity. More precisely, the sMINLP uses a relaxation of the feasible region of the compressor group computed as the convex hull of the union of the feasible regions of the different configurations. The reduced NLP uses the bounds (after preprocessing; see Section 5.4) of the operating range of the compressor group, specifically the computed bounds on throughput and pressure increase (in absolute and relative terms).

The relation of the power consumption of a compressor and its resulting pressure increase is only necessary for our models when we are trying to minimize the energy expenditure or to represent drives accurately. This is integrated into the MPEC model and the validation NLP.

5.1.7 • Control valves

Control valves can regulate the flow or the pressure, i.e., they allow specifying a flow rate or a target pressure and making sure that this flow or pressure is attained (see Section 2.3.4). In the networks considered here, all control valves are modeled as pressure regulating valves, i.e., the target pressure is specified. We do not consider flow regulating valves, where the target flow rate is specified. As with compressor stations, these valves are typically part of control valve stations; see Section 2.4.1 and Figure 2.7 for the typical network layout. These stations have an elaborate piping, which we model using an inlet resistor and an outlet resistor. As the control valve may sometimes be bypassed, we also add (normal) valves to model this bypass (see again Figure 2.7).

A control valve $a = (u,v) \in A_{cv}$ can be *active* or *closed*. Moreover, nonautomated control valves can be in *bypass*. We call a control valve *open* if it is *active* or in *bypass*.

If the valve is active, it is capable of inducing a pressure reduction $\Delta_a = p_u - p_v \geq 0$ within the interval $[\underline{\Delta}_a, \overline{\Delta}_a]$. In the case that the control valve cannot be remotely operated, an outflow pressure p_a^{set} is preset if the control valve is active.

The MILP model directly implements the full station using binary variables to represent the different operation modes. The sMINLP model uses a slight adaptation; additional constraints are needed as the sMINLP model uses only pressure squared variables. The reduced NLP model heuristically chooses *open* or *closed* mode and uses constraints for the open ones to decide active or bypass mode without representing the station network. In the case that the valve cannot be remotely operated, a relaxation is used. The MPEC model uses complementarity constraints to represent the different operation modes.

5.2 • Abstract problem statement

Chapters 6–10 deal with the problem of the *validation of nominations* (NoVa): Informally speaking, for a given network and nomination, we try to find configurations of the active elements such that the given nomination can be transported through the network. The approaches discussed below all deal with special cases of the following abstract problem.

We are given a directed network $G = (V, A)$ with the node set V partitioned as $V = V_+ \cup V_- \cup V_0$ and the arc set A partitioned into the arc sets given in Table 5.1. For each $a \in A$, the required parameters as specified in Table 5.1 are given. Furthermore, we are given pressure bounds at each node $u \in V$ that correspond to different technical requirements. Additionally flow bounds for each arc $a \in A$ are given, which correspond to technical or legal throughput restrictions. The specification of the network is completed by a list of subnetwork operation modes.

5.2. Abstract problem statement

In this book we consider a nomination to be a set of power values P_u for each node $u \in V_\pm$ such that the following balance condition holds:

$$\sum_{u \in V_+} P_u = \sum_{u \in V_-} P_u.$$

With the exception of the NLP model, our models cannot, however, directly deal with such nominations. This is further discussed in Section 5.3.3. We therefore consider in the following mainly nominations that are directly given in mass flow q_u^{nom}, where the following mass flow balance condition holds:

$$\sum_{u \in V_+} q_u^{\text{nom}} = \sum_{u \in V_-} q_u^{\text{nom}}.$$

In the following chapters, the term nomination is used to specify either a setting of power or mass flow values at the entries and exits. We only explicitly state whether the nominations are given in terms of power or in terms of mass flow when this is not clear from the context.

Technically, additional pressure requirements at the entries and exits are also part of the nomination. However, these pressure bounds can in our case be directly calculated from the nominated power/flow values. For ease of terminology we, therefore do not consider them part of the nomination.

The variables in our models can be classified into four types:

1. mass flow variable q_a for each arc $a \in A$,
2. pressure variable p_u for each node $u \in V$,
3. decision variables for all active elements and subnetwork operation modes, and
4. further model-specific variables.

The first two types of variables are used in all models or can be uniquely reconstructed from the solutions. The variables of the third type are fixed in the NLP model.

The constraints can be subdivided in five types:

1. the mass flow conservation condition for each $u \in V$,
2. a set of constraints modeling the corresponding element for each $a \in A$,
3. additional constraints that model further technical restrictions along certain arcs (e.g., maximal pressure, speed restrictions on pipes),
4. a set of constraints modeling the mixture conditions for each $u \in V_0$, and
5. constraints on q_u^{nom} and p_u for each $u \in V_\pm$ that are given directly through the nomination or are implied by it.

The mass flow conservation constraints are the basis of all our models:

$$\sum_{a \in \delta^+(u)} q_a - \sum_{a \in \delta^-(u)} q_a = q_u^{\text{nom}} \quad \text{for all } u \in V.$$

Additionally, $q_u^{\text{nom}} = 0$ holds for all $u \in V_0$. The main difference between our models is the choice of constraints of Type 2 and 3. Mixture constraints (Type 4) are used only by the NLP model. Constraints of Type 2 are local constraints in the sense that such a constraint for an arc $a = (u, v)$ depends only on the flow along this arc q_a and the pressure at its endpoints p_u, p_v (and additional model-specific variables).

A feasible solution for such a problem is, of course, given by values for the different variables such that all constraints are satisfied.

As our problem is first of all a feasibility problem, we do not prescribe an objective function. It can, however, be helpful for some models to introduce an artificial objective (e.g., the MILP model; see Section 6.1.9).

The main difference of this formulation compared to the planning problem considered by transmission system operators (TSOs) is that we directly specify the given supplies and demands as *flow* and not as *power* requirements (see the discussion in Section 5.3.3).

5.3 • Additional modeling considerations

Apart from the detailed modeling of the elements of the gas network, we have to make some global modeling decisions. We first discuss our decision to only use stationary models. We then discuss how to deal with the effects of gas temperature. Finally, we discuss how to deal with the problem in which the instances generated by the nomination generation approach of Chapter 14 do not necessarily satisfy mass flow balance. The reason for this is that the effects of gas mixing and the power needed by gas driven compressors cannot be determined a priori, but depend on the configuration of active elements. In this section we will outline the problems and explain some of our mitigation strategies. The consequences of these decisions will be discussed in more detail in Chapters 11 and 12.

5.3.1 • Assumption of stationarity

As already mentioned in Chapter 2, we are interested in solutions for mid- to long-term planning problems. From a planning point of view, this typically means using a stationary (steady state) model for gas transport. In practice, however, gas networks are operated/dispatched in a dynamic/transient manner, i.e., over a continuous time horizon. A transient model of a gas network, thus, would require as input both initial states and future nomination profiles in continuous time. Since we consider mid- to long-term planning, both are unknown. Thus, in this book we exclusively deal with stationary gas transportation. We will discuss this topic more extensively in Section 11.1.

One additional benefit of the restriction to stationary flows is that nominations are usually aggregated, e.g., over a day. This improves data quality. For a transient setting, our input data would have to be sliced on an hourly basis, which would lead to larger errors in the forecast (see Chapter 13). It is an open question whether the increased accuracy of a transient model yields better results when having to deal with worse data. In practice, as outlined in Chapter 4, planning questions are currently analyzed using stationary models, even though simulation tools for transient models are available. In fact, the methods are usually the same, and stationary solutions are computed by considering a very long time horizon in a transient simulation.

A general question, however, is to what extent transient gas transport allows for a larger set of transportable nominations. We have to leave this complex question for future research.

5.3.2 • Temperature effects

With the exception of the NLP model, we consider the *isothermal* case, i.e., we do not take into account changes in gas temperature. Phrased differently, we ignore the conservation of energy in all the other models. This decision results from the observation that models using the energy equation are much harder to solve. In fact, experiments based on the NLP approach in Chapter 10 showed that including temperature effects leads to models that are extremely difficult to solve. Moreover, this again is intractable for models that incorporate discrete decisions. For this reason, we have restricted the main discussion of this book to the isothermal case. A thorough discussion of the modeling of the temperature effects is given in Chapter 10.

There are some heuristic arguments as to why neglecting temperature effects can be acceptable in our planning setting. Clearly, taking the energy equation into account has the largest impact at places in the network where major changes in the temperature occur. There are two types of effects that lead to major changes in gas temperature which we consider in turn.

First, gas is heated or cooled when it is compressed and expanded, respectively. This can be observed, e.g., at compressors (see Section 10.1.10) and at control valves (see Section 10.1.8). To guarantee smooth operation, however, network operators have installed gas coolers and preheaters at these places. This means that if necessary the gas is typically cooled down before leaving the compressor group. Similarly, the gas is heated up in so-called preheaters before it passes an active control valve. This makes sure that local temperature differences are limited, which in turn makes the effects of the neglected energy equation less pronounced. With the exception of the NLP model, all coolers and preheaters are therefore modeled just as resistors and all temperature-related effects are ignored.

The second effect is the exchange of heat through pipe walls (see Section 10.1.4). The typical gas pipe we consider is buried underground. This allows us to assume that the surrounding temperature of the pipe is almost constant. This means that in large parts of the network temperature changes are negligible. The main exceptions, again, are parts that are close to compressor groups and control valves. Here the gas temperature and the temperature of the surrounding soil differ by a comparatively large amount. Nevertheless, we assume that this effect, as discussed above, is not large.

5.3.3 ▪ Flow balance and gas mixtures

Most of our models, again with the exception of the NLP model, assume that a specific form of flow balance holds for the given nomination: More precisely, they assume that the *normal volumetric* in- and outflows are balanced. This assumption, though mathematically convenient, is, in general, not true. There are two main reasons for this: One lies in the imperfection of the physical model when dealing with different types of natural gas. The other lies in the fact that gas powered compressors consume some of the gas flowing through, the amount of which is not known a priori. We have decided, for the purposes of this book, to treat only the case where we have one type of natural gas and assume that the gas needed to power gas driven compressors is supplied externally. Wherever the nominations assume that different types of gas are present, we compute an "average" gas composition to achieve this. The only exception to this is the NLP model of Chapter 10, which is able to consider the mixture effects that occur when using multiple types of gas. In the rest of this section we discuss these two decisions.

Real-life gas networks typically transport different types of natural gas (for a thorough discussion see Section 10.1.2). The natural gas of different suppliers has quite distinct characteristics, one of which is its calorific value. This is typically taken into account in real-life gas networks by using the delivered power values as the basis of delivery contracts instead of the normal volumetric flow. Thus, power values should also be the basis for optimization approaches. This also means that an accurate model has to consider the mixing of the different types of gas.

Our physical models, however, are not formulated in terms of power, but use conservation of mass. This translates into a conservation of mass flow in the network (which is the basis of the NLP model). The translation between mass flow and normal volumetric flow depends on the assumed equation of state, which also depends on the type of gas in the network; see Eq. (2.20) in Chapter 2. This means that only if we correctly take the mixing of gas into account at all points in the network and approximate the equation of state with reasonable accuracy, it holds that conservation of mass flow translates into conservation of normal volumetric flow. As a special case, this occurs when the network contains only one type of gas. All our models (except for the NLP), however, need to approximate the equation of state rather coarsely. Thus, the models cannot guarantee exact conservation of normal volumetric flow, unless we restrict ourselves to the case of one type of natural gas, which we will do in the rest of this book, with the exception of Chapter 10.

Realistic gas networks do not satisfy this assumption. The different entries supply gas with different calorific values. To get a benchmark set of nominations that allows us to compare different approaches on realistic networks, we decided to transform the nominations in such a way that the supplies/demands are given in terms of normal volumetric flow instead of power as follows.

We denote by P_u the power supply/demand at node u and assume that we are given a nomination that is power-balanced, i.e., $\sum_{u \in V_+} P_u = \sum_{u \in V_-} P_u$ holds. As we know the calorific value ($H_{c,u}$) and the power supply at each entry u, we can compute the mass flow of gas that is supplied. We then compute an average calorific value for the whole network by weighting the calorific values with the mass flow of gas supplied at each node. More formally, the mass flow q_u^{nom} of gas supplied or discharged at node u is given by $q_u^{\text{nom}} = P_u / H_{c,u}$. As an approximation, we now compute the average calorific value of the nomination as

$$H_c^m := \frac{\sum_{u \in V_+} H_{c,u} q_u^{\text{nom}}}{\sum_{u \in V_+} q_u^{\text{nom}}}.$$

The resulting mass flows at the exits are then computed as

$$q_u^{\text{nom}} := \frac{P_u}{H_c^m}.$$

A short computation shows that the resulting supply and demand values are now given as mass flow and fulfill the mass flow balance condition.

Another source of the inexactness is the above mentioned fact that gas driven compressors consume gas depending on their power needs. The amount of which is, of course, not known a priori and, thus, we need to work around this problem as well. Here our approach is to treat these compressors similarly to electrically driven compressors in the sense that we disregard the gas needed to power the turbine and assume that this gas is supplied externally.

5.4 ▪ Pre- and postprocessing

In addition to the main solution procedures described in the following chapters, we use some pre- and postprocessing techniques that are implemented independently of these procedures. These techniques are meant to reduce the size of the underlying problem and to strengthen bounds, but they have also been very useful to detect and fix problems with the real-world data given to us from Open Grid Europe GmbH (OGE). Often, infeasibilities detected in preprocessing led us directly to problems in the input data, e.g., inconsistent bounds and problems with inaccurate node heights.

The preprocessing steps can be classified according to different criteria: The major criterion is whether the preprocessing step only uses information about the network or whether it additionally uses the nomination at hand.

5.4. Pre- and postprocessing

The major preprocessing step only requiring network information is the aforementioned determination of flow and pressure bounds for compressors (see Section 5.1.6). For all models, we first compute bounds for each machine on the maximum absolute pressure increase ($\overline{\Delta p}$), maximum relative pressure increase ($\overline{\varepsilon}$), and the minimal and maximal volumetric flow ($\underline{Q}$ and $\overline{Q}$). For this we use the compressor model of the validation NLP (see Section 10.1.10). The resulting NLP model is small enough to be solved to global optimality.

The nomination-dependent preprocessing steps are mainly bound strengthening and variable fixing steps. The latter steps try to fix discrete decisions whenever flow and/or pressure bounds disallow certain valve/control valve/compressor configurations. As an example one can consider the case of a valve: If the pressure intervals at the end nodes of the valve are disjoint, the valve has to be closed.

We use three different kinds of bound strengthening:

1. bound propagation using interval arithmetic,
2. strengthening of flow bounds using min-cost-flow computations, and
3. additional bound strengthening for the pressure drop along resistors.

Basic bound propagation based on interval arithmetic is used to propagate bounds through the network. The underlying physical model used is a relaxation of the high resolution NLP model of Section 10.4. To additionally strengthen flow bounds, we use simple linear min-cost-flow computations to determine lower and upper bounds on the flow of all arcs of the network. The third step solves a small NLP model to compute bounds on the pressure drop along a resistor. The resulting problems are small enough to be solved to global optimality in a reasonable time. This third step is useful, since standard interval arithmetic techniques are not sufficient to get good bounds for the pressure decrease along the resistor, and resistors are often situated at crucial points in the network (e.g., at the inlet node of compressor groups).

The basic bound propagation is a standard procedure in MINLP solvers. The additional strengthening of flow bounds is a procedure specific for our problem type. It allows us to exploit the network structure of our problem fairly well. We observed difficulties in the bound strengthening of resistors. Due to the ubiquity of resistors at important points in the network (e.g., compressor groups), it is worthwhile to apply the rather costly step of computing tight bounds for the pressure drop.

Postprocessing is used for models that only use a compressor group model and do not explicitly take single compressors into account. We employ a heuristic to decide which configuration of the compressor group is used and how the flow is split inside the group. We assume that we are given a solution that contains values for in-/outflow and pressure at the boundary nodes of the compressor group. For each possible configuration we now estimate flows and pressure increases inside the group using the following rules:

1. If k machines are used in parallel, we split the flow equally between the k machines.
2. If k machines are used in series, we assume that each machine increases the pressure by $\Delta p / k$.

Using these rules, we get values for the flow through every compressor machine and the resulting pressure increase. We can then verify exactly whether the machine is able to satisfy these requirements. If all machines are able to satisfy these flow and pressure requirements, we report the resulting configuration. This heuristic is used for solutions of the reduced NLP and the sMINLP model and has proven very robust as can be seen in the results of Chapter 12.

5.5 ▪ Overview of the literature

A general survey over the application of optimization methods in the natural gas industry is given by Zheng et al. (2010). Closer to the problems we are studying are the surveys by Shaw (1994) and Ríos-Mercado and Borraz-Sánchez (2012). A monograph outlining the earlier state of the art is the book Osiadacz (1987).

One class of approaches uses MILP solvers as the main working horse. The main difficulty is then to get an adequate model for the nonlinearities. For the stationary case, Möller (2004) used piecewise-linear approximations; see also Martin, Möller, and Moritz (2006). This approach was then further extended to the transient case (Moritz (2007); Mahlke, Martin, and Moritz (2010); Fügenschuh et al. (2009); Geißler et al. (2011); Geißler et al. (2012); Domschke et al. (2011)).

A similar approach using different linearizations was used by Tomasgard et al. (2007) and Nørstebø, Rømo, and Hellemo (2010). The linearization chosen in those two papers is relatively simple: a local linearization of a variant of (2.24). As the focus of these two papers is an integrated model of the full gas value chain, the transportation aspect plays only a minor role.

Starting from a transient model, Strusberg and Engell (2002) solve a discretized version of an optimal control problem using MILP techniques.

Among the many NLP approaches, we mention Percell and Ryan (1987), who use a gradient-descent–based method and de Wolf and Smeers (2000), who consider sequential linear programming approaches. Based on a fine simulation model, Jeníček (1993) and Vostrý (1993) use subgradient-based methods. But also sequential quadratic programming (SQP) has been considered as well as interior point methods. See (Furey (1993); Ehrhardt and Steinbach (2005); and Steinbach (2007). Bonnans, Spiers, and Vie (2011) use interval analysis techniques to analyze the Belgian network. Similar models as in Chapter 9 are used in Baumrucker, Renfro, and Biegler (2008) and Baumrucker and Biegler (2010). Another approach to analyze the problem at hand is to treat it as a game and search for equilibrium solutions (Perdicoúlis and Fletcher (1999); Ramchandani (1993)). Li et al. (2011) formulate an MINLP which tackles the stochastic pooling problem and solve it by a decomposition approach. Approaches that have a similar basis as the approach outlined in Chapter 8 are discussed in Hamam and Brameller (1971), Mallinson et al. (1993), Wu et al. (2000), and Ríos-Mercado et al. (2002).

Another widely used approach is dynamic programming. The definitive survey for the work up to 1998 is that by Carter (1998). The first application were so-called gun-barrel networks, i.e., straight line networks with compressors (Wong and Larson (1968), (1968)). The method was later extended to more complex network topologies; see e.g., Lall and Percell (1990), Gilmour, Luongo, and Schroeder (1989), and Zimmer (1975). Further extensions of this approach were proposed by Borraz-Sánchez and Ríos-Mercado (2004).

Not surprisingly, many purely heuristic approaches have been developed. We mention applications of simulated annealing by Wright, Somani, and Ditzel (1998) (specifically for the optimization of compressor operations) and Mahlke, Martin, and Moritz (2007), tabu search by Borraz-Sánchez and Ríos-Mercado (2004), general expert systems by Sun et al. (2000), genetic algorithms by Li et al. (2011), and ant colony optimization by Chebouba et al. (2009). Kim, Ríos-Mercado, and Boyd (2000) and Ríos-Mercado, Kim, and Boyd (2006) present a two-stage iterative heuristic to minimize fuel cost of the compressors.

Lloyd et al. (2006) and Sekirnjak (1996) describe practical experience with using optimization methods in natural gas transport.

Of course, also more specialized problems have been discussed in the literature. We note especially the treatment of compressors discussed by multiple authors (Carter (1996); Osiadacz (1980); Wright, Somani, and Ditzel (1998). For a discussion of the discontinuities arising in the operation of compressor groups, see Carter, Schroeder, and Harbick (1993). When considering the transient problem, it is often difficult to decide whether the network is controllable at all. Questions of this type have been considered by various authors (Banda and Herty (2008); Banda, Herty, and Klar (2006); Brouwer, Gasser, and Herty (2011); Gugat (2003); Gugat and Leugering (2003); Leugering and Schmidt (2002); Gugat et al. (2001)), especially for the isothermal model. Another related topic is optimizing the topology of gas networks; see André, Bonnans, and Cornibert (2009), Babonneau, Nesterov, and Vial (2012), de Wolf and Smeers (1996), Hansen, Madsen, and Nielsen (1991), and Zhang and Zhu (1996).

In closing, we mention that overviews of the methods used here have appeared in the literature: For a discussion from an application point of view we refer the reader to Martin et al. (2011) and Fügenschuh et al. (2013). For a more detailed description of the mathematical approaches we refer the reader to Pfetsch et al. (2015).

5.6 • Overview of our approaches

In the following chapters a variety of approaches on how to validate nominations (i.e., solve NoVa) will be described. In the preceding sections we have discussed their differences mainly from the point of view of the physical model. We will now discuss their mathematical differences. We consider three types of approaches. The first type consists of approaches that try to solve (coarse) MINLP models of NoVa exactly: these are the MILP approach of Chapter 6 and the sMINLP approach of Chapter 7. Approaches of the second type try to construct solutions heuristically: these are the reduced NLP approach of Chapter 8 and the MPEC approach of Chapter 9. As mentioned in the introduction, we refer to the first two types as *decision approaches*. The third type uses a fixed setting of the discrete variables and then computes high accuracy solutions: a family of such approaches is described in Chapter 10. One of these models of the third type, which we call ValNLP, will be our reference model to determine the physical accuracy of all other models.

We also combine these approaches into a unified approach: In a first stage we compute a *solution candidate* using one or more approaches of the first two types. If we find a feasible solution in this way, in a second stage we then fix the discrete decisions and compute a solution using ValNLP. This combined approach, when successful, allows us to compute high accuracy solutions to our problem.

The approach of Chapter 6 starts with a relatively detailed MINLP model. This model is then transformed into an MILP model using a general technique given by Geißler et al. (2012). The resulting MILP is a relaxation of the underlying MINLP. The advantage of this technique is that we can give a priori guarantees on the error of the solution with respect to the underlying MINLP model. It is also able to certify the infeasibility of the underlying MINLP. The approach is also independent of the solver that is used for the resulting MILP.

The sMINLP approach of Chapter 7 works differently. It starts with a somewhat coarser MINLP formulation of our problem. This problem is then solved with a specialized algorithm implemented in the general-purpose MINLP solver framework SCIP (Achterberg (2009), (2013)). The main advantage of this approach is the possibility of specializing the solver to NoVa, which leads to short computation times.

In Chapter 8 we discuss a heuristic that is based on a coarse NLP formulation. This formulation employs a reduction of the number of variables and allows fast computation times, which in turn allows us to try a large number of different possible discrete settings of the active elements in our network. The number of settings to be tested is greatly reduced by first building a setting adapted to the general flow situation of the current nomination. For the analysis of the flow situation, a transshipment problem for the underlying nomination is solved on an aggregated graph. As in the case of the sMINLP, the approach is attractive as the computation times are short.

The MPEC heuristic of Chapter 9 starts with a very detailed MINLP model, where the discrete decisions are then rewritten using complementarity constraints. The resulting problems are then solved using standard regularization approaches from the literature. The advantage of this method is that it is able to yield high accuracy solutions comparatively quickly (see also Schmidt, Steinbach, and Willert (2013); Schmidt (2013)).

As described above, the models described so far use a rather coarse physical modeling. To validate the results of these approaches, we use each result as a starting point for a suitable NLP model as described in Chapter 10, and in particular for our reference model, the validation NLP (ValNLP) described in Section 10.4.

To do so, we fix the discrete decisions computed in the first-stage approaches. This leaves two types of continuous variables remaining: The first type are variables that occur in both the foregoing model and ValNLP; the values from the solution can directly be used as a starting point for the NLP. The second type are the variables that occur only in the validation NLP model; here heuristic initialization rules are used to set a starting point. We discuss the difficulties associated with this two-stage approach in Chapter 11 (especially in Section 11.3). It has, however, proven to be quite satisfactory in practice. This is discussed in Chapter 12, in particular in Section 12.6.

The approaches discussed here can, of course, also be combined. We use all decision approaches to generate discrete decisions and validate them using ValNLP. We then pick the best discrete decision with respect to the results of ValNLP in case the instance is feasible, or we report infeasible if the MILP model or the sMINLP model reports infeasibility. This approach cannot, in general, guarantee consistent results (see Chapter 11). Again, however, this approach gives, in general, good results (see the discussion in Section 12.7).

Chapter 6
The MILP-relaxation approach

Björn Geißler, Alexander Martin, Antonio Morsi, Lars Schewe

Abstract *We describe how to tackle the problem of validating nominations using mixed-integer programming methods. To this end we first give a mixed-integer nonlinear programming formulation of NoVa, from which we derive a mixed-integer linear relaxation. Our reformulation technique allows us to prescribe an a priori bound on the error we make when relaxing the nonlinear functions.*

Our approach for the solution of the problem of validating nominations (NoVa) consists of two steps: We first give an *MINLP formulation* of the problem. We then use the technique by Geißler et al. (2012) to derive a *MILP relaxation* of this mixed-integer nonlinear program (MINLP). The resulting mixed-integer linear program (MILP) is then solved using off-the-shelf MILP solvers. With this approach we are able to give *infeasibility results* for a relatively "exact" underlying MINLP model. As we are able to choose the acceptable relaxation error a priori, we can also produce solutions that are typically quite close to solutions of the precise NLP model described in Chapter 10.

6.1 ▪ An MINLP model for the validation of nominations

According to Section 5.3.1, our aim is to derive an MINLP-model for stationary state gas transport. The features of our model can be summarized as follows: We neglect *temperature differences* in general and assume *constant density* along the pipes in the network. This allows us to use an *algebraic model* for pressure loss along a pipe. Furthermore, we assume that we do not need to take the effect of *gas mixing* into account. All parameters depending on the gas composition are considered constant. *Compressor groups* are modeled in detail: We model each compressor machine as a single unit. We do not, however, model the compressor drives explicitly. Compressor groups are treated as subnetworks within our approach. As we are using an MILP solver, we are able to handle all *discrete decisions* with the aid of binary variables exactly.

The rest of this section deals with the details of deriving the model. The main objective is to set up an accurate MINLP model, which is then transformed into a "well-behaved" MILP model as described in Section 6.2.

6.1.1 ▪ Network model

As already stated in Section 2.5, we model a gas network by means of a directed finite graph $G = (V, A)$ and we assume lower and upper bounds $\underline{q}_a$ and $\overline{q}_a$ on the mass flow variables q_a to be given, i.e.,

$$\underline{q}_a \leq q_a \leq \overline{q}_a \quad \text{for all } a \in A. \tag{6.1}$$

For each node $u \in V$ we introduce a variable for the pressure p_u and assume minimal and maximal pressure values $\underline{p}_u$ and $\overline{p}_u$, i.e.,

$$\underline{p}_u \leq p_u \leq \overline{p}_u \quad \text{for all } u \in V,$$

to be given. Further, we have *conservation of mass*. This physical law ensures that the mass of gas flowing into a node u is equal to the mass of gas flowing away from u minus the amount of gas withdrawn from the network at u. Thus,

$$\sum_{a \in \delta^+(u)} q_a - \sum_{a \in \delta^-(u)} q_a = q_u^{\text{nom}} \quad \text{for all } u \in V, \tag{6.2}$$

see Eq. (2.8). We remark that q_u^{nom} is nonnegative for exits, nonpositive for entries and zero for junctions, i.e.,

$$q_u^{\text{nom}} \begin{cases} \geq 0 & \text{for all } u \in V_+, \\ = 0 & \text{for all } u \in V_0, \\ \leq 0 & \text{for all } u \in V_-. \end{cases}$$

We proceed with a detailed description of each type of network element starting with pipes.

6.1.2 ▪ Pipes

As already mentioned in Section 5.1.2, a pipe $a = (u, v) \in A_{\text{pi}}$ is specified by its length L_a, diameter D_a, and roughness k_a, and we assume all pipes to be straight and of cylindrical shape. Further, we assume *constant gas temperature*, i.e., $T_u = T_m$ for all $u \in V$; see Section 5.3.2.

To model pressure loss along a pipe, we use the algebraic model described in Eq. (2.24). The friction factor λ_a, is computed with Nikuradse's formula (2.19), which is suitable for large Reynolds numbers. Since our aim is to perform NoVa on large transport networks, where flows are typically highly turbulent, Nikuradse's formula matches our requirements (see Section 2.3.1). Moreover, we use the formula of Papay (2.4) from Section 2.3.1, with constant mean pressure,

$$p_{\text{m},a} = \frac{1}{2}(\max\{\underline{p}_u, \underline{p}_v\} + \min\{\overline{p}_u, \overline{p}_v\}) \quad \text{for all } a = (u,v) \in A_{\text{pi}}, \tag{6.3}$$

to compute the compressibility factor in our simplified model. Pressure changes along a pipe are thus modeled, according to Eq. (2.24), as

$$p_v^2 - c_{1,a} p_u^2 = c_{2,a} |q_a| q_a \quad \text{for all } a = (u,v) \in A_{\text{pi}}, \tag{6.4}$$

with nonnegative constants

$$c_{1,a} = e^{-S_a},$$

$$c_{2,a} = \begin{cases} -\Lambda_a \frac{e^{S_a}-1}{S_a e^{S_a}} & \text{for } s_a \neq 0, \\ -\Lambda_a & \text{otherwise,} \end{cases}$$

$$\Lambda_a = \frac{L_a \lambda_a R_s z_m T_m}{A_a^2 D_a},$$

$$S_a = 2g \frac{s_a L_a}{R_s z_m T_m},$$

$$z_m = z(p_{m,a}, T_m).$$

6.1.3 ▪ Short cuts

Another type of gas network element is a so-called *short cut*. As already discussed in Section 5.1.3, these do not exist in reality, but are widely used by practitioners to model, e.g., hybrid points, such as *gas storages*. To model a short cut, we only need to require that the pressure values at its endpoints are equal. Therefore, we add

$$p_u = p_v \quad \text{for all } (u,v) \in A_{sc} \tag{6.5}$$

to our model. Since short cuts have no physical properties, no further conditions are necessary.

6.1.4 ▪ Resistors

As already mentioned in Section 2.3.2, further network elements causing a pressure drop are represented by so-called *resistors*. We distinguish between two types of resistors: those with constant pressure drop in flow direction and those for which the pressure reduction additionally depends on the amount of flow. The set of resistors of the former type is denoted by $A_{\text{lin-rs}}$ and those of the second type belong to $A_{\text{nl-rs}}$; see also Table 5.1 in Section 5.1.1.

Since in both cases the pressure reduction depends on the flow direction, we introduce a binary variable $r_a \in \{0,1\}$ for the *direction of gas flow* along each resistor $a = (u,v) \in A_{rs}$, where $q_a > 0$ implies $r_a = 0$ and $q_a < 0$ implies $r_a = 1$. To reflect this property in our model, we add the constraints

$$\underline{q}_a r_a \leq q_a \leq \overline{q}_a (1-r_a) \quad \text{for all } a \in A_{rs}. \tag{6.6}$$

Each resistor $a = (u,v) \in A_{\text{nl-rs}}$ is specified by its drag factor ζ_a and diameter D_a. The pressure difference between the endpoints of a is given by Eq. (2.30). Using the equation of state for real gases (2.20) and the velocity equation (2.31), we can rewrite Eq. (2.30) in terms of our model variables as

$$p_u - p_v = \begin{cases} \frac{8\zeta_a R_s T_u}{\pi^2 D_a^4} \frac{q_a^2 z(p_u, T_u)}{p_u} & \text{for } q_a \geq 0 \\ -\frac{8\zeta_a R_s T_v}{\pi^2 D_a^4} \frac{q_a^2 z(p_v, T_v)}{p_v} & \text{for } q_a \leq 0 \end{cases} \quad \text{for all } a = (u,v) \in A_{\text{nl-rs}}. \tag{6.7}$$

Using our assumption of *constant temperature* T_m and a *mean pressure* $p_{m,a}$ computed according to Eq. (6.3), we can substitute the compressibility factors in (6.7) by a constant mean compressibility factor $z_{m,a} = z(p_{m,a}, T_m)$ such that we obtain for all $a = (u,v) \in A_{\text{nl-rs}}$

$$p_u - p_v = \begin{cases} c_a \frac{q_a^2}{p_u} & \text{for } q_a \geq 0, \\ -c_a \frac{q_a^2}{p_v} & \text{for } q_a \leq 0, \end{cases} \tag{6.8}$$

with $c_a = 8\zeta_a R_s T_m z_{m,a}/(\pi^2 D_a^4)$. Finally, we model Eq. (6.8) by adding the constraints

$$p_u^2 - p_v^2 + |\Delta_a|\Delta_a = 2c_a |q_a| q_a \qquad \text{for all } a = (u,v) \in A_{\text{nl-rs}}, \tag{6.9a}$$
$$\Delta_a = p_u - p_v \qquad \text{for all } a = (u,v) \in A_{\text{nl-rs}}, \tag{6.9b}$$
$$p_u - p_v \leq (\overline{p}_u - \underline{p}_v)(1 - r_a) \qquad \text{for all } a = (u,v) \in A_{\text{nl-rs}}, \tag{6.9c}$$
$$p_v - p_u \leq (\overline{p}_v - \underline{p}_u) r_a \qquad \text{for all } a = (u,v) \in A_{\text{nl-rs}}. \tag{6.9d}$$

A resistor $a = (u,v) \in A_{\text{lin-rs}}$ is specified by a fixed pressure drop ξ_a in flow direction. According to Eq. (2.32) in Section 2.3.2, the pressure values at nodes u and v are related by

$$p_u - p_v = \text{sgn}(q_a)\xi_a.$$

We approximate this *discontinuous* function for $a = (u,v) \in A_{\text{lin-rs}}$ by the continuous piecewise linear function

$$p_u - p_v = \begin{cases} -\xi_a & \text{for } q_a \leq -\varepsilon, \\ \frac{\xi_a}{\varepsilon} q_a & \text{for } -\varepsilon \leq q_a \leq \varepsilon, \\ \xi_a & \text{for } q_a \geq \varepsilon, \end{cases} \tag{6.10}$$

with some carefully chosen $\varepsilon > 0$. For the computational results presented in Chapter 12 we choose $\varepsilon = 1/3600$ m³/s ρ_0. This piecewise linear function is then modeled by means of mixed-integer linear constraints according to the *incremental method* (Markowitz and Manne (1957)). For further details on this technique we also refer to Section 6.2 below.

6.1.5 • Valves

As outlined in Section 2.3.3, a *valve* $a = (u,v) \in A_{\text{va}}$ may be either *closed* or *open*. In the former case it prevents gas from passing and pressures at nodes u and v are decoupled. In contrast, if a is open, we have $p_u = p_v$ and the mass flow q_a is only restricted by its bounds. In order to reflect this behavior in our model we introduce an additional binary variable $s_a \in \{0,1\}$ for the *switching state* of a valve. If $s_a = 1$, the valve a is open, and $s_a = 0$ means that a is closed. Our model for valves is

$$\overline{q}_a s_a \geq q_a \qquad \text{for all } a = (u,v) \in A_{\text{va}}, \tag{6.11a}$$
$$\underline{q}_a s_a \leq q_a \qquad \text{for all } a = (u,v) \in A_{\text{va}}, \tag{6.11b}$$
$$(\overline{p}_v - \underline{p}_u)s_a + p_v - p_u \leq \overline{p}_v - \underline{p}_u \qquad \text{for all } a = (u,v) \in A_{\text{va}}, \tag{6.11c}$$
$$(\overline{p}_u - \underline{p}_v)s_a + p_u - p_v \leq \overline{p}_u - \underline{p}_v \qquad \text{for all } a = (u,v) \in A_{\text{va}}. \tag{6.11d}$$

6.1.6 • Control valves

It is necessary to reduce pressure, when gas is transported from a large conveyor pipeline into a regional subnetwork. For that purpose, *control valves* are usually located at such transition points. We have to distinguish between *remotely controllable, fully automated control valves* and those with only *manually adjustable preset pressure*. Thus, according to

Section 5.1.1, we actually have $A_{cv} = A_{cv}^{aut} \dot\cup A_{cv}^{man}$, where A_{cv}^{aut} denotes the set of automated control valves, and the set A_{cv}^{man} contains all control valves with preset pressure. For a more detailed discussion of the technical properties of both types of control valves we refer to Section 2.3.4.

Just like a valve, a control valve can be either *closed* or *open*. When it is closed, gas cannot pass and the pressure values at its incident nodes are decoupled, and with an open control valve, pressure can be reduced. Again, we introduce a binary variable $s_a \in \{0, 1\}$ for each control valve $a = (u, v) \in A_{cv}$ to model whether it is open ($s_a = 1$) or closed ($s_a = 0$).

For an automated control valve $a = (u, v) \in A_{cv}^{aut}$, we assume a lower bound $\underline{\Delta}_a \geq 0$ and an upper bound $\overline{\Delta}_a \geq \underline{\Delta}_a$ on the attainable pressure reduction to be given. Gas flow against the direction of the arc is impossible, i.e., $q_a \geq 0$. For control valves of this type we add constraints

$$\overline{q}_a s_a \geq q_a \qquad \text{for all } a = (u, v) \in A_{cv}^{aut},$$
$$\underline{q}_a s_a \leq q_a \qquad \text{for all } a = (u, v) \in A_{cv}^{aut},$$
$$(\overline{p}_v - \underline{p}_u + \underline{\Delta}_a) s_a + p_v - p_u \leq \overline{p}_v - \underline{p}_u \qquad \text{for all } a = (u, v) \in A_{cv}^{aut},$$
$$(\overline{p}_u - \underline{p}_v - \overline{\Delta}_a) s_a + p_u - p_v \leq \overline{p}_u - \underline{p}_v \qquad \text{for all } a = (u, v) \in A_{cv}^{aut},$$

to our model.

In contrast, a control valve $a = (u, v) \in A_{cv}^{man}$ with *preset pressure* might be in three different operation modes. If pressure at node u exceeds the preset pressure p_a^{set}, the control valve is *active*, and the pressure at node v has to be equal to p_a^{set}. If p_v is below p_a^{set}, the control valve is in *bypass* mode, and thus $p_u = p_v$. Finally, whenever the pressure at node v exceeds the preset pressure, the control valve must be *closed*. In order to model all three possible states we introduce two binary variables $s_a^{ac}, s_a^{bp} \in \{0, 1\}$ in addition to s_a. Here, $s_a^{ac} = 1$ if and only if the control valve is active and $s_a^{bp} = 1$ if and only if a is in bypass mode. Altogether this results in the following model:

$$s_a^{ac} + s_a^{bp} = s_a \qquad \text{for all } a = (u, v) \in A_{cv}^{man}, \quad (6.12a)$$
$$p_u - p_v \leq (1 - s_a^{bp})(\overline{p}_u - \underline{p}_v) \qquad \text{for all } a = (u, v) \in A_{cv}^{man}, \quad (6.12b)$$
$$p_u - p_v \geq (1 - s_a^{bp})(\underline{p}_u - \overline{p}_v) \qquad \text{for all } a = (u, v) \in A_{cv}^{man}, \quad (6.12c)$$
$$p_v + s_a^{bp}(\overline{p}_v - p_a^{set}) \leq \overline{p}_v \qquad \text{for all } a = (u, v) \in A_{cv}^{man}, \quad (6.12d)$$
$$p_v + s_a^{ac}(\overline{p}_v - p_a^{set}) \leq \overline{p}_v \qquad \text{for all } a = (u, v) \in A_{cv}^{man}, \quad (6.12e)$$
$$p_v + s_a^{ac}(\underline{p}_v - p_a^{set}) \geq \underline{p}_v \qquad \text{for all } a = (u, v) \in A_{cv}^{man}, \quad (6.12f)$$
$$p_u - p_v \geq (1 - s_a^{ac})(\underline{p}_u - \overline{p}_v) \qquad \text{for all } a = (u, v) \in A_{cv}^{man}, \quad (6.12g)$$
$$q_a \geq (1 - s_a^{ac}) \underline{q}_a \qquad \text{for all } a = (u, v) \in A_{cv}^{man}, \quad (6.12h)$$
$$p_v + s_a(p_a^{set} - \underline{p}_v) \geq p_a^{set} \qquad \text{for all } a = (u, v) \in A_{cv}^{man}, \quad (6.12i)$$
$$q_a \leq \overline{q}_a s_a \qquad \text{for all } a = (u, v) \in A_{cv}^{man}, \quad (6.12j)$$
$$q_a \geq \underline{q}_a s_a \qquad \text{for all } a = (u, v) \in A_{cv}^{man}. \quad (6.12k)$$

Equation (6.12a) states that a control valve has to be either closed or active or in bypass. Inequalities (6.12b)–(6.12d) guarantee $p_u = p_v \leq p_a^{set}$ when the control valve is in bypass mode. Due to the presence of inequalities (6.12e)–(6.12h), we have $p_u \geq p_v = p_a^{set}$ and $q_a \geq 0$ for an active control valve. Constraints (6.12i)–(6.12k) model the closed state of the control valve. In this case, we have $p_v \geq p_a^{set}$ and $q_a = 0$. Note that it is easily possible to eliminate s_a from (6.12) by substituting $s_a = s_a^{ac} + s_a^{bp}$ in (6.12j)–(6.12k) and replacing Eq. (6.12a) by $s_a^{ac} + s_a^{bp} \leq 1$. However, we decided for the above presentation, especially in order to simplify the description of our subnetwork operation mode model in Section 6.1.8.

We further remark that pressure happens to be reduced when gas enters or leaves a control valve due to viscous drag of fittings or due to the presence of facilities like, e.g., gas preheaters. To reflect these typically small pressure reductions in our model, a control valve is usually located between two suitably dimensioned resistors (see Section 6.1.4).

Further, automated control valves are *unidirectional* elements, i.e., $\underline{q}_a \geq 0$. It is therefore only possible to reduce pressure in flow direction. However, for a given control valve $a \in A_{cv}^{aut}$ there always is a distinguished valve, the so-called *bypass valve*, connected in *parallel* to the control valve in order to allow backward flow. Usually it is not possible to open a control valve and its associated bypass valve simultaneously. We refer to such restrictions, which describe dependencies between switching states of various controllable network elements, as subnetwork operation modes; see Section 2.4.3. How subnetwork operation modes are represented within our model is described in Section 6.1.8.

In the following, we refer to the serial connection of a control valve, the two resistors, and, in case of an automated control valve, the parallel bypass valve, as a *control valve station*; see also Section 2.4.1. A schematic plot of a control valve station is given in Figure 2.7.

6.1.7 ▪ Compressors

Since on the one hand NoVa is a pure feasibility problem, the power consumption of the compressors and thus the resulting costs for electricity or fuel gas are only of minor importance. Therefore, we decided not to model these quantities at all. On the other hand, the operating range of a compressor is of significant importance. It is given by its in general non-convex characteristic diagram. As explained in Section 2.3.5, a detailed modeling of compressors significantly increases the complexity of the overall model. Thus, in order to keep our model tractable, we apply a couple of simplifications, which are discussed in the remainder of this section. To this end, each compressor $a = (u, v) \in A_{cm}$ is assumed to be given together with a constant *mean adiabatic efficiency* $\eta_{ad,m,a}$, an *isentropic exponent* x_a, bounds for the manageable *mass flow rate* $\underline{q}_a, \overline{q}_a$, *compression ratio* $\underline{\varepsilon}, \overline{\varepsilon}$, *pressure increase* $\underline{\Delta p}, \overline{\Delta p}$, specific change in *adiabatic enthalpy* $\underline{H}_{ad,a}, \overline{H}_{ad,a}$, and *volumetric flow rate* $\underline{Q}_a, \overline{Q}_a$. We remark that each of these bounds can be computed by solving a non-convex but relatively small nonlinear program (NLP), with its feasible range given by the characteristic diagram of the respective compressor to global optimality, see Section 2.3.5.

In order to decide whether a compressor is running, we introduce a binary variable $s_a \in \{0, 1\}$, which is equal to one if and only if a is operating. In our model, the specific change in adiabatic enthalpy $H_{ad,a}$ is, according to Eq. (2.42), modeled as

$$H_{ad,a} = c_{H_{ad,a}} \left(\left(\frac{p_v}{p_u} \right)^{\gamma_a} - 1 \right) s_a \quad \text{for all } a = (u,v) \in A_{cm}, \tag{6.13}$$

with constants

$$c_{H_{ad,a}} = \frac{z_m T_m R_s}{\gamma_a} \quad \text{and} \quad \gamma_a = \frac{x_a - 1}{x_a},$$

where the temperature T_m is again considered constant, and thus so is

$$z_m = z\left(\frac{1}{2}(\underline{p}_u + \overline{p}_u), T_m\right).$$

6.1. An MINLP model for the validation of nominations

In order to obtain a first approximation of the feasible range of a compressor we add

$$q_a \geq s_a \underline{q}_a \quad \text{for all } a = (u,v) \in A_{\text{cm}}, \quad (6.14\text{a})$$
$$q_a \leq s_a \overline{q}_a \quad \text{for all } a = (u,v) \in A_{\text{cm}}, \quad (6.14\text{b})$$
$$p_v - p_u \geq \underline{\Delta p}\, s_a + (\underline{p}_v - \overline{p}_u)(1 - s_a) \quad \text{for all } a = (u,v) \in A_{\text{cm}}, \quad (6.14\text{c})$$
$$p_v - p_u \leq \overline{\Delta p}\, s_a + (\overline{p}_v - \underline{p}_u)(1 - s_a) \quad \text{for all } a = (u,v) \in A_{\text{cm}}, \quad (6.14\text{d})$$
$$p_v \geq \underline{\varepsilon}\, p_u - (1 - s_a)(\underline{\varepsilon}\,\overline{p}_u + \underline{p}_v) \quad \text{for all } a = (u,v) \in A_{\text{cm}}, \quad (6.14\text{e})$$
$$p_v \leq \overline{\varepsilon}\, p_u - (1 - s_a)(\overline{\varepsilon}\, \underline{p}_u - \overline{p}_v) \quad \text{for all } a = (u,v) \in A_{\text{cm}} \quad (6.14\text{f})$$

to our model.

6.1.7.1 ▪ Turbo compressors

Again, as outlined in Section 2.3.5, the feasible operating range of a *turbo compressor* is mainly described by a set of (bi)quadratic least-squares fits of measured data points. In our model, we substitute this, in general nonconvex, operating range by a convex outer approximation. According to Eq. (2.47), the operating range of a turbo compressor is bounded from below and above. These bounds are quadratic polynomials in the volumetric flow rate Q_a, whose coefficients are determined by measured data points, the *minimum speed* $\underline{n}_a$ for the lower bound, and the *maximum speed* $\overline{n}_a$ for the upper bound. The *surgeline* and *chokeline* constraints are also quadratic polynomials in Q_a; see (2.49). Therefore, we have two constraints of the form

$$H_{\text{ad},a} \geq \alpha_2 Q_a^2 + \alpha_1 Q_a + \alpha_0, \quad (6.15)$$

namely for the minimum speed and chokeline restrictions and two constraints of type

$$H_{\text{ad},a} \leq \alpha_2 Q_a^2 + \alpha_1 Q_a + \alpha_0, \quad (6.16)$$

for the maximum speed and surgeline restrictions for each turbo compressor $a \in A_{\text{cm}}$. Here, $\alpha_0, \alpha_1, \alpha_2$ are real-valued coefficients. Note the we actually have *different coefficients* $\alpha_1, \alpha_2, \alpha_3$ for each of the four restrictions and for each compressor. But since the general modeling approach is similar for all four constraints, we do not differentiate between these coefficients by means of notation in order to improve readability.

Before we actually derive convex relaxations of Eq. (6.15) and Eq. (6.16), we have to define the notions *convex underestimator*, *concave overestimator*, and *convex (concave) envelope*. We follow the definition given by Jach, Michaels, and Weismantel (2008), but also refer to Rockafellar (1970) for an equivalent definition.

Definition 6.1.

(a) *Let $S \subseteq \mathbb{R}^n$ be a convex set and $f : S \to \mathbb{R} \cup \{\pm\infty\}$. Then a function $g : \mathbb{R}^n \to \mathbb{R} \cup \{\pm\infty\}$ is called* underestimator *(*overestimator*) of f on S if $g(x) \leq f(x)$ ($g(x) \geq f(x)$) for all $x \in S$. If g is convex (concave) on S, then we call g a convex underestimator (concave overestimator) of f on S.*

(b) *The pointwise supremum of all convex underestimators of f on S is called convex envelope of f on S and is denoted by $\text{vex}_S[f](x)$. The pointwise infimum of all concave overes-*

timators of f on S is called *concave envelope* of f on S and is denoted by $\text{cave}_S[f](x)$, i.e.,

$$\text{vex}_S[f](x) := \sup\{g(x) \mid g : S \to \mathbb{R} \cup \{\pm\infty\} \text{ with}$$
$$g(x) \leq f(x) \text{ for all } x \in S, \text{ and } g \text{ convex}\},$$
$$\text{cave}_S[f](x) := \inf\{g(x) \mid g : S \to \mathbb{R} \cup \{\pm\infty\} \text{ with}$$
$$g(x) \geq f(x) \text{ for all } x \in S, \text{ and } g \text{ concave}\}.$$

The function $\text{vex}_S[f](x)$ is the tightest convex underestimator, and $\text{cave}_S[f](x)$ is the tightest concave overestimator of f on S.

Assuming again a *constant temperature* T_m and a *constant mean compressibility factor* z_m, the equation of state for real gases (2.20) together with Eq. (2.38) yields

$$Q_a = \zeta_a \frac{q_a}{p_u}, \quad \text{with} \quad \zeta_a = R_s T_m z_m. \tag{6.17}$$

Using Eq. (6.13) with $s_a = 1$ and (6.17), we can rewrite inequality (6.15) in terms of our model variables p_u, p_v, and q_a as

$$g_1(p_u, p_v) + \alpha_2 g_2(p_u, q_a) + \alpha_1 \zeta_a q_a + (c_{H_{ad,a}} + \alpha_0) p_u \leq 0 \tag{6.18}$$

and Eq. (6.16) now reads

$$g_1(p_u, p_v) + \alpha_2 g_2(p_u, q_a) + \alpha_1 \zeta_a q_a + (c_{H_{ad,a}} + \alpha_0) p_u \geq 0, \tag{6.19}$$

with

$$g_1(p_u, p_v) = -c_{H_{ad,a}} p_u^{1-\gamma_a} p_v^{\gamma_a}$$

and

$$g_2(p_u, q_a) = \zeta_a^2 \frac{q_a^2}{p_u}.$$

A short calculation of the eigenvalues of the Hessians of g_1 and g_2 shows that both functions are convex for $c_{H_{ad,a}}, \zeta_a \geq 0$, and $0 < \gamma_a < 1$, which is always the case in practice; see Section 2.3.5.1. Thus, for $\alpha_2 \geq 0$, constraint (6.18) is convex and

$$\text{cave}_{D_1}[g_1](p_u, p_v) + \alpha_2 \text{cave}_{D_2}[g_2](p_u, q_a) + \alpha_1 \zeta_a q_a + (c_{H_{ad,a}} + \alpha_0) p_u \geq 0$$

is a convex relaxation of constraint (6.19). In the other case, i.e., $\alpha_2 < 0$,

$$g_1(p_u, p_v) + \alpha_2 \text{cave}_{D_2}[g_2](p_u, q_a) + \alpha_1 \zeta_a q_a + (c_{H_{ad,a}} + \alpha_0) p_u \leq 0$$

is a convex relaxation of (6.18) and

$$\text{cave}_{D_1}[g_1](p_u, p_v) + \alpha_2 g_2(p_u, q_a) + \alpha_1 \zeta_a q_a + (c_{H_{ad,a}} + \alpha_0) p_u \geq 0$$

is a convex relaxation of (6.19).

Therefore, we have to determine the (linear) concave envelopes of g_1 and g_2 over suitable domains D_1 and D_2. To this end we transform the bounds on $H_{ad,a}$ and Q_a into the (p_u, p_v, q_a)-space, yielding

$$p_u \left(c_{H_{ad,a}}^{-1} \underline{H}_{ad,a} + 1 \right)^{\frac{1}{\gamma_a}} \leq p_v \leq p_u \left(c_{H_{ad,a}}^{-1} \overline{H}_{ad,a} + 1 \right)^{\frac{1}{\gamma_a}}, \tag{6.20}$$

$$\zeta_a^{-1} \underline{Q}_a p_u \leq q_a \leq \zeta_a^{-1} \overline{Q}_a p_u. \tag{6.21}$$

6.1. An MINLP model for the validation of nominations

To simplify notation we define

$$\underline{\mathcal{H}} := c_{H_{ad,a}}^{-1} \underline{H}_{ad,a} + 1 \quad \text{and} \quad \overline{\mathcal{H}} := c_{H_{ad,a}}^{-1} \overline{H}_{ad,a} + 1.$$

Making use also of the bounds on the pressure at node u we can thus define the polytope

$$D := \operatorname{conv}\left\{ \left(\underline{p}_u, \underline{p}_u \underline{\mathcal{H}}^{\frac{1}{\gamma_a}}, \zeta_a^{-1} \underline{p}_u \underline{Q}_a\right), \left(\underline{p}_u, \underline{p}_u \overline{\mathcal{H}}^{\frac{1}{\gamma_a}}, \zeta_a^{-1} \underline{p}_u \underline{Q}_a\right), \right.$$
$$\left(\overline{p}_u, \overline{p}_u \underline{\mathcal{H}}^{\frac{1}{\gamma_a}}, \zeta_a^{-1} \overline{p}_u \underline{Q}_a\right), \left(\overline{p}_u, \overline{p}_u \overline{\mathcal{H}}^{\frac{1}{\gamma_a}}, \zeta_a^{-1} \overline{p}_u \underline{Q}_a\right),$$
$$\left(\underline{p}_u, \underline{p}_u \underline{\mathcal{H}}^{\frac{1}{\gamma_a}}, \zeta_a^{-1} \underline{p}_u \overline{Q}_a\right), \left(\underline{p}_u, \underline{p}_u \overline{\mathcal{H}}^{\frac{1}{\gamma_a}}, \zeta_a^{-1} \underline{p}_u \overline{Q}_a\right),$$
$$\left. \left(\overline{p}_u, \overline{p}_u \underline{\mathcal{H}}^{\frac{1}{\gamma_a}}, \zeta_a^{-1} \overline{p}_u \overline{Q}_a\right), \left(\overline{p}_u, \overline{p}_u \overline{\mathcal{H}}^{\frac{1}{\gamma_a}}, \zeta_a^{-1} \overline{p}_u \overline{Q}_a\right) \right\}$$

that is the convex hull of the bounds on $H_{ad,a}$ and Q_a transformed into the (p_u, p_v, q_a)-space. The projection of D on the (p_u, p_v)-space yields

$$D_1 := \operatorname{conv}\left\{ \left(\underline{p}_u, \underline{p}_u \underline{\mathcal{H}}^{\frac{1}{\gamma_a}}\right), \left(\underline{p}_u, \underline{p}_u \overline{\mathcal{H}}^{\frac{1}{\gamma_a}}\right), \left(\overline{p}_u, \overline{p}_u \underline{\mathcal{H}}^{\frac{1}{\gamma_a}}\right), \left(\overline{p}_u, \overline{p}_u \overline{\mathcal{H}}^{\frac{1}{\gamma_a}}\right) \right\}$$

and the projection of D on the (p_u, q_a)-space gives us

$$D_2 := \operatorname{conv}\left\{ \left(\underline{p}_u, \zeta_a^{-1} \underline{p}_u \underline{Q}_a\right), \left(\underline{p}_u, \zeta_a^{-1} \underline{p}_u \overline{Q}_a\right), \left(\overline{p}_u, \zeta_a^{-1} \overline{p}_u \underline{Q}_a\right), \left(\overline{p}_u, \zeta_a^{-1} \overline{p}_u \overline{Q}_a\right) \right\}.$$

Since g_i is convex on D_i, the concave envelope of g_i is polyhedral and $\operatorname{cave}_{D_i}[g_i](v_i^j) = g_i(v_i^j)$ for all extreme points $\mathbf{v}_{i_j}$ of D_i, for $i = 1, 2$ and $j = 1, \ldots, 4$. Solving these two systems of linear equations, we obtain

$$\operatorname{cave}_{D_1}[g_1](p_u, p_v) = \frac{c_{H_{ad,a}}}{\Delta \mathcal{H}} \left(\overline{\mathcal{H}} \underline{\mathcal{H}}^{\frac{1}{\gamma}} - \underline{\mathcal{H}} \overline{\mathcal{H}}^{\frac{1}{\gamma}} \right) p_u + \frac{c_{H_{ad,a}}}{\Delta \mathcal{H}} (\underline{\mathcal{H}} - \overline{\mathcal{H}}),$$

$$\operatorname{cave}_{D_2}[g_2](p_u, q_a) = - \underline{Q}_a \overline{Q}_a p_u + \zeta_a (\underline{Q}_a + \overline{Q}_a) q_a,$$

where

$$\Delta \mathcal{H} = \overline{\mathcal{H}}^{\frac{1}{\gamma_a}} - \underline{\mathcal{H}}^{\frac{1}{\gamma_a}}.$$

Finally, we add the *convex relaxations* of the minimal and maximal speed line, the surgeline, and the chokeline constraints to our model. In cases where $\alpha_2 \geq 0$, these are

$$s_a \left(g_1(p_u, p_v) + \alpha_2 g_2(p_u, q_a) + \alpha_1 \zeta_a q_a + (c_{H_{ad,a}} + \alpha_0) p_u \right) \leq 0 \tag{6.22}$$

for the minimal speed and chokeline and

$$s_a \left(\operatorname{cave}_{D_1}[g_1](p_u, p_v) + \alpha_2 \operatorname{cave}_{D_2}[g_2](p_u, q_a) \right.$$
$$\left. + \alpha_1 \zeta_a q_a + (c_{H_{ad,a}} + \alpha_0) p_u \right) \geq 0$$

for the maximal speed and surgeline. In the other case, i.e., $\alpha_2 < 0$, the relaxed minimal speed and chokeline constraints are

$$s_a \left(g_1(p_u, p_v) + \alpha_2 \operatorname{cave}_{D_2}[g_2](p_u, q_a) + \alpha_1 \zeta_a q_a + (c_{H_{ad,a}} + \alpha_0) p_u \right) \leq 0,$$

and the relaxed maximal speed and surgeline constraints are given by

$$s_a\left(\text{cave}_{D_1}[g_1](p_u, p_v) + \alpha_2 g_2(p_u, q_a) + \alpha_1 \zeta_a q_a + (c_{H_{\text{ad},a}} + \alpha_0)p_u\right) \geq 0. \tag{6.23}$$

Note that multiplying the binary switching variable on the left-hand side of inequalities (6.22)–(6.23) is necessary in order to obtain valid constraints even for cases where the respective compressor is inactive. Further note that due to the multiplication of s_a, inequalities (6.22)–(6.23) are nonconvex. A convexification of these constraints is discussed in Section 6.2.3.

6.1.7.2 ▪ Piston compressors

As described in Section 2.3.5, the characteristic diagram of a *piston compressor* $a = (u, v) \in A_{\text{cm}}$ is given as a box in the (Q_a, M_a)-space, where the *shaft torque* M_a is defined as in Eq. (2.50). We model the bounds on the operating volume flow by including Eq. (6.21) to our model, i.e.,

$$\zeta_a^{-1} \underline{Q}_a p_u \leq q_a \leq \zeta_a^{-1} \overline{Q}_a p_u. \tag{6.24}$$

In cases where explicit bounds on the pressure increase or the compression ratio are given, shaft torque bounds are implicitly incorporated via constraints (6.14c)–(6.14f). In the remaining cases, where an explicit bound $\overline{M}_a$ on the shaft torque is given, valid bounds on pressure increase, compression ratio, and volumetric flow rate are computed in a preprocessing step as described in Section 5.1.6. Thus, constraints (6.14c)–(6.14f), together with constraint (6.24), yield a relaxation of the operating range.

6.1.7.3 ▪ Compressor groups

As outlined in Section 2.4.2, a number of compressors are usually assembled in a *compressor group*, where the network operator has to choose one out of a number of different *configurations*, i.e., serial connections of parallel compressors. Additionally, as in case of control valve stations, the operator has also the possibility to *bypass* the whole compressor group or to switch it off, i.e., *shut down* all participating compressors. Again, the bypass is usually modeled by introducing a valve connected in *parallel* to the entire compressor group, and exactly as in the case of control valves, resistors are located at both ends of a compressor group to account for pressure reductions due to inner station piping or, e.g., gas coolers. We further remark that we treat multiple occurrences of the same compressor within different configurations as multiple compressors of identical type. Figure 6.1 shows an example of how a compressor group with two compressors and two configurations is reflected in our model.

Similar to the case of control valve stations, the fact that one has to choose among bypass mode, closing the compressor group, or selecting one of its configurations, can be encoded in terms of subnetwork operation modes. For the precise modeling of subnetwork operation modes we refer to Section 6.1.8.

6.1.8 ▪ Subnetwork operation modes

In the following, we describe how subnetwork operation modes, see Section 2.4.3, are treated within our MINLP model. To this end, we denote by $\mathscr{S}$ the set of subnetworks with given operation modes, where each subnetwork $S \in \mathscr{S}$ is a triple $S = (A_S, \mathscr{M}_S, f_S)$ of the subnetwork elements $A_S \subseteq A_{\text{active}}$, possible operation modes $\mathscr{M}_S \subseteq \{0, 1\}^{A_S}$, and a

6.1. An MINLP model for the validation of nominations

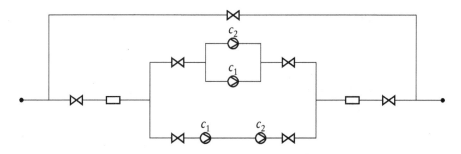

Figure 6.1. *A compressor group with two compressors $c_1, c_2 \in A_{cm}$, and two possible configurations. In the first configuration the two compressors are connected in parallel. The second configuration consists of a serial connection of the same compressors. The bypass valve connected in parallel to the rest of the compressor group allows (uncompressed) flow in both directions.*

function $f_S : A_S \times \mathcal{M}_S \to \{-1, 0, 1\}$ with

$$f_S(a, m) = \begin{cases} -1 & \text{if gas flows from } v \text{ to } u \text{ in operation mode } m, \\ 0 & \text{if the flow direction on } a \text{ is undefined in operation mode } m, \\ 1 & \text{if gas flows from } u \text{ to } v \text{ in operation mode } m, \end{cases}$$

for $a = (u, v)$ and $m \in \mathcal{M}_S$. As introduced in Section 2.5, the set A_{active} consists of all active elements of the network. In our model we introduce binary decision variables $s_m \in \{0, 1\}$ for all $m \in \mathcal{M}_S$ and $S \in \mathcal{S}$. With these variables we are able to formulate that for each subnetwork *exactly one* operation mode has to be chosen, i.e.,

$$\sum_{m \in \mathcal{M}_S} s_m = 1 \qquad \text{for all } S \in \mathcal{S}. \tag{6.25a}$$

Further, a subnetwork element $a \in A_S$ may only be open (closed) if it has been declared to be open (closed) in the selected operation mode:

$$\sum_{m \in \mathcal{M}_S} m_a s_m \geq s_a \qquad \text{for all } S \in \mathcal{S}, a \in A_S, \tag{6.25b}$$

$$\sum_{m \in \mathcal{M}_S} (1 - m_a) s_m \geq (1 - s_a) \qquad \text{for all } S \in \mathcal{S}, a \in A_S. \tag{6.25c}$$

Note that $m_a \in \{0, 1\}$ denotes the (open/closed) state of arc $a \in A_S$ in decision $m \in \{0, 1\}^{A_S}$.

Finally, if an operation mode that specifies flow directions for some of the subnetwork elements is chosen, we have to guarantee that these flow direction requirements are satisfied. This is achieved by adding the following inequalities:

$$\left(1 + \sum_{(a,m) \in A_S \times \mathcal{M}_S} f_S(a, m) s_m\right) \overline{q}_a \geq q_a \qquad \text{for all } S \in \mathcal{S}, a \in A_S, \tag{6.25d}$$

$$\left(1 - \sum_{(a,m) \in A_S \times \mathcal{M}_S} f_S(a, m) s_m\right) \underline{q}_a \leq q_a \qquad \text{for all } S \in \mathcal{S}, a \in A_S. \tag{6.25e}$$

6.1.9 • Objective function

As stated earlier, for the validation of a nomination it is sufficient to find a *feasible* point, i.e., a control for all switchable elements such that gas is transported according to the nomination and according to our stationary model of gas flow. Therefore, we decided to chose an *objective function* that is likely to even out at least some of the model simplifications outlined above.

We particularly want to employ an objective function that tends to produce solutions, where all running compressors are operated within their (nonconvex) feasible operating range, although our model only contains a convex outer approximation of the respective set of constraints.

To this end, we a priori compute a point $(Q_a^*, \varepsilon_a^*) \in [\underline{Q}_a, \overline{Q}_a] \times [\underline{\varepsilon}_a, \overline{\varepsilon}_a]$ in the interior of the feasible operating range of every compressor $a \in A_{\mathrm{cm}}$ in terms of operating volume flow Q_a and pressure ratio $\varepsilon_a = p_v/p_u$. If we would obtain a solution that satisfies $\zeta_a q_a = Q_a^* p_u$, see (6.17), and $p_v = \varepsilon_a^* p_u$ for all operating compressors $a = (u,v) \in A_{\mathrm{cm}}$, we would safely satisfy any of the (nonconvex) constraints that specify the feasible working ranges. In almost all cases, however, such a solution does not exist, and thus we introduce slack variables $\sigma_{Q_a}^+, \sigma_{Q_a}^-, \sigma_{\varepsilon_a}^+, \sigma_{\varepsilon_a}^- \in \mathbb{R}_{\geq 0}$ to formulate relaxations,

$$\zeta_a q_a - Q_a^* p_u \leq M_{1,a}(1 - s_a) + \sigma_{Q_a}^+ \quad \text{for all } a = (u,v) \in A_{\mathrm{cm}}, \quad (6.26a)$$

$$\zeta_a q_a - Q_a^* p_u \geq M_{2,a}(1 - s_a) - \sigma_{Q_a}^- \quad \text{for all } a = (u,v) \in A_{\mathrm{cm}}, \quad (6.26b)$$

$$p_v - \varepsilon_a^* p_u \leq M_{3,a}(1 - s_a) + \sigma_{\varepsilon_a}^+ \quad \text{for all } a = (u,v) \in A_{\mathrm{cm}}, \quad (6.26c)$$

$$p_v - \varepsilon_a^* p_u \geq M_{4,a}(1 - s_a) - \sigma_{\varepsilon_a}^- \quad \text{for all } a = (u,v) \in A_{\mathrm{cm}}, \quad (6.26d)$$

of these requirements, with appropriate constants $M_{i,a}$ for $i = 1, \ldots, 4$, and for all $a \in A_{\mathrm{cm}}$. Our objective function is then chosen to be

$$\min \|\sigma\|_p, \quad p \in \{1, \infty\}, \quad (6.27)$$

where σ denotes the vector of slack variables $\sigma_{Q_a}^+, \sigma_{Q_a}^-, \sigma_{\varepsilon_a}^+, \sigma_{\varepsilon_a}^-$ for all $a \in A_{\mathrm{cm}}$.

6.2 • An MILP relaxation of the MINLP model

Most constraints of the MINLP model introduced in Section 6.1 consist of (mixed-integer) linear expressions only. The nonlinearities only show up in the constraints describing pressure drop along pipes and resistors, as well as in the description of the convexified feasible operating range of turbo compressors. The aim of this section is to construct MILP relaxations of a priori given precision of the MINLP model from Section 6.1. We proceed with Section 6.2.1, where we describe how the nonlinear equality constraints within the MINLP model from Section 6.1 are treated within our MILP models. Next, in Section 6.2.2 we show how to reduce the number of these nonlinearities in order to obtain MILP formulations with fewer (binary) variables and constraints. Finally, in Section 6.2.3 we present a cutting plane algorithm for the constraints defining the operating ranges of turbo compressors.

6.2.1 • MILP formulations of pressure drop

In this section we show how the pressure loss equations for pipes (6.4) and for resistors (6.9a) are represented in our MILP model. For convenience we repeat these

6.2. An MILP relaxation of the MINLP model

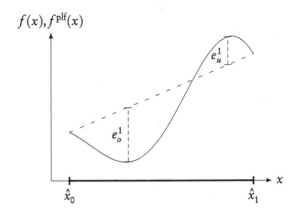

Figure 6.2. *The maximum underestimation e_u^1 and the maximum overestimation e_o^1 of f by f^{plf} over $[\hat{x}_0, \hat{x}_1]$.*

equations:

$$p_v^2 - c_{1,a} p_u^2 = c_{2,a} |q_a| q_a \quad \text{for all } a = (u,v) \in A_{\mathrm{pi}}, \tag{6.28}$$

$$p_u^2 - p_v^2 + |\Delta_a| \Delta_a = 2c_a |q_a| q_a \quad \text{for all } a = (u,v) \in A_{\mathrm{nl\text{-}rs}}. \tag{6.29}$$

In order to arrive at a mixed-integer linear model for pressure drop, we introduce variables π_u, π_v for the *squared pressures* at nodes u and v, a variable ϕ_a for the term $|q_a|q_a$ for all $a = (u,v) \in A_{\mathrm{pi}} \cup A_{\mathrm{nl\text{-}rs}}$, and a variable ω_a for the term $|\Delta_a|\Delta_a$ for all $a \in A_{\mathrm{nl\text{-}rs}}$. Using these variables, Eq. (6.28) and (6.29) become linear,

$$\pi_v - c_{1,a} \pi_u = c_{2,a} \phi_a \quad \text{for all } a = (u,v) \in A_{\mathrm{pi}},$$

$$\pi_u - \pi_v + \omega_a = 2c_a \phi_a \quad \text{for all } a = (u,v) \in A_{\mathrm{nl\text{-}rs}},$$

and we have to make sure that $\pi_u = p_u^2$, $\pi_v = p_v^2$, $\phi_a = |q_a|q_a$ for all $a = (u,v) \in A_{\mathrm{pi}} \cup A_{\mathrm{nl\text{-}rs}}$ and $\omega_a = |\Delta_a|\Delta_a$ for all $a \in A_{\mathrm{nl\text{-}rs}}$. These nonlinear equations can of course only be modeled approximately within an MILP model. In order to do this with an a priori specified approximation accuracy, we apply the approximation algorithm proposed in Geißler (2011) and Geißler et al. (2012). With this approach it is, in particular, possible to compute a piecewise linear interpolation f^{plf} of a univariate (nonlinear) function $f(x)$ over an interval $[\underline{x}, \overline{x}]$ that satisfies any a priori given upper bound $\tau_{\mathrm{abs}}^{\mathrm{plf}} > 0$ on the absolute approximation error. In addition to the actual piecewise linear interpolation f^{plf} that is represented by a set of automatically chosen breakpoints $\underline{x} = \hat{x}_0 \leq \cdots \leq \hat{x}_k = \overline{x}$, the algorithm also yields the *maximum overestimation* e_o^i and the *maximum underestimation* e_u^i of f by f^{plf} over each interval $[\hat{x}_{i-1}, \hat{x}_i]$ for $i = 1, \ldots, k$; see also Figure 6.2. In a second step, we use the piecewise linear interpolation f^{plf}, together with the knowledge about their maximum underestimation and maximum overestimation of f, to construct a set of mixed-integer linear constraints that are satisfied by all points on the graph of $f(x)$ on $[\underline{x}, \overline{x}]$.

To this end, we introduce a variable $y \approx f(x)$ for the approximate function value and apply the *extended incremental method* for piecewise polyhedral envelopes from

Geißler (2011) and Geißler et al. (2012):

$$x = \hat{x}_0 + \sum_{i=1}^{k}(\hat{x}_i - \hat{x}_{i-1})\delta_i, \tag{6.30a}$$

$$y = f(\hat{x}_0) + \sum_{i=1}^{k}(f(\hat{x}_i) - f(\hat{x}_{i-1}))\delta_i + e, \tag{6.30b}$$

$$e \leq e_u^1 + \sum_{i=1}^{n-1} z_i(e_u^{i+1} - e_u^i), \tag{6.30c}$$

$$e \geq -e_o^1 - \sum_{i=1}^{n-1} z_i(e_o^{i+1} - e_o^i), \tag{6.30d}$$

$$\delta_{i+1} \leq z_i, \qquad \text{for } i = 1,\ldots,k-1, \tag{6.30e}$$
$$z_i \leq \delta_i, \qquad \text{for } i = 1,\ldots,k-1, \tag{6.30f}$$
$$z_i \in \{0,1\}, \qquad \text{for } i = 1,\ldots,k-1. \tag{6.30g}$$
$$\delta_1 \leq 1, \tag{6.30h}$$
$$\delta_k \geq 0. \tag{6.30i}$$

Assuming $e = 0$ in Eq. (6.30b) and deleting inequalities (6.30c)–(6.30d), we obtain the well-known (unextended) incremental method (Markowitz and Manne (1957)) for a piecewise linear function f^{plf} from (6.30). Then, a point $(x, f^{\text{plf}}(x))$ on the graph of the piecewise linear function is described in terms of Eq. (6.30a)–(6.30b), since the so-called *filling condition*, i.e., $1 \geq \delta_1 \geq z_1 \geq \cdots \geq \delta_{k-1} \geq z_{k-1} \geq \delta_k \geq 0$, is enforced by constraints (6.30e)–(6.30i). Due to the filling condition, every point feasible for (6.30) satisfies $\delta_i = z_i = 1$ for all $i < j$ and $\delta_i = z_i = 0$ for all $i > j$ and some index $j \in \{1,\ldots,k\}$ (assuming $z_k = 0$). Thus, the value of x is located in the interval $[\hat{x}_{j-1}, \hat{x}_j]$ and can be expressed as $x = \hat{x}_{j-1} + (\hat{x}_j - \hat{x}_{j-1})\delta_j$. The value of y is determined accordingly. The incremental method has the nice property of being a locally ideal formulation (Padberg (2000)) meaning the polytope given through constraints (6.30) has only integer vertices.

In the extended incremental method (6.30), we add an additional continuous variable e to the right-hand side of (6.30b) such that, for given x, the value of y can be arbitrarily chosen from $[f^{\text{plf}}(x) + \underline{e}, f^{\text{plf}}(x) + \overline{e}]$, where $\underline{e}$ and $\overline{e}$ denote the upper and lower bound of e. Eventually, inequalities (6.30c)–(6.30d) adjust these bounds according to minus the maximum overestimation $e_o^i \leq \tau_{\text{abs}}^{\text{plf}}$ and the maximum underestimation $e_u^i \leq \tau_{\text{abs}}^{\text{plf}}$ of f by f^{plf} on the interval containing x.

Thus, the projection of the set of points feasible for constraints (6.30) onto the (x,y) plane properly contains the graph of f on $[\underline{x}, \overline{x}]$. In addition, if (6.30) is satisfied by a point $(x^*, y^*, \delta_1^*, \ldots, \delta_k^*, z_1^*, \ldots, z_{k-1}^*, e^*)$, then $|f(x^*) - y^*| \leq \tau_{\text{abs}}^{\text{plf}}$. For further details on this technique we again refer to Geißler (2011) and Geißler et al. (2012).

Finally, in order to arrive at an MILP-formulation for pressure drop, we apply this technique to the nonlinear functions $x \mapsto x^2 \approx y$ with $(x,y) \in \{(p_u, \pi_u), (p_v, \pi_v)\}$ for all $(u,v) \in A_{\text{pi}} \cup A_{\text{nl-rs}}$ and to the function $x \mapsto |x|x \approx y$ with $(x,y) = (q_a, \phi_a)$ for all $a \in A_{\text{pi}} \cup A_{\text{nl-rs}}$ and with $(x,y) = (\Delta_a, \omega_a)$ for all $a \in A_{\text{nl-rs}}$.

As already mentioned in Section 6.1.4, we model the piecewise linear pressure drop function (6.10) of each resistor $a \in A_{\text{lin-rs}}$ in terms of the (unmodified) incremental method (Markowitz and Manne (1957)).

To further strengthen the MILP-relaxation obtained this way, especially in cases where $\tau_{\text{abs}}^{\text{plf}} \gg 0$, we add the linear parts of the convex and concave envelopes of the approximated

6.2. An MILP relaxation of the MINLP model

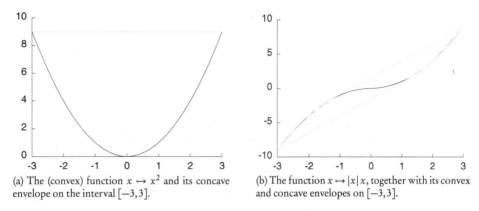

(a) The (convex) function $x \mapsto x^2$ and its concave envelope on the interval $[-3,3]$.

(b) The function $x \mapsto |x|x$, together with its convex and concave envelopes on $[-3,3]$.

Figure 6.3. *Two simple functions with their convex and concave envelopes.*

functions as constraints to our MILP model. For approximations of squared pressure variables we have to deal with functions of type $x \mapsto x^2 \approx y$. In this case, the concave envelope is linear and we add the constraints

$$y \leq (\underline{x} + \overline{x})x - \underline{x}\,\overline{x}, \tag{6.31}$$

with $(x, y) \in \{(p_u, \pi_u), (p_v, \pi_v)\}$ for all $(u, v) \in A_{\text{pi}} \cup A_{\text{nl-rs}}$ to our model; see Figure 6.3(a). In case of an approximation of a function of type $f : x \mapsto |x|x \approx y$, the convex envelope $\text{vex}_{[\underline{x},\overline{x}]}[f](x)$ of f on $[\underline{x},\overline{x}] \neq \emptyset$ is given by (see Geißler (2011))

$$\text{vex}_{[\underline{x},\overline{x}]}[f](x) = \begin{cases} -(\underline{x}+\overline{x})x + \underline{x}\,\overline{x} & \text{for } \overline{x} < 0, \\ \underline{x}(2-\sqrt{8})x + \underline{x}^2(\sqrt{8}-3) & \text{for } 0 \in [\underline{x},\overline{x}],\, x \in [\underline{x}, \underline{x}(1-\sqrt{2})], \\ |x|x & \text{otherwise.} \end{cases}$$

The concave envelope $\text{cave}_{[\underline{x},\overline{x}]}[f](x)$ of f on $[\underline{x},\overline{x}]$ is given by

$$\text{cave}_{[\underline{x},\overline{x}]}[f](x) = \begin{cases} (\underline{x}+\overline{x})x - \underline{x}\,\overline{x} & \text{for } \underline{x} > 0, \\ \overline{x}(\sqrt{8}-2)x + \overline{x}^2(3-\sqrt{8}) & \text{for } 0 \in [\underline{x},\overline{x}],\, x \in [\overline{x}(1-\sqrt{2}), \overline{x}], \\ |x|x & \text{otherwise.} \end{cases}$$

An illustration of the convex and concave envelopes of $|x|x$ is given in Figure 6.3(b). Since we only make use of the linear parts of the envelopes, we have to distinguish between three cases. Whenever $\overline{x} < 0$, we add the constraints

$$y \geq -(\underline{x}+\overline{x})x + \underline{x}\,\overline{x},$$

whenever $\underline{x} > 0$, we add the constraints

$$y \leq (\underline{x}+\overline{x})x - \underline{x}\,\overline{x},$$

and whenever $0 \in [\underline{x},\overline{x}]$ we add the constraints

$$y \geq \underline{x}(2-\sqrt{8})x + \underline{x}^2(\sqrt{8}-3),$$
$$y \leq \overline{x}(\sqrt{8}-2)x + \overline{x}^2(3-\sqrt{8}),$$

with $(x,y) = (q_a, \phi_a)$ for all $a \in A_{\text{pi}} \cup A_{\text{nl-rs}}$ and with $(x,y) = (\Delta_a, \omega_a)$ for all $a \in A_{\text{nl-rs}}$, to our MILP-relaxation model.

Finally, we remark that we do not yet incorporate tangential cuts to the nonlinear parts of the convex and concave envelopes, neither a priori nor via a cutting plane algorithm. However, this would further improve the tightness of the relaxation, especially in case of large $\tau_{\text{abs}}^{\text{plf}}$, and can therefore be regarded as a possible extension.

6.2.2 ▪ Reducing the number of nonlinear equations

As it has been mentioned several times, the vast majority of elements in a gas transport network are pipes. Thus, the major part of nonlinear equality constraints in the MINLP model derived in Section 6.1 consists of the corresponding pressure loss equations (6.4). For convenience we repeat these equations for a pipe $a = (u,v) \in A_{\text{pi}}$ again here,

$$p_v^2 - c_{1,a} p_u^2 = c_{2,a} |q_a| q_a. \qquad (6.32)$$

As one might have noted while reading Section 6.2.1, the size, especially in terms of the number of binary variables, of a MILP-relaxation model therefore crucially depends on the number of approximations of the nonlinear terms in Eq. (6.32) and their complexity. Roughly speaking, we have to incorporate three piecewise polyhedral envelopes for each pipe: two for the approximations of the squared pressures and one for the right-hand side expression $|q_a| q_a$. Since only squared pressures show up in Eq. (6.32) and since all pressure variables are bounded from below by a nonnegative value, there is no need to explicitly introduce variables for the (nonsquared) pressure at nodes that are incident to pipes only.

Further, the single constraint (6.5) modeling a short cut $(u,v) \in A_{\text{sc}}$ can be equivalently formulated as

$$\pi_u = \pi_v, \qquad (6.33)$$

with variables π_u and π_v for the squared pressures at nodes u and v, respectively. Moreover, constraints (6.11a)–(6.11d) specifying the properties of a valve $(u,v) \in A_{\text{va}}$ can also be formulated in terms of squared pressure variables:

$$\overline{q}_a s_a \geq q_a, \qquad (6.34a)$$
$$\underline{q}_a s_a \leq q_a, \qquad (6.34b)$$
$$(\overline{p}_v^2 - \underline{p}_u^2) s_a + \pi_v - \pi_u \leq \overline{p}_v^2 - \underline{p}_u^2, \qquad (6.34c)$$
$$(\overline{p}_u^2 - \underline{p}_v^2) s_a + \pi_u - \pi_v \leq \overline{p}_u^2 - \underline{p}_v^2. \qquad (6.34d)$$

However, we were not able to reformulate the constraints modeling compressors, control valves, and resistors in terms of squared pressure variables only. In particular, for resistors $(u,v) \in A_{\text{nl-rs}}$ we need *both* types of variables at nodes u and v.

Nonetheless, there is still place for tremendous reductions of the size of the MILP relaxations by introducing a piecewise polyhedral envelope of the relation $\pi_u = p_u^2$ at node u only if actually needed. In order to identify a set of nodes, where variables for pressure and squared pressure have to coexist that leads to a minimum size MILP-model, we solve a binary integer program with variables $\alpha_u^p, \alpha_u^\pi \in \{0,1\}$ for all nodes $u \in V$. A solution to this problem has to be interpreted as follows. If $\alpha_u^p = 1$ then we have to introduce the pressure variable p_u, and if $\alpha_u^\pi = 1$ we have to introduce the variable π_u for the squared pressure at node u to our model. Both types of variables have to be introduced at node u if and only if $\alpha_u^p = \alpha_u^\pi = 1$. Of course many of these variables have to be fixed in advance in order to obtain the desired result. To this end, we define the set $A_p :=$

6.2. An MILP relaxation of the MINLP model

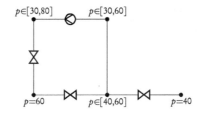

Figure 6.4. *A small network consisting of 5 nodes with given pressure bounds, 3 valves, 1 pipe, and 1 compressor. Solving model (6.35) for this network shows that it is sufficient to introduce both pressure and squared pressure variables, only at the upper right and lower left nodes. Since pressure is fixed at the lower left node, only a single piecewise polyhedral relaxation of the relation $\pi = p^2$ has to be constructed this way.*

$A_{cv} \cup A_{cm} \cup A_{rs}$ of arcs which can only be modeled with the aid of pressure variables, the set $A_\pi := A_{pi} \cup A_{nl\text{-}rs}$ of arcs that can only be modeled with variables for squared pressure at its incident nodes. The resulting model reads as follows:

$$\min \sum_{u \in V} n_{x^2}(\underline{p}_u, \overline{p}_u, \tau^{\text{plf}}_{\text{abs}})(\alpha^p_u + \alpha^\pi_u) \tag{6.35a}$$

$$\text{s.t.} \quad \alpha^p_u + \alpha^\pi_v \geq 1 \quad \text{for all } (u,v) \in A \setminus (A_p \cup A_\pi), \tag{6.35b}$$

$$\alpha^\pi_u + \alpha^p_v \geq 1 \quad \text{for all } (u,v) \in A \setminus (A_p \cup A_\pi), \tag{6.35c}$$

$$\alpha^p_u + \alpha^\pi_u \geq 1 \quad \text{for all } u \in V, \tag{6.35d}$$

$$\alpha^p_u = \alpha^p_v = 1 \quad \text{for all } (u,v) \in A_p, \tag{6.35e}$$

$$\alpha^\pi_u = \alpha^\pi_v = 1 \quad \text{for all } (u,v) \in A_\pi, \tag{6.35f}$$

$$\alpha^p_u, \alpha^\pi_u \in \{0,1\} \quad \text{for all } u \in V, \tag{6.35g}$$

where $n_f(\underline{x}, \overline{x}, \tau)$ denotes the minimum number of subdivisions of an interval $[\underline{x}, \overline{x}]$, necessary to interpolate f with approximation error at most τ. Subtracting $\sum_{u \in V} n_{x^2}(\underline{p}_u, \overline{p}_u, \tau^{\text{plf}}_{\text{abs}})$ from the objective function (6.35a) yields the number of polyhedral pieces needed to achieve the desired accuracy. Thus the number of additional binary variables introduced to our MILP-relaxation model by the technique described in Section 6.2.1 is minimized this way. For an illustrative example, we refer to Figure 6.4.

While it is in general hard to compute $n_f(\underline{x}, \overline{x}, \tau)$, we are able to give a closed form expression for the case $f(x) = x^2$ as needed in (6.35):

Lemma 6.2. *For $f : \mathbb{R} \to \mathbb{R}$ with $f(x) = x^2$, $\underline{x} \leq x \leq \overline{x}$, and $\tau > 0$ we have*

$$n_f(\underline{x}, \overline{x}, \tau) = \left\lceil \frac{\overline{x} - \underline{x}}{2\sqrt{\tau}} \right\rceil.$$

Proof. Since f is convex, its concave envelope, see (6.31), is equal to its linear interpolation on $[\underline{x}, \overline{x}]$. We thus define the error function $e(x) := \text{cave}_{[\underline{x}, \overline{x}]}[f](x) - f(x) = (\underline{x} + \overline{x})x - \underline{x}\,\overline{x} - x^2$. The function e is concave as a difference of a linear and a convex function.

Further we have $e(x) \geq 0$ for $x \in [\underline{x}, \overline{x}]$, $e(x) < 0$ for $x \in \mathbb{R} \setminus [\underline{x}, \overline{x}]$, and $e(\underline{x}) = e(\overline{x}) = 0$. Thus the maximal approximation error on $[\underline{x}, \overline{x}]$ is attained at the unique solution x^*

of $\frac{\partial e(x)}{\partial x} = 0$. A short calculation yields

$$\frac{\partial e(x)}{\partial x} = \underline{x} + \overline{x} - 2x,$$

and thus $x^* = (\underline{x} + \overline{x})/2$. Plugging x^* into e then gives the maximum approximation error

$$e(x^*) = \frac{1}{4}(\overline{x} - \underline{x})^2 = \frac{w^2}{4},$$

with $w := \overline{x} - \underline{x}$. Since the maximum approximation error only depends on the width w of the interval, the problem of minimizing the maximum error using $n+1$ breakpoints is equivalent to finding a subdivision of $[\underline{x}, \overline{x}]$ into n subintervals $[\hat{x}_{i-1}, \hat{x}_i]$ for $i = 1, \ldots, n$, with $\hat{x}_0 = \underline{x}$ and $\hat{x}_n = \overline{x}$ such that $\max\{w_i = \hat{x}_i - \hat{x}_{i-1} \mid i = 1, \ldots, n\}$ is minimized. This is done by an equidistant subdivision with $w_i = (\overline{x} - \underline{x})/n$ for $i = 1, \ldots, n$ and maximum approximation error

$$e_n^{\max} := \frac{1}{4}\left(\frac{\overline{x} - \underline{x}}{n}\right)^2.$$

In order to find the minimal $n^* \in \mathbb{R}$ with $e_{n^*}^{\max} \leq \tau$, we obtain

$$n^* = \frac{\overline{x} - \underline{x}}{2\sqrt{\tau}},$$

and thus

$$n_f(\underline{x}, \overline{x}, \tau) = \left\lceil \frac{\overline{x} - \underline{x}}{2\sqrt{\tau}} \right\rceil,$$

which finishes the proof. □

6.2.3 ▪ Characteristic diagrams of turbo compressors

In this section we describe how the *convex relaxation* of the feasible operating range of a turbo compressor $a = (u, v) \in A_{\text{cm}}$ is treated within our MILP-relaxation model. Our aim is to fulfill the constraints in question by successively adding linear inequalities during the solution of a MILP-relaxation model with a branch-and-bound algorithm.

The bounding constraints for the operating range of a turbo compressor have already been derived in Section 6.1.7.1; see constraints (6.22)–(6.23). Each of these constraints is of type

$$s_a(f_1(p_u, p_v) + f_2(p_u, q_a) + c\, q_a + d\, p_u) \leq 0, \tag{6.36}$$

with continuously differentiable convex functions $f_1, f_2 : \mathbb{R}^2 \to \mathbb{R}$ and constant multipliers $c, d \in \mathbb{R}$. The binary variable s_a indicates whether a is running or not; see Section 6.1.7. Due to their nonlinearity, constraints of type (6.36) can not be directly incorporated in a MILP-relaxation model. Thus, in a first step, we linearize the product of the binary variable and the rest of the left-hand side:

$$f_1(p_u, p_v) + f_2(p_u, q_a) + c\, q_a + d\, p_u + M s_a \leq M, \tag{6.37}$$

with some

$$M \geq \max f_1(p_u, p_v) + f_2(p_u, q_a) + c\, q_a + d\, p_u$$
$$\text{s.t.}\ \underline{p}_u \leq p_u \leq \overline{p}_u,$$
$$\underline{p}_v \leq p_v \leq \overline{p}_v,$$
$$\underline{q}_a \leq q_a \leq \overline{q}_a.$$

6.2. An MILP relaxation of the MINLP model

Next, we apply a simple *cutting plane algorithm*. Whenever a mixed-integer feasible solution $(p_u^*, p_v^*, q_a^*, s_a^*)$ to our MILP-relaxation model with $p_u = p_u^*$, $p_v = p_v^*$, $q_a = q_a^*$, and $s_a = s_a^* = 1$ such that $f_1(p_u^*, p_v^*) + f_2(p_u^*, q_a^*) + c\, q_a^* + d\, p_u^* > 0$ has been found in a node of the branch-and-bound tree, we add the globally valid inequality

$$f_1(p_u^*, p_v^*) + \nabla f_1(p_u^*, p_v^*) \begin{pmatrix} p_u - p_u^* \\ p_v - p_v^* \end{pmatrix} \\ + f_2(p_u^*, q_a^*) + \nabla f_2(p_u^*, q_a^*) \begin{pmatrix} p_u - p_u^* \\ q_a - q_a^* \end{pmatrix} + c\, q_a + d\, p_u + M'\, s_a \leq M', \quad (6.38)$$

with

$$M' := \max f_1(p_u^*, p_v^*) + \nabla f_1(p_u^*, p_v^*) \begin{pmatrix} p_u - p_u^* \\ p_v - p_v^* \end{pmatrix} \\ + f_2(p_u^*, q_a^*) + \nabla f_2(p_u^*, q_a^*) \begin{pmatrix} p_u - p_u^* \\ q_a - q_a^* \end{pmatrix} + c\, q_a + d\, p_u \quad (6.39a)$$

$$\text{s.t. } \underline{p}_u \leq p_u \leq \overline{p}_u, \quad (6.39b)$$

$$\underline{p}_v \leq p_v \leq \overline{p}_v, \quad (6.39c)$$

$$\underline{q}_a \leq q_a \leq \overline{q}_a. \quad (6.39d)$$

Inequality (6.38) is valid for $s_a = 0$ due to the definition of M'; see (6.39). For $s_a = 1$, the left-hand side of inequality (6.38) is equal to the first-order Taylor approximation of the left-hand side of inequality (6.37). Since inequality (6.37) is convex in this case, (6.39a) is globally valid. Those points with $s_a = 1$ that satisfy the cut (6.38) with equality, lie on a hyperplane tangential to our convex relaxation of the operating range of the compressor at (p_u^*, p_v^*, q_a^*).

6.2.4 ▪ Conclusions

In Section 6.1, we presented an MINLP model for NoVa with objective function (6.27) that is to be minimized subject to the constraints (6.1)–(6.2), (6.4)–(6.6), (6.9), (6.10)–(6.12), (6.14), and (6.22)–(6.26). In Section 6.2, we have shown how an MILP relaxation of this MINLP is constructed. We introduced a technique to replace the nonlinearities that show up in constraints (6.4) and (6.9) with piecewise polyhedral envelopes of arbitrary tightness. In the course of this, we presented the extended incremental method for modeling piecewise polyhedral envelopes in terms of mixed-integer linear constraints; see constraint (6.30). The number of additional (binary) variables and constraints that are introduced to the MILP relaxation model via constraint (6.30) has then been reduced by the techniques presented in Section 6.2.2. As part of this, the short cut constraints (6.5) have partly been replaced by Eq. (6.33), and the valve model (6.11) has partly been replaced by (6.34). Finally, in Section 6.2.3, we presented an outer approximation scheme for convex relaxations of the operating ranges of turbo compressors. This outer approximation algorithm cuts off infeasible solutions via the addition of constraints of type (6.38) during branch-and-bound. These constraints are mixed-integer linear reformulations of inequalities (6.22)–(6.23) that were already part of our MINLP model.

Due to the application of outer approximation and piecewise polyhedral envelopes, the resulting MILP model is indeed a relaxation of the MINLP from Section 6.1. However, it is not a relaxation of the NLP model used for the computations in Chapter 12. This is due to the assumption of elementwise constant compressibility factors and the smooth approximation of the friction term in the momentum equation (2.13) used in the NLP model from Section 10. All other simplifications that were made for the derivation of our MINLP model, i.e., the convexification of the operating ranges of compressors, the nonconsideration of drives, and the neglect of temperature changes do not affect the relaxation property.

However, that these simplifications are reasonable in the sense that they result in computationally tractable MILP relaxation models with physically meaningful solutions to NoVa can be seen from the computational studies presented in Chapter 12.

Chapter 7
The specialized MINLP approach

Jesco Humpola, Armin Fügenschuh, Benjamin Hiller,
Thorsten Koch, Thomas Lehmann, Ralf Lenz,
Robert Schwarz, Jonas Schweiger

Abstract *We propose an approach to solve the validation of nominations problem using mixed-integer nonlinear programming (MINLP) methods. Our approach handles both the discrete settings and the nonlinear aspects of gas physics. Our main contribution is an innovative coupling of mixed-integer (linear) programming (MILP) methods with nonlinear programming (NLP) that exploits the special structure of a suitable approximation of gas physics, resulting in a global optimization method for this type of problem.*

In this chapter, we describe an approach to solve the nomination validation problem using mixed-integer nonlinear programming (mixed-integer nonlinear program (MINLP)) methods. Just like the approach presented in the preceding chapter, we model the pressure drop in a pipe using the nonlinear approximations for the stationary case from Section 2.3.1. However, instead of linearizing the nonlinearities to obtain a mixed-integer linear program (mixed-integer linear program (MILP)), we carefully construct a model that contains only nonlinearities exhibiting a certain structure. This enables us to decide on the feasibility of a nomination without having to relax or approximate the nonlinearities. Our approach handles both the discrete settings which correspond to active elements in the network and the nonlinear aspects of gas physics.

Our method uses standard MINLP methods like outer approximations and branch-and-bound for the discrete decisions of the nomination validation problem. Usually, the resulting relaxation needs to be refined by *spatial branching* (see Section 7.1.1). We avoid this spatial branching by employing nonlinear program (NLP) techniques that exploit the special structure of the chosen approximation of gas flow physics. The result is an implementation of a specialized solver based on the framework SCIP (Achterberg (2009), (2013)). The MINLP model can be solved to global optimality, implying that if no feasible solution is found, the given nomination cannot be fulfilled.

Let us briefly recall the setup from the preceding chapters. We are given a directed graph $G = (V, A)$ that models a gas network. The arcs represent the various elements of a gas network. We distinguish between passive elements, namely pipes and resistors, whose behavior cannot be influenced, and active elements, namely valves, control valves, and compressors, which allow controlling the network. The active and passive elements are collected in the arc sets A_{active} and A_{passive}, respectively. In addition to the network, there is a nomination q^{nom} that specifies for each $u \in V$ the amount of flow that leaves ($q_u^{\text{nom}} \leq 0$) or enters ($q_u^{\text{nom}} \geq 0$) the network at u. We note that $q_u^{\text{nom}} = 0$ holds at inner nodes of the network. The nomination q^{nom} is balanced, i.e., we have

$$\sum_{u \in V} q_u^{\text{nom}} = 0.$$

All flows are specified as mass flow rates, which are, given our model assumptions, equivalent to volumetric flow rates under normal conditions. The task is to decide whether or not the nomination q^{nom} can be technically realized by the network.

Apart from the element-specific models described later, we use the following generic variables and constraints in our model. As in the previous chapter, we have variables for the flow through each element a (given as mass flow rate) that are bounded by $\underline{q}_a$ and $\overline{q}_a$:

$$q_a \in [\underline{q}_a, \overline{q}_a] \quad \text{for all } a \in A.$$

Since we are using a directed graph, we follow the convention that the sign of the flow indicates the direction: positive means flow in direction of the arc, negative means flow in the opposite direction. To ensure a consistent flow model, flow conservation constraints

$$\sum_{a \in \delta^+(u)} q_a - \sum_{a \in \delta^-(u)} q_a = q_u^{\text{nom}} \quad \text{for all } u \in V \tag{7.1}$$

state that the balance of entering and leaving flows has to match exactly the nominated amount q_u^{nom} at each node $u \in V$. Moreover, we need to track the pressure (in bar) at each node u in the network. Instead of using pressures, it is more convenient for our approach to use squared pressure variables

$$\pi_u \in [\underline{p}_u^2, \overline{p}_u^2] \quad \text{for all } u \in V,$$

where $\underline{p}_u$ and $\overline{p}_u$ are the specified pressure bounds at node $u \in V$. Only when absolutely necessary we introduce pressure variables p_u and the coupling constraint $\pi_u = p_u^2$. Note that the pressure in the system is never smaller than the atmospheric pressure and therefore nonnegative, so using squared pressure causes no ambiguities.

The remainder of this chapter is structured as follows. First, we study the special case of passive networks consisting of pipes only in Section 7.1. We review results from the literature and establish that the existence problem of validating a nomination can be equivalently cast as a convex optimization problem. Section 7.2 presents our main theoretical contribution, namely a framework and solution approach that allows the use of convex optimization methods to decide the feasibility of nonconvex subproblems that arise in the presence of active elements. We use these ideas in Section 7.3 to model the nomination validation problem in the framework of Section 7.2.

7.1 ▪ Passive pipe networks

Pipes are the most frequent components in gas networks. In this section, we therefore investigate networks consisting of pipes only, which we call *passive pipe networks*.

We use the model established in Section 2.3.1 (in particular the approximations for the stationary case given by (2.24), (2.25), (2.28), and (2.29)) for the relationship between flow and pressure difference in a pipe a. After rearranging terms, this relationship reads

$$\alpha_a |q_a| q_a = p_u^2 - \beta_a p_v^2. \tag{7.2}$$

7.1. Passive pipe networks

The parameters α_a and β_a are constants determined by the properties of the pipe $a \in A$, i.e.,

$$\alpha_a := \begin{cases} \Lambda \frac{e^S - 1}{S} & S \neq 0 \\ \Lambda & \text{otherwise} \end{cases} \quad \text{and} \quad \beta_a := e^S. \tag{7.3}$$

Note that the definitions of the specific constants Λ and S are given by (2.25) and (2.28). Thus, the so-called Weymouth constant α_a depends on the diameter and the length of pipe a. The second constant β_a reflects the impact of the slope of the pipe.

To simplify Eq. (7.2), we introduce a variable for the squared pressure at a node:

$$\pi_u := p_u^2 \quad \text{for all } u \in V.$$

The pressure loss equation (7.2) then becomes

$$\alpha_a |q_a| q_a = \pi_u - \beta_a \pi_v. \tag{7.4}$$

The problem of validating a nomination on a passive pipe network corresponds to an existence problem with a nonlinear, nonconvex constraint set:

$$\exists \, q, \pi \tag{7.5}$$

$$\text{s.t.} \quad \sum_{a \in \delta^+(u)} q_a - \sum_{a \in \delta^-(u)} q_a = q_u^{\text{nom}} \quad \text{for all } u \in V,$$

$$\alpha_a |q_a| q_a = \pi_u - \beta_a \pi_v \quad \text{for all } a = (u, v) \in A,$$

$$\underline{\pi}_u \leq \pi_u \leq \overline{\pi}_u \quad \text{for all } u \in V,$$

$$\underline{q}_a \leq q_a \leq \overline{q}_a \quad \text{for all } a \in A.$$

As any nonconvex NLP, the validation of nomination problem (NoVa) for passive pipe networks can in principle be solved by spatial branching. However, also sophisticated techniques based on convex reformulations are available. We discuss both approaches next.

7.1.1 ▪ Spatial branching for nonconvex problems

Spatial branching is a technique to handle nonconvex optimization problems. In conjunction with outer approximation and branching on integer variables, spatial branching is part of state-of-the-art MINLP solvers (see Belotti et al. (2009); Tawarmalani and Sahinidis (2002), (2004), (2005); Smith and Pantelides (1999); Vigerske (2012)).

Let us give a brief review about the concept of outer approximation and spatial branching. Let $S \subseteq \mathbb{R}^n$ be the (nonconvex) feasible set of the problem. A linear outer approximation of the feasible set is computed such that

$$S \subseteq \{x \,|\, Dx \leq d\}$$

for suitable $D \in \mathbb{Q}^{m \times n}$ and $d \in \mathbb{Q}^m$. This relaxation is used during the branch-and-bound algorithm and is successively refined by additional cutting planes. Branching on integer variables deals with potential integrality requirements. When all integer variables take integral values, the relaxation is strengthened by spatial branching, i.e., branching on continuous variables appearing in nonlinear terms. Spatial branching on the continuous variable x_i of the solution x^* to the linear relaxation refers to subdividing the previous linear relaxation into two parts,

$$S \subseteq \{x \,|\, Dx \leq d, \, x_i \leq x_i^*\} \cup \{x \,|\, Dx \leq d, \, x_i \geq x_i^*\}.$$

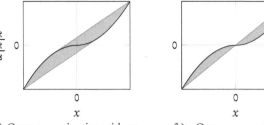
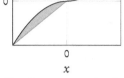
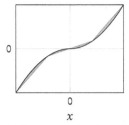

(a) Outer approximation without branching.
(b) Outer approximation with branching on zero.
(c) Outer approximation with several branchings on x.

Figure 7.1. *Successive improvement of the approximation of $x \mapsto \alpha x|x|$ by spatial branching and outer approximation. (Source: Pfetsch et al. (2015).)*

For each part of the relaxation, a subproblem is created, and a tighter outer approximation can be computed due to tighter variable bounds; see Figure 7.1 for an example. Spatial branching thus improves the convex relaxation, in particular, at places where the describing functions are nonconvex. Branching is pursued until all integral variables take integral values and the outer approximation is "close enough" to the feasible region. This way, global bounds on the objective function can be computed and the problem can be solved to global optimality up to a certain accuracy.

7.1.2 ▪ Solution via convex reformulation

It was shown by Collins et al. (1978) and Maugis (1977) that NoVa on a passive pipe network is always feasible if no bounds on the variables are to be respected. Moreover, the admissible squared pressure vectors form a line in $\mathbb{R}^V$.

Theorem 7.1. *(Collins et al. (1978); Maugis (1977).) Let $q^{\mathrm{nom}} \in \mathbb{R}^V$ be a nomination and Φ_a a strictly increasing function for each pipe $a \in A$. Then the solution space of the problem*

$$\exists \quad q, \pi \tag{7.6}$$
$$\text{s.t.} \quad \sum_{a \in \delta^+(u)} q_a - \sum_{a \in \delta^-(u)} q_a = q_u^{\mathrm{nom}} \quad \text{for all } u \in V,$$
$$\Phi_a(q_a) = \pi_u - \pi_v \quad \text{for all } a = (u,v) \in A,$$

fulfills the following conditions:

1. *The projection on the flow component q contains only one point, i.e., the flow is unique.*
2. *The projection on the squared pressure component π has the form*

$$\{\pi^* + \eta \mathbb{1} \mid \eta \in \mathbb{R}\}$$

for some π^*, i.e., the squared pressure is unique up to a constant shift.

7.1. Passive pipe networks

Proof. Collins et al. (1978) show the feasibility of system (7.6). To this end, they consider the following auxiliary nonlinear program, where q_a^0 is a root of Φ_a, i.e., $\Phi(q_a^0) = 0$:

$$\min \sum_{a \in A} \int_{q_a^0}^{q_a} \Phi_a(t) \, dt \tag{7.7}$$

$$\text{s.t.} \sum_{a \in \delta^+(u)} q_a - \sum_{a \in \delta^-(u)} q_a - q_u^{\text{nom}} = 0 \quad \text{for all } u \in V.$$

Let $q^* \in \mathbb{R}^A$ denote an optimal point of this system. The Karush-Kuhn-Tucker conditions for this optimization problem imply the existence of $\pi^* \in \mathbb{R}^V$ with

$$\Phi_a(q_a^*) = \pi_u^* - \pi_v^* \quad \text{for all } a = (u, v) \in A.$$

The pair (q^*, π^*) is a feasible solution to (7.6).

Next, we prove uniqueness of the flow. In fact, uniqueness of the flow follows from the strong convexity of the objective function of (7.7). Nevertheless, we give another proof. Therefore we assume to have two different solutions (q', π') and (q'', π''). For each arc we change the orientation if necessary such that $q' - q'' \geq 0$. Note that changing the direction of an arc goes along with replacing the constraint $\Phi_a(q_a) = \pi_u - \pi_v$ by $-\Phi_a(-q_a) = \pi_v - \pi_u$. As Φ_a is a strictly increasing function, the same holds for $q_a \mapsto -\Phi_a(-q_a)$. Hence the characteristics of (7.6) remain valid after reorientation.

Classical network flow theory shows that the difference $q' - q''$ consists of circuit flows only (see, e.g., Korte and Vygen (2007)). Thus, we obtain from $q' \geq q''$ that there exists a set of circuits $C_1, \ldots, C_\ell$ and corresponding flow values $q_{C_i} > 0, i = 1, \ldots, \ell$, such that

$$q_a' - q_a'' = \sum_{\substack{i = 1, \ldots, \ell \\ a \in C_i}} q_{C_i} \quad \text{for all } a \in A.$$

Note that the sum can be empty for an arc $a \in A$. We consider a single circuit C out of $C_1, \ldots, C_\ell$. For each arc $a \in C$ it holds $q_a' > q_a''$. Since Φ_a is strictly increasing we get

$$\pi_u' - \pi_v' > \pi_u'' - \pi_v'' \quad \text{for all } (u, v) \in C.$$

Summing up these inequalities gives

$$\sum_{(u,v) \in C} \pi_u' - \pi_v' > \sum_{(u,v) \in C} \pi_u'' - \pi_v''.$$

On the other hand, the summands cancel each other out, and we have

$$\sum_{(u,v) \in C} \pi_u' - \pi_v' = 0 = \sum_{(u,v) \in C} \pi_u'' - \pi_v'',$$

which is a contradiction. This proves $q' = q''$.

Since the flow is unique and Φ_a is strictly increasing, the squared pressure differences $\pi_u - \pi_v$ are uniquely determined for all arcs. Thus $\{\pi^* + \eta \mathbb{1} \mid \eta \in \mathbb{R}\}$ forms the set of the feasible squared pressure values.

Shifting all values by a constant is feasible, since the shift cancels out in the difference $\pi_u - \pi_v$, which is the only place π occurs in the problem. Constant shift, however, is the only possible source of difference. □

This theorem can be adapted to the case of bounded variables.

Theorem 7.2. *Let $q^{\mathrm{nom}} \in \mathbb{R}^V$ be a balanced nomination and Φ_a a strictly increasing function. Then the solution space of the problem*

$$
\begin{aligned}
\exists \quad & q, \pi & & (7.8)\\
\text{s.t.} \quad & \sum_{a \in \delta^-(u)} q_a - \sum_{a \in \delta^+(u)} q_a = q_u^{\mathrm{nom}} & & \text{for all } u \in V, \\
& \Phi_a(q_a) = \pi_u - \pi_v & & \text{for all } a = (u,v) \in A, \\
& \underline{\pi}_u \leq \pi_u \leq \overline{\pi}_u & & \text{for all } u \in V, \\
& \underline{q}_a \leq q_a \leq \overline{q}_a & & \text{for all } a \in A,
\end{aligned}
$$

is either empty or fulfills the following conditions:

1. *The projection on the flow component q consists of only one point, i.e., the flow is unique.*
2. *The projection on the squared pressure component π has the form*

$$\{\pi^* + \eta \mathbb{1} \mid \underline{\eta} \leq \eta \leq \overline{\eta}\}$$

for some π^*, $\underline{\eta}$ and $\overline{\eta}$, i.e., the squared pressure is unique up to a constant shift within some interval.

Proof. To prove this statement we make use of Theorem 7.1. Problem (7.6) is a relaxation of (7.8). Thus, in the case when (7.8) is feasible, the arc flow q is unique. Furthermore, the solution space for π is a straight line that is additionally bounded by the constraints $\underline{\pi}_u \leq \pi_u \leq \overline{\pi}_u$ for each node $u \in V$. The result is a line segment which can be described by

$$\{\pi^* + \eta \mathbb{1} \mid \underline{\eta} \leq \eta \leq \overline{\eta}\},$$

for some π^*, and $\underline{\eta} := \max_{u \in v}\{\underline{\pi}_u - \pi_u^*\}$ and $\overline{\eta} := \min_{u \in v}\{\overline{\pi}_u - \pi_u^*\}$. □

In particular, these theorems show that the respective feasible sets of the constraints are convex. This observation can be exploited by applying efficient local solution methods to obtain global information. The corresponding solution approach is based on the reformulation of the existence problem, which motivates the subsequent definition.

Definition 7.3. *Let Π be some existence problem and Π' be a minimization problem with a nonnegative objective value that is equivalent to Π in the following sense: If the optimal solution value for Π' is positive, then Π is infeasible and if, on the contrary, there is an optimal solution of Problem Π' with value 0, this solution is feasible for Problem Π. Then we call Π' a* slack reformulation *of Π.*

We remark that the idea of a slack reformulation is well known in other contexts. The following theorem presents a convex slack reformulation for Problem (7.8).

7.1. Passive pipe networks

Theorem 7.4. *The feasible set of the problem*

$$\min \quad \sum_{u \in V} \Delta_u + \sum_{a \in A} \Delta_a \qquad (7.9)$$

$$\begin{aligned}
\text{s.t.} \quad & \sum_{a \in \delta^+(u)} q_a - \sum_{a \in \delta^-(u)} q_a = q_u^{\text{nom}} & & \text{for all } u \in V, \\
& \Phi_a(q_a) = \pi_u - \pi_v & & \text{for all } a = (u,v) \in A, \\
& \pi_u - \Delta_u \leq \overline{\pi}_u & & \text{for all } u \in V, \\
& \pi_u + \Delta_u \geq \underline{\pi}_u & & \text{for all } u \in V, \\
& \Delta_u \geq 0 & & \text{for all } u \in V, \\
& q_a - \Delta_a \leq \overline{q}_a & & \text{for all } a \in A, \\
& q_a + \Delta_a \geq \underline{q}_a & & \text{for all } a \in A, \\
& \Delta_a \geq 0 & & \text{for all } a \in A
\end{aligned}$$

is nonempty and convex. Moreover, Problem (7.9) is a slack reformulation of Problem (7.8).

Proof. By Theorem 7.1, the constraint system

$$\sum_{a \in \delta^+(u)} q_a - \sum_{a \in \delta^-(u)} q_a = q_u^{\text{nom}} \qquad \text{for all } u \in V, \qquad (7.10)$$

$$\Phi_a(q_a) = \pi_u - \pi_v \qquad \text{for all } a = (u,v) \in A,$$

is feasible and has a unique solution flow q while the feasible vectors π form a straight line. The additional inequalities of Problem (7.9) in comparison to (7.10) are linear. Thus, the feasible region of Problem (7.9) is convex, which implies that the problem is a convex optimization problem. For every feasible solution (q, π) of (7.10),

$$\begin{aligned}
\Delta_u &:= \max\{0, \pi_u - \overline{\pi}_u, \underline{\pi}_u - \pi_u\} & & \text{for all } u \in V, \\
\Delta_a &:= \max\{0, q_a - \overline{q}_a, \underline{q}_a - q_a\} & & \text{for all } a \in A,
\end{aligned}$$

gives values for Δ such that (Δ, q, π) is feasible for Problem (7.9). Moreover, $\Delta = 0$ if and only if (q, π) does not violate any bounds. Thus Problem (7.9) is a slack reformulation of Problem (7.8), which proves the theorem. □

The factor β_a in (7.3) is 1 if and only if the slope of pipe a equals zero. In this case, we can therefore solve a convex optimization problem to determine the feasibility of the nonconvex existence Problem (7.5). This is due to the fact that

$$\Phi_a(q_a) = \alpha_a |q_a| q_a$$

is strictly increasing, which yields a convex solution space of (7.10), so that Theorem 7.4 provides a convex slack reformulation of (7.5). Therefore, a global solution can be determined efficiently without time-consuming spatial branching. Since for convex problems, local optima are global optima, efficient local solvers can be applied.

Approximation of slopes To be able to use this machinery, the pressure loss has to be described by an equation of the type

$$\Phi_a(q_a) = \pi_u - \pi_v. \qquad (7.11)$$

To this end, we approximate the term $\pi_u - \beta_a \pi_v$ from our original formulation of pipe model (7.4) by $\varepsilon_a (\pi_u - \pi_v)$. The approximation factor ε_a can then be moved to the left-hand side. We choose an approximation factor which minimizes the error in the squared ℓ_2-norm over the domain of the pressure variables, i.e., a minimizer of

$$\min_{\varepsilon \in \mathbb{R}} \left(\int_{\underline{\pi}_u}^{\overline{\pi}_u} \int_{\underline{\pi}_v}^{\overline{\pi}_v} ((\pi_u - \beta_a \pi_v) - \varepsilon (\pi_u - \pi_v))^2 \, d\pi_v \, d\pi_u \right).$$

In case that the endnodes u and v have identical pressure bounds, we can *analytically* evaluate this expression and obtain

$$\varepsilon_a = \frac{1}{2}(1 + \beta_a).$$

With this model adjustment, we can determine the feasibility of a nomination by solving just one convex optimization problem. This eliminates the need for time-consuming spatial branching.

7.2 • From passive pipe networks to gas networks with active devices

Real gas networks also incorporate active elements, which allow controlling the gas flow through the network. Depending on the type of the element, certain discrete and/or continuous control decisions have to be taken. We thus face a mixed-integer, nonlinear, nonconvex existence problem. Of course, the combination of integer and spatial branching described in Section 7.1.1 can be applied. With a careful element modeling, however, we can treat the subproblem remaining after fixing all discrete decisions very similarly to the case of a passive pipe network, i.e., we can decide the feasibility of the search tree node by solving a single or at least a small number of convex NLPs.

All discrete control decisions for the gas network elements described in Section 2.3 can be reduced to opening or closing a valve (see Section 2.3.3). For instance, a compressor machine may only operate if the valve on the pipe to that machine is open. If a valve is open, the pressures at both ends have to be equal, which is essentially the same as identifying the end nodes in the graph model. If a valve is closed, there must be no flow through it and the pressures at its ends are decoupled, which corresponds to removing the arc from the model. Thus, once the states of all valves are fixed, the remaining problem is a continuous existence problem. In the special case that all active elements are valves, we face a passive pipe network problem. Hence, each complete setting of all valves corresponds to a unique (flow, squared pressure difference) combination; however, the actual pressure level may be shifted (see Theorem 7.2). Validating a nomination on such a network corresponds to "selecting" a setting of the valves for which the flow is feasible and a feasible pressure distribution across the network can be found. This is illustrated in Figure 7.2.

For a passive pipe network, the relation between the flow q_a over a pipe $a = (u,v)$ and the squared inlet and outlet pressures π_u and π_v, respectively, is given by a strictly increasing function of the flow, yielding the squared pressure difference $\pi_u - \pi_v$ for a given flow q_a (see Eq. (7.4) with $\beta_a = 1$). In contrast, the characteristic of active elements is that for a given flow q_a, the squared pressure π_v at the outlet may be controlled in a certain range, depending on the inlet squared pressure π_u.

7.2. From passive pipe networks to gas networks with active devices

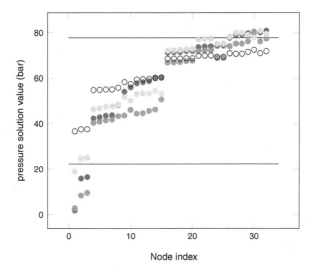

Figure 7.2. *Pressure vectors for four valve settings of an example network consisting of 37 pipes and 2 valves. Horizontal lines mark the squared pressure bounds $\underline{p}_u = 22$ and $\overline{p}_u = 78$. The flows are feasible in all cases. The settings marked by rings are feasible, i.e., they fulfill the pressure bounds, whereas the others are infeasible, i.e., the objective of Problem (7.9) was positive. Therefore, the pressure values cannot be shifted such that the bounds are respected.*

Let us denote by $\mathscr{F}_a \subseteq \mathbb{R}^3$ the feasible operating range of an active element $a \in A_{\text{active}}$, i.e., the set of (q_a, π_u, π_v)-tuples that can be realized by a. To obtain the desired functional relation between the flow and the squared pressure difference which is required in order to apply Theorem 7.4, we discretize the set $\mathscr{F}_a$ by introducing parallel hyperplanes $H_{(r_a, \gamma_{a,0})}, \ldots, H_{(r_a, \gamma_{a,n})}$ relating q_a and $\pi_u - \pi_v$ via

$$q_a = r_a(\pi_u - \pi_v) + \gamma_{a,h} \quad \text{for all } h \in \{0, \ldots, n\}, \tag{7.12}$$

where $r_a > 0$ is a parameter controlling the direction of the hyperplanes and $\gamma_{a,h} \in [\underline{\gamma}_{a,h}, \overline{\gamma}_{a,h}]$ are suitably chosen parameters "selecting" the discretization hyperplane. Assuming all discrete variables (i.e., binary switching variables and variables selecting a single $\gamma_{a,h}$ for each active element $a \in A_{\text{active}}$) have been fixed, the remaining problem can now be stated in the form of Theorem 7.4 with $\Phi(q_a) := (q_a - \gamma_{a,h})/r_a$ for each active element $a \in A_{\text{active}}$ plus some additional continuous constraints modeling $\mathscr{F}_a$. Abstractly, these can be thought of as being given by a vector-valued function $g_a(q_a, \pi_u, \pi_v)$ for each active element $a \in A_{\text{active}}$ and the constraints

$$g_a(q_a, \pi_u, \pi_v) \leq 0 \quad \text{for all } a \in A_{\text{active}}. \tag{7.13}$$

If, for each active element a, the set described by (7.13) is convex, the following theorem provides a means to determine the feasibility of the corresponding node in the branch-and-bound tree by solving just a single convex NLP.

Theorem 7.5. *Let $q^{\text{nom}} \in \mathbb{R}^V$ be a nomination and Φ_a a strictly increasing function. If, for each active element $a \in A_{\text{active}}$, the function g_a has the property that $\{(\pi_u, \pi_v) \mid g_a(q_a, \pi_u, \pi_v) \leq 0\}$ is a convex set for fixed q_a, a convex slack reformulation for the problem*

$$\exists \quad q, \pi \tag{7.14}$$
$$\text{s.t.} \quad \sum_{a \in \delta^+(u)} q_a - \sum_{a \in \delta^-(u)} q_a = q_u^{\text{nom}} \quad \text{for all } u \in V,$$
$$\Phi_a(q_a) = \pi_u - \pi_v \quad \text{for all } a = (u,v) \in A,$$
$$g_a(q_a, \pi_u, \pi_v) \leq 0 \quad \text{for all } a \in A_{\text{active}},$$
$$\underline{\pi}_u \leq \pi_u \leq \overline{\pi}_u \quad \text{for all } u \in V,$$
$$\underline{q}_a \leq q_a \leq \overline{q}_a \quad \text{for all } a \in A$$

is given by

$$\min \quad \sum_{u \in V} \Delta_u + \sum_{a \in A} \Delta_a \tag{7.15}$$
$$\text{s.t.} \quad \sum_{a \in \delta^+(u)} q_a - \sum_{a \in \delta^-(u)} q_a = q_u^{\text{nom}} \quad \text{for all } u \in V,$$
$$\Phi_a(q_a) = \pi_u - \pi_v \quad \text{for all } a = (u,v) \in A,$$
$$g_a(q_a, \pi_u, \pi_v) - \Delta_a \leq 0 \quad \text{for all } a \in A_{\text{active}},$$
$$q_a - \Delta_a \leq \overline{q}_a \quad \text{for all } a \in A,$$
$$q_a + \Delta_a \geq \underline{q}_a \quad \text{for all } a \in A,$$
$$\Delta_a \geq 0 \quad \text{for all } a \in A,$$
$$\pi_u - \Delta_u \leq \overline{\pi}_u \quad \text{for all } u \in V,$$
$$\pi_u + \Delta_u \geq \underline{\pi}_u \quad \text{for all } u \in V,$$
$$\Delta_u \geq 0 \quad \text{for all } u \in V.$$

Proof. To show that (7.15) is a convex optimization problem, the crucial observation is that, by Theorem 7.1, the flow vector q is uniquely determined by the flow conservation constraints and the relationships between flow and squared pressure differences given by the strictly increasing functions Φ_a. Denote this fixed flow vector by q^*. The solution space of Problem (7.15) is that of Problem (7.9) (which by Theorem 7.4 is convex), intersected with the feasible set of

$$g_a(q_a^*, \pi_u, \pi_v) - \Delta_a \leq 0 \quad \text{for all } a \in A,$$

which by assumption is convex.

Establishing that Problem (7.15) is a slack reformulation of Problem (7.14) works similarly to the proof of Theorem 7.4. We extend Problem (7.9) by the additional constraints $g_a(q_a, \pi_u, \pi_v) - \Delta_a \leq 0$ for all $a \in A_{\text{active}}$, and define Δ_a for $a \in A_{\text{active}}$ by

$$\Delta_a := \max\{0, g_a(q_a, \pi_u, \pi_v), q_a - \overline{q}_a, \underline{q}_a - q_a\}.$$

Obviously, $\Delta = 0$ if and only if no bound and no active element constraint is violated. □

In general, the set described by (7.13) for a fixed flow vector q need not be convex. We can deal with this situation by just solving Problem (7.15) and interpreting its solution

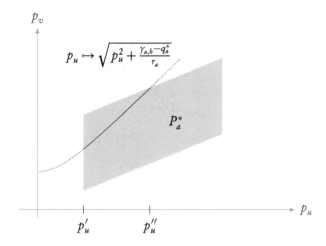

Figure 7.3. *Intersection of the polytope P_a^* describing the feasible operating range for $q_a = q_a^*$ with the curve determined by the selected hyperplane. In this case, the intersection corresponds to an interval $[p_u', p_u'']$, and we have the relation $\pi_u \in [{p_u'}^2, {p_u''}^2]$ if and only if $(q_a^*, \pi_u, \pi_v) \in \mathscr{F}_a$.*

properly. In the case $\Delta = 0$, we have been lucky and found a feasible solution. If $\Delta \neq 0$ we cannot reliably claim infeasibility yet. However, the flow q^* is available which is unique by Theorem 7.5.

The feasible operating range $\mathscr{F}_a \subseteq \mathbb{R}^3$ (as a subset of the (q_a, π_u, π_v)-space) of an active element $a \in A_{\text{active}}$ is naturally modeled as a polytope P_a in the (q_a, p_u, p_v)-space plus constraints coupling pressures and squared pressures, i.e.,

$$\pi_u = p_u^2 \quad \text{and} \quad \pi_v = p_v^2.$$

After fixing the flow q^*, the polytope P_a^* arising from restricting P_a to $q_a = q_a^*$ describes the remaining freedom to choose the pressures p_u and p_v. From the selected hyperplane

$$q_a = r_a(\pi_u - \pi_v) + \gamma_{a,h}$$

we obtain the outlet pressure p_v as a function of the inlet pressure p_u as

$$p_v(p_u) = \sqrt{p_u^2 + \frac{\gamma_{a,h} - q_a^*}{r_a}}. \tag{7.16}$$

We can now compute the set of inlet pressures p_u, for which $(q_a^*, p_u, p_v(p_u))$ as given by (7.16) is contained in P_a^*. In general, the corresponding set in the (q_a^*, π_u, π_v)-space is given by a union of intervals $I_{a,u}^{(i)}, I_{a,v}^{(i)}$ for π_u and π_v, i.e., $(\pi_u, \pi_v) \in \bigcup_i I_{a,u}^{(i)} \times I_{a,v}^{(i)}$. This method is illustrated in Figure 7.3. Finally we solve the optimization problem

$$\exists \pi \text{ s.t.} \quad \pi_u - \pi_v = \Phi_a(q_a^*) \quad \text{for all } a = (u,v) \in A, \tag{7.17}$$

$$\bigcup_i (\pi_u, \pi_v) \in I_{a,u}^{(i)} \times I_{a,v}^{(i)} \quad \text{for all } a = (u,v) \in A.$$

If it is feasible, the optimal solution π_u^* yields a primal solution to (7.14) given by (q_a^*, π_u^*). Otherwise, if the above problem is infeasible, then (7.14) is infeasible.

7.3 • Element modeling

We are now prepared to present the model for the remaining network elements. For each element, we will describe the model which can be used to formulate a general MINLP to be solved by spatial branching. Let k_a be a constant for each $a \in A$, x_a be a binary variable that indicates whether flow is allowed on the arc a, y_a a discretization variable, $X \subseteq \{0,1\}^A$ and $\gamma_a y_a = \gamma_{a,h}$. Then the general MINLP model is written as follows:

$$\exists q, \pi, p, x, y \tag{7.18}$$

$$\text{s.t. } x_a = 1 \Rightarrow \alpha_a q_a |q_a|^{k_a} + \gamma_a y_a - (\pi_u - \beta_a \pi_v) = 0 \quad \text{for all } a = (u,v) \in A,$$

$$x_a = 1 \Rightarrow g_a(q_a, \pi_u, \pi_v) \leq 0 \quad \text{for all } a = (u,v) \in A,$$

$$x_a = 0 \Rightarrow q_a = 0 \quad \text{for all } a \in A,$$

$$\sum_{a \in \delta^+(u)} q_a - \sum_{a \in \delta^-(u)} q_a = q_u^{\text{nom}} \quad \text{for all } u \in V,$$

$$\underline{\pi}_u \leq \pi_u \leq \overline{\pi}_u \quad \text{for all } u \in V,$$

$$\underline{q}_a \leq q_a \leq \overline{q}_a \quad \text{for all } a \in A,$$

$$x_a \in \{0,1\} \quad \text{for all } a \in A,$$

$$y_a \in \{0,\ldots,n\} \quad \text{for all } a \in A,$$

$$x \in X.$$

For ease of notation we write $\Phi_a(q_a) := \alpha_a q_a |q_a|^{k_a} + \gamma_a y_a$ in the following. The model is solved by SCIP. All implications are modeled by indicator constraints. We set branching priorities to ensure that no spatial branching is performed before all discrete decisions are fixed. Additionally, we solve (7.15) whenever the LP relaxation of a node of the branching tree yields integer values for the integer variables. We either obtain a feasible solution or solve (7.17) to obtain a feasible solution or to detect the infeasibility of the current integer decisions. In the last case, we cut off the node from the branch-and-bound tree corresponding to the integer decisions.

7.3.1 • Resistors

We model two distinct types of resistors: those with a constant pressure loss and those with a flow dependent pressure loss. The "exact" formulas for resistors from Section 2.3.2 do not fit in our framework. We cope with this by using best approximating formulas of the required type.

Constant pressure loss The first type is modeled by (2.32). Typically this type of resistor represents a measuring system that causes a fixed pressure decrease by $\xi_a \in \mathbb{R}_{\geq 0}$ in the direction of the flow. Originally, such a resistor $a = (u,v) \in A_{\text{rs}}$ is described by

$$p_u - p_v = s_a^{\text{fwd}} \xi_a - s_a^{\text{bwd}} \xi_a,$$

$$s_a^{\text{fwd}} = 1 \implies q_a > 0,$$

$$s_a^{\text{bwd}} = 1 \implies q_a < 0,$$

$$s_a^{\text{zero}} = 1 \implies q_a = 0,$$

$$s_a^{\text{fwd}} + s_a^{\text{bwd}} + s_a^{\text{zero}} = 1,$$

7.3. Element modeling

where s_a^{fwd}, s_a^{bwd}, and s_a^{zero} indicate forward flow, backward flow, and no flow, respectively. Observe that although resistors are passive elements, we need to introduce binary variables for modeling reasons.

We express the pressure loss $p_u - p_v$ in squared pressure variables π. To this end the following approximation is employed:

$$\varepsilon_a (\pi_u - \pi_v) = \xi_a.$$

The linearization factor ε_a is a minimizer of

$$\min_{\varepsilon \in \mathbb{R}} \int_{\underline{p}_u}^{\overline{p}_u} \int_{\underline{p}_v}^{\overline{p}_v} \left[(p_u - p_v) - \varepsilon \cdot (p_u^2 - p_v^2) \right]^2 dp_v \, dp_u,$$

and ε_a can be computed analytically, but we forgo to show the formula here. Then, the following model conforms with Problem (7.18):

$$\varepsilon_a (\pi_u - \pi_v) = s_a^{\text{fwd}} \xi_a - s_a^{\text{bwd}} \xi_a,$$
$$s_a^{\text{fwd}} = 1 \implies q_a > 0,$$
$$s_a^{\text{bwd}} = 1 \implies q_a < 0,$$
$$s_a^{\text{zero}} = 1 \implies q_a = 0,$$
$$s_a^{\text{fwd}} + s_a^{\text{bwd}} + s_a^{\text{zero}} = 1.$$

Flow dependent pressure loss The second type of resistors models a pressure difference that depends on the flow q_a for $a = (u,v) \in A_{\text{rs}}$. Its model is given by the constraints (2.30) and (2.31). Reformulating these equations with density

$$\rho = \frac{\rho_0 z_0 T_0}{p_0} \frac{p}{z(p,T) T}$$

and compressibility factor as by the AGA equation (2.5), we obtain

$$\frac{p_u^2 - p_u p_v}{1 + \zeta_a p_u} = \xi_a q_a |q_a|$$

as model for the resistor with $p_u \geq p_v$. The constants ζ_a and ξ_a are resistor specific.

To deal with this nonlinearity, we compute ε_a to achieve the following approximation in terms of a Weymouth equation:

$$\varepsilon_a (p_u^2 - p_v^2) \approx \begin{cases} \dfrac{p_u^2 - p_u p_v}{1 + \zeta_a p_u} & \text{if } p_u \geq p_v, \\ -\dfrac{p_v^2 - p_u p_v}{1 + \zeta_a p_v} & \text{if } p_u \leq p_v. \end{cases}$$

Let $\ell := \min\{\underline{p}_u, \underline{p}_v\}$ and $u := \max\{\overline{p}_u, \overline{p}_v\}$. Our goal is define ε_a such that the approximation error is minimal. Hence ε_a is a solution of this minimization problem:

$$\min_{\varepsilon \in \mathbb{R}} \int_\ell^u \int_\ell^{p_u} \left(\frac{p_u^2 - p_u p_v}{1 + \zeta_a p_u} - \varepsilon(p_u^2 - p_v^2) \right)^2 dp_v \, dp_u$$
$$+ \int_\ell^u \int_\ell^{p_v} \left(-\frac{p_v^2 - p_u p_v}{1 + \zeta_a p_v} - \varepsilon(p_u^2 - p_v^2) \right)^2 dp_u \, dp_v.$$

Again, symbolic algebra packages derive a closed form expression for ε_a. Using this solution we model the second type of resistors by a Weymouth constraint:

$$q_a |q_a| = \frac{\varepsilon_a}{\xi_a}(\pi_u - \pi_v).$$

Setting

$$\Phi_a(q_a) := \frac{\xi_a}{\varepsilon_a} q_a |q_a|,$$

$$g_a(q_a, \pi_u, \pi_v) := 0,$$

and fixing $x_a = 1$ yields a model conforming with Problem (7.18).

7.3.2 ▪ Valves

Following the description in Section 2.3.3, a valve $a = (u,v) \in A_{va}$ has two physical states: open or closed. When closed, no gas flow along a is allowed, so $q_a = 0$. When open, the pressures at the end nodes u and v must be equal, i.e., $p_u = p_v$, which is equivalent to shrinking u and v to a single node.

A binary variable s_a distinguishes between these states. With this variable we use the following constraints for a valve:

$$s_a = 0 \implies q_a = 0,$$
$$s_a = 1 \implies \pi_u = \pi_v.$$

Thus $g_a(q_a, \pi_u, \pi_v) := 0$ and $\Phi_a(q_a) := 0$ yields a model conforming with Problem (7.18).

7.3.3 ▪ Control valves

A control valve $a = (u,v) \in A_{cv}$ is used to reduce the pressure from u to v. Its technical details are described in Section 2.3.4. Three binary variables represent the states of a control valve: s_a indicates whether the control valve is open or closed, s_a^{ac} indicates whether it is active, and s_a^{bp} whether it is bypassed.

$$s_a = 0 \implies q_a = 0, \tag{7.19a}$$
$$s_a^{ac} = 1 \implies \underline{\Delta}_a \leq p_u - p_v \leq \overline{\Delta}_a, \tag{7.19b}$$
$$s_a^{ac} = 1 \implies q_a \geq 0, \tag{7.19c}$$
$$s_a^{bp} = 1 \implies p_u = p_v, \tag{7.19d}$$
$$s_a^{ac} + s_a^{bp} = s_a, \tag{7.19e}$$
$$\pi_u = p_u^2 \quad \text{and} \quad \pi_v = p_v^2. \tag{7.19f}$$

Usually, bounds on the pressure differences $\underline{\Delta}_a, \overline{\Delta}_a$ are given in terms of pressure instead of squared pressure; see Eq. (2.37). Hence, the explicit coupling constraints (7.19f) are needed. To obtain a model which conforms with Problem (7.18), we define $\Phi_a(q_a) := 0$ for the bypass case. For the active case we define $g_a(q_a, \pi_u, \pi_v)$ by constraints (7.19b), (7.19c), and (7.19f).

When the control valve has a preset downstream pressure p_a^{set}, the states closed, active, and bypassed of the control valve are modeled using the corresponding binary variables

7.3. Element modeling

leading to additional constraints:

$$s_a = 0 \implies p_v \geq p_a^{\text{set}},$$
$$s_a^{\text{ac}} = 1 \implies p_v = p_a^{\text{set}},$$
$$s_a^{\text{bp}} = 1 \implies p_v \leq p_a^{\text{set}}.$$

Again, these constraints complement the definition of $g_a(q_a, \pi_u, \pi_v)$.

Discretization of the feasible operating range According to Section 7.2, the feasible operating range of control valve $a = (u, v) \in A_{\text{cv}}$ is described by a set $\mathscr{F}_a$ within the (q_a, π_u, π_v)-space, which is modeled by a polytope P_a in the (q_a, p_u, p_v)-space and the additional pressure coupling constraints (see (7.19)):

$$\pi_u = p_u^2 \quad \text{and} \quad \pi_v = p_v^2.$$

It is sufficient to model $\mathscr{F}_a$ for the active state (i.e., $s_a^{\text{ac}} = 1$) only, since in the remaining states the correct behavior is already enforced by the constraints (7.19a) and (7.19d). For an active control valve, the set $\mathscr{F}_a$ is

$$\mathscr{F}_a = \{(q_a, \pi_u, \pi_v) \in \mathbb{R}^3 \mid \underline{p}_v^2 \leq \pi_u \leq \pi_v \leq \overline{p}_u^2,$$
$$\underline{\Delta}_a \leq p_u - p_v \leq \overline{\Delta}_a,$$
$$\pi_u = p_u^2, \pi_v = p_v^2,$$
$$0 \leq q_a \leq \overline{q}_a\}.$$

Since pressure difference bounds are given, the model for the constraints g_a in Problem (7.14) requires additional variables p_u and p_v; the function g_a then describes the set

$$\{(q_a, \pi_u, \pi_v, p_u, p_v) \in \mathbb{R}^5 \mid \underline{\Delta}_a \leq p_u - p_v \leq \overline{\Delta}_a, \pi_u = p_u^2, \pi_v = p_v^2\},$$

as the remaining constraints can be expressed as variable bounds.

To account for the relationship between q_a and $\pi_u - \pi_v$, we discretize the set $\mathscr{F}_a$ by parallel hyperplanes. The intersection of these hyperplanes with $\mathscr{F}_a$ then represents the discretized feasible operating range of the control valve. Note that the pressure p_v can be fixed to p_a^{set} which has no affect on this discretization. We generate two hyperplanes that enclose $\mathscr{F}_a$:

$$\underline{q}_a = r_a \overline{\delta}_a + \gamma_{a,0} \quad \text{and} \quad \overline{q}_a = r_a \underline{\delta}_a + \gamma_{a,n}, \tag{7.20}$$

with $\underline{\delta}_a \leq \pi_u - \pi_v \leq \overline{\delta}_a$ and $\underline{\delta}_a := \underline{\pi}_u - \overline{\pi}_v$, $\overline{\delta}_a := \overline{\pi}_u - \underline{\pi}_v$. We model the rotation parameter as $r_a = (\overline{q}_a - \underline{q}_a)/(\overline{\delta}_a - \underline{\delta}_a)$ to map the shape of $\mathscr{F}_a$ to some extent. For a fixed parameter r_a, the two parameters $\gamma_{a,0}$ and $\gamma_{a,n}$ can be derived from (7.20). The discretization of $\mathscr{F}_a$ by hyperplanes $H_{(r_a, \gamma_{a,0})}, \ldots, H_{(r_a, \gamma_{a,n})}$ is then given by

$$q_a = r_a(\pi_u - \pi_v) + \frac{h(\gamma_{a,n} - \gamma_{a,0})}{n} + \gamma_{a,0},$$

with $h \in \{0, \ldots, n\}$. This relation between the difference of squared pressures and the flow yields the definition of $\Phi_a(q_a)$ for the active case.

7.3.4 ▪ Compressor machines and groups

Our model for a single compressor machine uses the convex hull of the characteristic diagram as the feasible operating range of the machine and relaxes restrictions that may be imposed by the drive. As described in Section 2.4.2, often a number of individual machines (typically between 1 and 4) constitute a compressor group and are treated as a single entity in the network model. This group may be run in parallel, serial, or as a mixture of both. Each possible mode of operating is called a configuration. As in the approach presented in Chapter 6, a configuration may be represented by a series-parallel subnetwork specifying the gas flow through the compressor group, determining which compressors are used in which order. Configurations may be directly modeled by including the subnetwork for each configuration in the network and requiring that only one of them be used, as shown in Figure 6.1. However, this introduces complexity in form of additional constraints and variables. Instead, we propose an approach based on projections of polytopes that does not increase the model size. Wu et al. (2000) describe an aggregation method for homogeneous compressors running in parallel. Our approach can also be used for inhomogeneous compressors and any combination of serial and parallel piping.

We use a single polytope in the (q_a, p_u, p_v)-space to describe the feasible operating range of an entire compressor group $a = (u, v) \in A_{cg}$ of the network model. This polytope is computed from polytopes modeling the feasible operating ranges of the compressor machines, which are combined according to the allowed configuration to derive an outer approximation of the operating range of the compressor group.

As in Section 7.3.3, three binary variables represent the states of the compressor group: s_a indicates whether the compressor group is open or closed, s_a^{ac} indicates whether it is active, and s_a^{bp} whether it is bypassed. This yields the model

$$s_a = 0 \implies q_a = 0, \tag{7.21a}$$
$$s_a^{ac} = 1 \implies (q_a, p_u, p_v) \in P_a, \tag{7.21b}$$
$$s_a^{ac} = 1 \implies q_a \geq 0, \tag{7.21c}$$
$$s_a^{bp} = 1 \implies p_u = p_v, \tag{7.21d}$$
$$s_a^{ac} + s_a^{bp} = s_a, \tag{7.21e}$$
$$\pi_u = p_u^2 \quad \text{and} \quad \pi_v = p_v^2. \tag{7.21f}$$

To obtain a model which conforms with Problem (7.18), we define $\Phi_a(q_a) := 0$ for the bypass case and $g_a(q_a, \pi_u, \pi_v)$ by constraints (7.21b), (7.21c), and (7.21f) for the active case.

The remainder of this section describes how to determine and treat the approximation to feasible operating ranges of the active state, i.e., the case $s_a^{ac} = 1$.

Modeling compressor machines Besides the nonlinear pressure loss equations of the pipelines, compressors are the other central source of nonlinear behavior of gas networks. The discrete decisions on activating and bypassing a compressor introduce a further source of discontinuity. However, for a correct model of the gas flow it is important to have a proper representation of their capabilities. More detailed discussions on compressor modeling can be found in Carter, Schroeder, and Harbick (1993), Carter (1996), and Wu et al. (2000), for instance.

In general, the feasible operating range of a compressor machine is a bounded nonconvex subset of the three-dimensional space. Usually, it is described as a projection to a two-dimensional space, which is called the characteristic diagram of the compressor (see

7.3. Element modeling

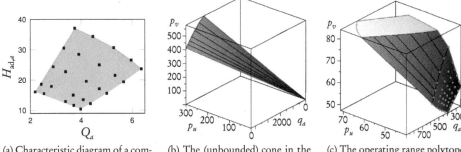

(a) Characteristic diagram of a compressor machine. (b) The (unbounded) cone in the transformed space. (c) The operating range polytope.

Figure 7.4. *From the characteristic diagram of a compressor machine to the operating range polytope. (Source: Pfetsch et al. (2015).)*

Section 2.3.5.1), which we assume to be given for all compressor machines of the network. In practice, it is gathered from measurements of various different flow and input/output pressure combinations. Each point in Figure 7.4(a) represents one such measurement. Below we present the mathematical details of the transformation from this input data to the feasible operating range. In our model we use a linear outer approximation for the feasible operating range.

Consider a compressor with inlet u and outlet v. The characteristic diagram is given by measurement points $\{(Q^{(k)}, H_{ad}^{(k)}) \mid k \in I\}$ where I is a finite index set. In order to use these quantities in the model directly, we need a map from the (Q, H_{ad})-space into the (q, p_u, p_v)-space of mass flow rate through the compressor, and inlet and outlet pressure, respectively. Using Eq. (2.20), (2.38), and (2.42), this mapping is given by

$$\begin{pmatrix} Q \\ H_{ad} \end{pmatrix} \mapsto \begin{pmatrix} q \\ p_u \\ p_v \end{pmatrix} = p_u \begin{pmatrix} c_1 Q \\ 1 \\ (c_2 H_{ad} + 1)^{\frac{x}{x-1}} \end{pmatrix}. \quad (7.22)$$

According to our model assumptions, $c_1 = \frac{1}{R_s T z}$ and $c_2 = \frac{x-1}{x R_s T z}$ are constant. The computation of a (constant) value for $z(p, T)$ necessitates the choice of a reference value for p_u. This choice and its impact is discussed in Section 12.3.4.

Each point $(Q^{(k)}, H_{ad}^{(k)})$ from the characteristic diagram can thus be translated into a ray of feasible combinations $(q^{(k)}, p_u^{(k)}, p_v^{(k)})$ for the compressor. The convex hull of all these rays, which is a cone C pointed in the origin for nonnegative p_u, gives us a linear approximation to the feasible operating range of the compressor. We intersect the cone C with

$$\{(q, p_u, p_v) \in \mathbb{R}^3 \mid \underline{p}_u \leq p_u \leq p_v \leq \overline{p}_v, \underline{q} \leq q \leq \overline{q}\},$$

corresponding to the technical limitations of the compressor in terms of minimal and maximal entry and exit pressure and throughput flow. This construction is visualized for an example in Figure 7.4.

Modeling compressor groups For a given configuration of a compressor group $a = (u, v) \in A_{cg}$, we consider the subnetwork representing this configuration. This subnetwork has exactly one entry u and exactly one exit v. Within the network, the gas flows in

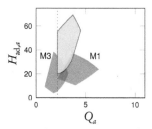 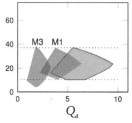 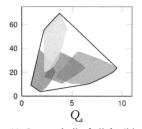

(a) Serial compression, i.e., first M3 then M1.

(b) Parallel compression, i.e., M1 parallel to M3.

(c) Convex hull of all feasible points.

Figure 7.5. *Resulting operating ranges for different combinations of two machines. (Source: Pfetsch et al. (2015).)*

serial and/or in parallel through the individual compressor machines. We set up a gas flow submodel for this subnetwork with the same constraints as described above: the flow conservation constraints (7.1) and constraints describing the feasible operating range of the compressor machines, i.e., linear constraints describing the polytope of each compressor machine. Note that this model does not contain binary variables, since we assume that the machines chosen in the configuration are set active (and not bypassed). By q_a we denote the flow through the subnetwork and by p_u and p_v the pressures at the entry and exit of the subnetwork, respectively. We introduce pressure variables $p_1, p_2, \ldots, p_n$ for the inner nodes and flow variables $q_1, q_2, \ldots, q_m$ for the arcs. This submodel describes a polytope with variables $(q_a, p_u, p_v, q_1, q_2, \ldots, q_m, p_1, p_2, \ldots, p_n)$. We apply Fourier–Motzkin elimination to project this polytope into the three-dimensional subspace of the variables (q_a, p_u, p_v). This gives a linear description of the feasible operating range, i.e., the set of feasible pressure and flow combinations, for the configuration under consideration. As an example, consider the characteristic diagrams in Figure 7.5. Here two characteristic diagrams for compressor machines M1 and M3 are shown. They can operate either in a serial compression (first M3 then M1) or in a parallel compression mode. The resulting characteristic diagram for serial compression is shown in Figure 7.5(a). The resulting characteristic diagram for parallel compression is shown in Figure 7.5(b).

Applying this procedure to all configurations of compressor group $a \in A_{cg}$, gives a family of polytopes in the same (q_a, p_u, p_v)-space, one for each configuration. The convex hull of the union of these polytopes constitutes an outer approximation to the feasible operating range of the whole compressor group. This convex hull is shown as a black line in Figure 7.5(c). The outer description of the resulting polytope yields a system of linear inequalities:

$$A_a \begin{pmatrix} q_a \\ p_u \\ p_v \end{pmatrix} \leq b_a.$$

All polyhedral calculations are carried out by the Parma Polyhedra Library (see Bagnara, Hill, and Zaffanella (2008)).

Note that the operating range polytope uses pressure variables, not squared pressure variables. Hence, explicit coupling constraints are needed at the inlet and the outlet of each compressor group.

Discretization of the feasible operating range As described in Section 7.2, the feasible operating range of an active element is described by a set $\mathscr{F}_a$ within the

7.3. Element modeling

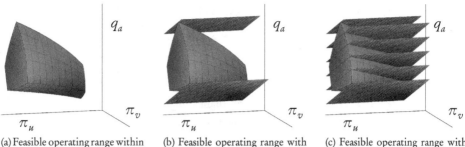

(a) Feasible operating range within the (q_a, π_u, π_v)-space.

(b) Feasible operating range with enclosing planes.

(c) Feasible operating range with discretization for $n = 6$.

Figure 7.6. *Feasible operating range of a compressor group within the (q_a, π_u, π_v)-space with discretization.*

(q_a, π_u, π_v)-space. According to the model above, $\mathscr{F}_a$ is given by

$$\mathscr{F}_a = \left\{ (q_a, \pi_u, \pi_v) \in \mathbb{R}^3 \,\Big|\, A_a \begin{pmatrix} q_a \\ p_u \\ p_v \end{pmatrix} \leq b_a,\, \pi_u = p_u^2, \pi_v = p_v^2, \right. $$

$$\left. \underline{\pi}_u \leq \pi_u \leq \pi_v \leq \overline{\pi}_v,\, \underline{q}_a \leq q_a \leq \overline{q}_a \right\}.$$

Similarly as for control valves, we discretize $\mathscr{F}_a$ by hyperplanes

$$q_a = r_a(\pi_u - \pi_v) + \gamma_{a,h} \quad \text{for all } a = (u,v) \in A_{cg},\, h \in \{0, \ldots, n\}.$$

Again, this relation between the difference of squared pressures and the flow yields the definition of $\Phi_a(q_a)$ for the active case.

An example of $\mathscr{F}_a$ and its discretization is shown in Figure 7.6. The intersection of these hyperplanes with $\mathscr{F}_a$ represents the discretized feasible operating range of the compressor group as described in Section 7.2.

We start the discretization process with the adiabatic characteristic diagram $\{(Q^{(k)}, H_{\text{ad}}^{(k)}) \mid k \in I\}$ where I is a finite index set. We calculate the convex hull of these points and sort them clockwise: $(Q^{(1)}, H_{\text{ad}}^{(1)}), \ldots, (Q^{(m)}, H_{\text{ad}}^{(m)})$. Each of these points corresponds to a curve in the (q_a, π_u, π_v)-space, defined by the following mapping:

$$f_k : \mathbb{R} \to \mathbb{R}^3$$

$$\pi_u \mapsto \begin{pmatrix} c_1 Q^{(k)} \sqrt{\pi_u} \\ \pi_u \\ \left(c_2 H_{\text{ad}}^{(k)} + 1\right)^{\frac{2x}{x-1}} \pi_u \end{pmatrix} = \begin{pmatrix} q_a \\ \pi_u \\ \pi_v \end{pmatrix}.$$

Differing from Eq. (7.22), the image is given as π, not p. We then approximate the feasible operating range $\mathscr{F}_a$ by

$$\tilde{\mathscr{F}}_a := \bigcup_{\pi_u \in [\underline{\pi}_u, \overline{\pi}_u]} \text{conv}\{f_k(\pi_u) \mid k \in \{1, \ldots, m\}\},$$

where the boundary of $\tilde{\mathscr{F}}_a$ is given by the union of the surfaces

$$S_k := \{\lambda f_k(\pi_u) + (1-\lambda) f_{k+1}(\pi_u) \mid \lambda \in [0,1],\ \underline{\pi}_u \leq \pi_u \leq \overline{\pi}_u\},$$

with $k \in \{1, \ldots, m\}$ and $m+1$ is identified with the first point. This means that we discretize $\tilde{\mathscr{F}}_a$ instead of $\mathscr{F}_a$.

To keep the number of hyperplanes and thus discretization variables small, we account for the shape of $\tilde{\mathscr{F}}_a$. We therefore calculate first a pair of hyperplanes $H_{(r_a, \gamma_{a,0})}, H_{(r_a, \gamma_{a,n})}$ that minimally enclose $\tilde{\mathscr{F}}_a$. The normal vector of these hyperplanes is given by $(1, -r_a, r_a^T)$, hence the parameter r_a is associated with the rotation or rather the tilt of the hyperplane, while the translation parameter $\gamma_{a,b}$ enables one to shift the hyperplane. The discretization level is then determined by the number of hyperplanes between the enclosing ones.

Let us describe the discretization process in more detail: The idea is to generate a candidate set of possible values for the parameter r_a. More precisely, we compute a possible value r_a^k for each of the two-dimensional manifolds S_k. This is done by forcing a hyperplane of the form (7.12) to intersect S_k at suitable points $(q_a^k, \pi_u^k, \pi_v^k), (q_a^{k+1}, \pi_u^{k+1}, \pi_v^{k+1}) \in \{f_k(\underline{\pi}_u), f_k(\overline{\pi}_u), f_{k+1}(\underline{\pi}_u), f_{k+1}(\overline{\pi}_u)\}$, which gives rise to the following system of equations:

$$-r_a^k \pi_u^k + r_a^k \pi_v^k + q_a^k = \gamma_a,$$
$$-r_a^k \pi_u^{k+1} + r_a^k \pi_v^{k+1} + q_a^{k+1} = \gamma_a,$$

yielding

$$r_a^k = \frac{q_a^k - q_a^{k+1}}{\pi_u^k - \pi_v^k + \pi_u^{k+1} - \pi_v^{k+1}}.$$

For each value r_a^k, we subsequently calculate the translation parameters $\gamma_{a,0}^k, \gamma_{a,n}^k$ such that the corresponding parallel hyperplanes $H_{(r_a^k, \gamma_{a,0}^k)}, H_{(r_a^k, \gamma_{a,n}^k)}$ enclose $\tilde{\mathscr{F}}_a$ with minimal distance.

We then select the tuple $(r_a, \gamma_{a,0}, \gamma_{a,n}) \in \{(r_a^1, \gamma_{a,0}^1, \gamma_{a,n}^1), \ldots, (r_a^m, \gamma_{a,0}^m, \gamma_{a,n}^m)\}$ that minimizes the distance between the enclosing hyperplanes, which is given by $(\gamma_{a,n} - \gamma_{a,0})/\sqrt{1 + r_a^2}$. Finally, the discretization of $\tilde{\mathscr{F}}_a$ is represented by the hyperplanes $H_{(r_a, \gamma_{a,0})}, \ldots, H_{(r_a, \gamma_{a,n})}$.

7.3.5 ▪ Combinatorics of subnetwork operation modes

A subnetwork operation mode is the designated way to describe a set of discrete states for the active arcs of some subnetwork; see Section 2.4.3. In addition to the choice between opening and closing elements, the flow direction can optionally be specified.

For every mode $m \in \mathscr{M}_S$ of a subnetwork, we introduce a binary variable $s_m \in \{0, 1\}$. Exactly one mode is to be selected:

$$\sum_{m \in \mathscr{M}_S} s_m = 1.$$

The state of individual elements is enforced by inequalities. Let $A^{\text{open}}(m)$ ($A^{\text{closed}}(m)$) denote the subset of arcs that are set to open (closed) in mode m. Then the state of arcs is enforced by the inequalities

$$s_m \leq s_a \quad \text{for all } a \in A^{\text{open}}(m),\ m \in \mathcal{M}_S,$$
$$s_m \leq 1 - s_a \quad \text{for all } a \in A^{\text{closed}}(m),\ m \in \mathcal{M}_S.$$

Note that these inequalities could be strengthened to equations

$$s_a = \sum_{m \in \mathcal{M}_S(a)} s_m, \qquad (7.23)$$

where $\mathcal{M}_S(a)$ denotes all modes where a is open. We could not, however, observe improvements for the dual bound, and even worse performance of primal heuristic plugins occurs in SCIP when using this strengthening, and thus we refrain from using it. Note that the above formulation with (7.23) is equivalent to Eq. (6.25).

Let mode m prescribe flow in forward direction on arc a. We then add the constraint

$$s_m = 1 \quad \Longrightarrow \quad q_a \geq 0,$$

which forces the flow direction.

7.4 · Conclusion

In Section 7.3, we presented an MINLP model (7.18) for NoVa. Furthermore, we presented a convex reformulation for NoVa for passive gas transmission networks in Section 7.1. This reformulation allows one to either compute a feasible solution or to detect the infeasibility of the nomination by solving a convex optimization problem. In the successive Section 7.2 we extended this approach to handle gas networks with active devices. Our solution approach to solve model (7.18) presented in Section 7.3 is then as follows: We use SCIP to solve model (7.18). Inside SCIP the nonlinearities are tackled by separation and spatial branching. We set branching priorities such that no spatial branching is applied before all integer variables are fixed. Whenever the LP relaxation of a node of the branching tree yields integer values for the integer variables we solve (7.15). We either obtain a feasible solution or solve (7.17) to obtain a feasible solution or to detect the infeasibility of the current integer decisions. In the last case, we cut off the node from the branch-and-bound tree corresponding to the integer decisions.

Characteristic of this approach is that we use an accurate model for pipes, while the feasible operating range of compressor groups is enlarged due to our convexification described in Section 7.3.4.

We will use our model in different variants for computational experiments in Chapter 12. We evaluate variants of the model that do not conform to the theory underlying the approach. To this end, we analyze the influence of the number of hyperplanes used to discretize the feasible operating ranges of active elements on the solvability and feasibility of the models. We show that the algorithm works better and has (almost) consistent results if the elements are not discretized at all. The same line of argumentation is used for the consideration of slopes in the pressure loss equations. Again, our approach is only in line with the theory if pipe slopes are ignored, but we show that it is practically justified to incorporate the respective coefficients in the model. Finally, we compare our problem specific approach to plain spatial branch-and-bound as described in Section 7.1.1.

Chapter 8
The reduced NLP heuristic

Ralf Gollmer, Rüdiger Schultz, Claudia Stangl

Abstract *We introduce a heuristic for the validation of nominations with two basic features: reducing the algebraic equations modeling conservation of mass and momentum as well as heuristically fixing binary decisions on active network elements.*

In the present chapter we develop a heuristic approach to validate nominations crucially relying on (i) model reduction in systems of algebraic equations resulting from Kirchhoff's laws, and (ii) (heuristic) procedures for fixing the binary decisions concerning the active network elements.

In stationarity, the continuity and momentum equations of gas dynamics can be reduced to algebraic equations whose unknowns are the pressure values at the nodes and flow values on the arcs of the gas transportation network. For further details of the physics within a gas pipeline see Section 2.3.1. This setting corresponds to what is known as Kirchhoff's current and voltage laws in electrical circuits or what is covered by the content and co-content models in water distribution networks. The counterparts to pressure and flow in gas networks are voltage and current in electrical circuits as well as hydraulic head and flow in water networks.

In all three areas, water, electricity, and gas, it more or less belongs to the folklore, but seems to have been "rediscovered" independently several times, that from the mentioned system of algebraic equations a substantial number of variables can be eliminated (Hamam and Brameller (1971); Mallinson et al. (1993); Ríos-Mercado et al. (2002)). More precisely, the elimination would lead to a system of nonlinear equations with as many unknowns as there are equations if we would take into consideration passive networks consisting of pipes only. This number coincides with the number of fundamental cycles in the network, i.e., the minimal number of arcs whose removal transforms the network into a tree. The number of unknowns also coincides with the number of fundamental cycles if we add compressor stations and control valves to the network with fixed pressure change.

Therefore, in one or another fashion, an initial system of equations

$$\mathscr{A} q = q^{\text{nom}},$$
$$\mathscr{A}^T p^2 = \alpha |q| q \tag{8.1}$$

arises for a gas network represented by a graph. Here, q^{nom} corresponds to the given nomination with positive components for supply and negative ones for demand. The network is represented by the node-arc-incidence matrix $\mathscr{A}$ with

$$\mathscr{A}_{ua} = \begin{cases} 1 & \text{if } a \in \delta^+(u), \\ -1 & \text{if } a \in \delta^-(u), \\ 0 & \text{otherwise.} \end{cases}$$

The symbol p^2 represents a vector, each component of which is the square of the pressure at the corresponding node. Similarly, $\alpha|q|q$ stands for a vector with a component for each arc of the form $\alpha_a|q_a|q_a$ with coefficients

$$\alpha_a = \begin{cases} \Lambda_a & \text{if arc } a \text{ is a pipe,} \\ 0 & \text{otherwise,} \end{cases} \quad (8.2)$$

where Λ_a is the Weymouth coefficient of pipe a; see Section 2.3.1 and the first part of Eq. (2.25) for a detailed description.

The two groups of equations correspond to Kirchhoff's first and second laws in a gas networks setting. Indeed, the first law states flow is conserved at any node. The second law, saying that the total pressure change around any closed loop in the network sums up to zero, is equivalent to the second group. The derivation of this equivalence is shown in Section 8.1 with an intuitive example at the end.

So far, the equations concern passive elements of the network, only. Active elements are modeled by additional constraints. Switching states for network elements, which could be in *closed* state, are handled separately. *Switching state* is a superordinate word for the two operation modes *closed* and *opened*. For any individual nomination, a custom-made aggregation of the network is followed by the solution of a custom-made transshipment problem on the condensed network. Based on its solutions, and, if available, expert knowledge, switching states controlling the active elements are chosen. A nonlinear (in)equality system remains, for which feasible solutions are sought by making use of standard NLP solvers.

8.1 ▪ Reduction of variables

One of the first works dealing with the reduction of variables in flow problems dates back to Hamam and Brameller (1971), where the authors presented two methods for network analysis for any fluid-flow problem. They distinguish between the nodal and the loop method, where the last one corresponds to the method that will be presented here. In Osiadacz and Pienkosz (1988) combinations of nodal and loop representations in methods for steady-state simulation of gas networks were investigated. Mallinson et al. (1993) presented another comparison between the nodal and the loop method. Ríos-Mercado et al. (2002) again used the reduction of variables to solve gas transportation problems. The authors also presented a proof for the existence and uniqueness of solutions of the system Eq. (8.1) for a purely passive network with fixed pressure at one node.

The gas network is modeled as a simple directed graph $G = (V, A)$. The set V of all nodes comprises the disjoint sets V_-, V_+, V_0 of nodes with type exit, entry, and inner node. The set A of all arcs is the union of the disjoint sets $A_{pi}, A_{sc}, A_{rs}, A_{va}, A_{cv}, A_{cg}$ of pipes, short cuts, resistors, valves, control valves, and compressor groups.

In this section we assume the state *closed* to be assigned a priori to a subset of the elements in A_{va}, A_{cv}, and A_{cg}. A heuristic for choosing such subset is presented below in Section 8.5.1.

8.1. Reduction of variables

The elements in *closed* mode are deleted from the graph. In case the graph is not connected, we add a minimal number of artificial arcs connecting components, which have flow fixed to zero and decoupled pressures at end nodes. Let the remaining graph $\bar{G} = (V, \bar{A})$ have $n+1$ nodes and $n+k$, $k \geq 0$ arcs. We assume that there exists at least one pipe, short cut, or valve in the network.

We thus work on a connected graph $\bar{G}$, having k fundamental cycles. To simplify the notation in the following, we denote the vector of supplies and demands in the total network by $q^{\mathrm{nom}+}$. In comparison, the vector q^{nom} only differs from $q^{\mathrm{nom}+}$ in the deleted first component. A nomination $q^{\mathrm{nom}+}$ has to be balanced, i.e., $\mathbb{1}^T q^{\mathrm{nom}+} = 0$. The flow rates belonging to the individual arcs of the network are assembled into the vector q with $n+k$ components. The definition of q can be found in the beginning of Section 2.2. Let furthermore $\mathscr{A}^+$ denote the corresponding node-arc-incidence matrix. The flow balance at the nodes now reads:

$$\mathscr{A}^+ q = q^{\mathrm{nom}+}.$$

It is well known that the rank of this system is n. To get a system with full rank, we delete the first row corresponding to the node with index 0 (without loss of generality node 0 is not incident to an active element). This node is chosen as a slack node and the root of the spanning tree used in what follows. This leads to

$$\mathscr{A} q = q^{\mathrm{nom}}$$

with an $(n, n+k)$-matrix $\mathscr{A}$ and n-vector q^{nom}.

For $a = (u, v) \in A_{\mathrm{pi}}$, the pressure loss equation couples the variables p_u, p_v, and q_a, as Eq. (2.24) derived in Section 2.3.1. For our purpose we simplify this equation further by assuming horizontal pipes as in Eq. (2.27). The pressure loss equation is nonlinear and thus defines a nonconvex set. With the coefficients defined in Eq. (8.2), we represent this approximation for the pressure loss in pipes within a unified relation for the squares of the pressures at the endpoints of any passive arc $a = (u, v)$:

$$p_u^2 - p_v^2 = \alpha_a |q_a| q_a.$$

For pipes it holds that $\alpha_a > 0$. Using the transpose of the node-arc-incidence matrix $\mathscr{A}^+$ with each row corresponding to one arc, we can combine all pressure loss equations into one system

$$(\mathscr{A}^+)^T (p^+)^2 - \alpha |q| q = 0,$$

where $(p^+)^2$ is an $(n+1)$-vector with $p_i^2, i = 1, \ldots, n+1$ as components for all nodes $u \in V$. Each component of $(\mathscr{A}^+)^T (p^+)^2$ is the difference of the squares of the pressures at tail and head of the corresponding arc. The pressure variable that corresponds to the chosen root node is denoted $p_r := p_0$, its square as p_r^2. We reformulate the system as

$$\mathscr{A}^T p^2 - \alpha |q| q = p_r^2 \mathscr{A}^T \mathbb{1}, \qquad (8.3)$$

where $\mathscr{A}$ is the node-arc-incidence matrix with deleted first row that corresponds to the node with index 0. The entry of $\mathscr{A}^T \mathbb{1}$ is zero if the corresponding arc is not incident to the fixed root node, the entry is -1 if the fixed node is the tail of the arc, and 1 if it is the head.

If the gas flows through a control valve (see Section 8.2.2) or a compressor group (see Section 8.2.1) the pressure can be changed actively. If it flows through a resistor (see Section 8.2.4) the relation of the pressures at its endpoints is defined by an equation. To model these changes of pressure, we subtract a $(n+k)$-vector Δ containing differences of squares of pressure variables from the left hand side of Eq. (8.3) with the following components:

$$\Delta_a = \begin{cases} p_u^2 - p_v^2 & \text{if } a = (u,v) \in A_{cg} \cup A_{cv} \cup A_{rs}, \\ 0 & \text{otherwise.} \end{cases}$$

We will present the constraints modeling compressor groups, control valves, and resistors in Sections 8.2.1, 8.2.2, 8.2.3 and 8.2.4 respectively. We will go on with the description for the system of pressure change equations along arcs of arbitrary type:

$$-\Delta + \mathcal{A}^T p^2 - \alpha |q| q = p_r^2 \mathcal{A}^T \mathbb{1}. \tag{8.4}$$

The flow conservation, the pressure change, and the limits on the pressure of the nodes $0 < \underline{p} \leq \overline{p}$ build the system of equations with added box constraints that is our main point of interest:

$$\begin{aligned} \mathcal{A} q &= q^{\text{nom}}, \\ -\Delta + \mathcal{A}^T p^2 - \alpha |q| q &= p_r^2 \mathcal{A}^T \mathbb{1}, \\ \underline{p} \leq p &\leq \overline{p}. \end{aligned} \tag{8.5}$$

The stationary equations in Eq. (8.5) have their physical origins in Kirchhoff's laws. From a broader perspective, including the dynamics of gas flow they are rooted in conservation laws of mass and momentum, and, still broader, in networked systems of hyperbolic partial differential equations; here, isothermal Euler equations with friction. The latter is an active field of contemporary mathematical research; see Gugat, Dick, and Leugering (2011), for instance.

The system Eq. (8.5) is well known in the literature to permit a substantial problem reduction by elimination of variables; see for instance Osiadacz and Pienkosz (1988). In the present setting this leads to the following. Since $\mathcal{A}$ has full rank, there exists a subset of the columns forming a regular matrix, and we can use the description $\mathcal{A} = (\mathcal{A}_B, \mathcal{A}_N)$, where $\mathcal{A}_B$ is regular. It holds that $B \cup N = \bar{A}$ and $B \cap N = \emptyset$.

Our notation at this point has been inspired by the classical simplex method with its basis and nonbasis variables. The set B comprises the arcs of a spanning tree, N the arcs of the corresponding cotree. In the fluid-networks literature, e.g., Osiadacz and Pienkosz (1988), these are called the tree and cotree variables, because of the correspondence between invertible submatrices of a node-arc-incidence matrix and spanning trees in networks.

For the construction of a spanning tree we use a breadth-first search and start at the chosen root node denoted with index 0. This root node can be chosen among all nodes that are not incident to a resistor, a control valve, or a compressor group. The vector q is split according to B and N: $q = (q_B, q_N)$.

From the first equation in Eq. (8.5) we get

$$q_B = \mathcal{A}_B^{-1} q^{\text{nom}} - \mathcal{A}_B^{-1} \mathcal{A}_N q_N. \tag{8.6}$$

From the system Eq. (8.4), we arrive at two systems for the arcs in B and N:

$$-\Delta_B + \mathcal{A}_B^T p^2 - \alpha_B |q_B| q_B = p_r^2 \mathcal{A}_B^T \mathbb{1}, \tag{8.7}$$

$$-\Delta_N + \mathcal{A}_N^T p^2 - \alpha_N |q_N| q_N = p_r^2 \mathcal{A}_N^T \mathbb{1}. \tag{8.8}$$

8.1. Reduction of variables

We multiply Eq. (8.7) with the inverse of $\mathcal{A}_B^T$ and get

$$-(\mathcal{A}_B^T)^{-1}\Delta_B + p^2 - (\mathcal{A}_B^T)^{-1}\alpha_B|q_B|q_B = (\mathcal{A}_B^T)^{-1}p_r^2\mathcal{A}_B^T\mathbb{1}$$
$$\Leftrightarrow \quad p_r^2\mathbb{1} + (\mathcal{A}_B^T)^{-1}(\Delta_B + \alpha_B|q_B|q_B) = p^2. \tag{8.9}$$

Applying Eq. (8.9) in Eq. (8.8) yields the following equation:

$$-\Delta_N + \mathcal{A}_N^T\big[p_r^2\mathbb{1} + (\mathcal{A}_B^T)^{-1}(\Delta_B + \alpha_B|q_B|q_B)\big]$$
$$-\alpha_N|q_N|q_N = p_r^2\mathcal{A}_N^T\mathbb{1}$$
$$\Leftrightarrow \quad \mathcal{A}_N^T(\mathcal{A}_B^T)^{-1}\big[\Delta_B + \alpha_B|q_B|q_B\big] = \alpha_N|q_N|q_N + \Delta_N. \tag{8.10}$$

And finally using Eq. (8.6) in Eq. (8.10) leads to

$$\mathcal{A}_N^T(\mathcal{A}_B^T)^{-1}\Big[\Delta_B + \alpha_B\big|\mathcal{A}_B^{-1}q^{\text{nom}} - \mathcal{A}_B^{-1}\mathcal{A}_Nq_N\big|(\mathcal{A}_B^{-1}q^{\text{nom}} - \mathcal{A}_B^{-1}\mathcal{A}_Nq_N)\Big]$$
$$= \alpha_N|q_N|q_N + \Delta_N. \tag{8.11}$$

Equations (8.6), (8.9), and (8.11) are the key equations above. They capture the basic relations between pressure and flow in a way particularly amenable to computations. Indeed, Eq. (8.6) provides explicit relations for the flows over all arcs in the tree and Eq. (8.9), again explicit, expressions for all node pressures.

Equation (8.11) is a system of $|N|$ specially structured nonlinear equations. Leaving aside the dependence of α on flow and pressure, the defining functions are formed by multiplication of linear terms and absolute values of linear terms and can be seen as piecewise quadratic polynomials, i.e., multivariate polynomials with monomials of degree at most 2. Equations (8.6), (8.9), and (8.11) can nicely be illustrated: Eq. (8.6), as in the simplex method of linear programming, allows for eliminating $|B|$ flow variables from the system. The explicit nonlinear equations (8.9) permit the elimination of most pressure variables. The only pressure variables remaining are the one at the root node and those at nodes incident to an active element in order to express Δ. Every row of Eq. (8.9) yields the pressure in a node with index $u > 0$ as resulting from the pressure change along the unique path from the node to the root node in the tree corresponding to B.

Equation (8.11) ensures for each fundamental cycle that the pressure change over the cotree arc is the same as the pressure change over the tree arcs of the corresponding fundamental cycle.

The matrix $\mathcal{A}_B^{-1}$ can be interpreted very well. Each row corresponds to a tree arc of the network and each column corresponds to a node except for the root. The entries in a column define the unique path from the node that corresponds to the column to the root. If the arc corresponding to a row appears in the forward direction on that path, the matrix entry is 1. If the arc appears in the backward direction, the matrix entry is -1, and if the arc cannot be found on the unique path the matrix entry is 0.

Pressure equation Consider the example of a small graph with the tree structure displayed in Figure 8.1. The arrows indicate the direction of the arc in the graph. As an example we consider the computation of the pressure in node 10. The row for this node in Eq. (8.9) defines the squared pressure for node 10. Assume that the gas flows from the root node to node 10. In the first term of Eq. (8.9) the pressure loss over the pipes is subtracted. Notice that in this example we have $q_{8,6} < 0$ in the terms expressing the pressure loss along pipes. The final term adds the pressure change by the compressor group.

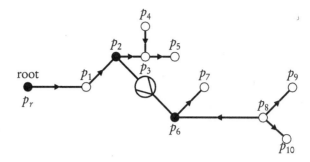

Figure 8.1. *Pressure as difference of the pressure in the root node and pressure change over the path.*

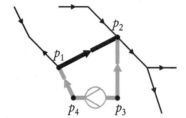

Figure 8.2. *Pressure loss over the thick black cotree arc must be identical to pressure loss over the thick gray tree arcs.*

Shortened pressure change equation Suppose there are active elements on the unique path from some node u to the root r, and let v be the first node incident to an active element along this path. Then the pressure p_u is fully determined by p_v, and the pressure change along the arcs between v and u.

Cycle equations Consider the example of a small graph with one cycle shown in Figure 8.2. Suppose the tree arcs are the gray colored ones and the cotree arc is the black one. The tree arc $(3,4)$ corresponds to a compressor group. The cotree arc closing the cycle is $(1,2)$. The resulting cycle equation (see Eq. (8.11)) then makes sure that the pressure loss over the three arcs on the left-hand side is equal to the pressure loss along the cotree arc on the right-hand side.

8.2 • Constraints for active elements

Until now we presented constraints for the pressure at each node and for the pressure balance in cycles. Now we turn to the constraints modeling special types of arcs.

It is possible that the amount of flow over an arc is limited due to technical or contractual reasons. These limitations are bounds on the flow variables, modeled with the use of Eq. (8.6) for the eliminated variables q_B. In this section u means the tail of the active element and v corresponds to its head.

8.2.1 • Compressor groups

A compressor group ($\mathscr{C}$) mostly consists of more than one compressor. Each compressor group contains some compressors and their drives, and there exist several possibilities to combine them. If some machines are combined serially, it is possible to reach a higher

8.2. Constraints for active elements

pressure than with each of the compressors alone. In contrast, if some machines are combined in parallel, the pressure can only be increased to the capabilities of each compressor, but the amount of gas being compressed can be higher. This level of detail is modeled only approximately here. All details of machines, drives, and compressor groups can be found in Section 2.3.5. In the following $q_\mathscr{C}$ is a variable for the gas flow over the compressor group, u is the tail of the group, and v is the head.

1. **Pressure increase**
   When passing through a compressor group, the pressure does not decrease:
   $$p_v - p_u \geq 0. \tag{8.12}$$

2. **Reverse direction bypass**
   The gas can flow through some compressor groups in reverse direction when in the *bypass* mode. In this case, pressure must not change:
   $$(p_v - p_u) \cdot q_\mathscr{C} \geq 0.$$

   ▷ If there is reverse flow through the group, it holds that $q_\mathscr{C} < 0$, and the inequality yields $(p_v - p_u) \leq 0$. Together with Eq. (8.12) this implies $p_v = p_u$.
   ▷ If there is forward flow through the group, it holds that $q_\mathscr{C} > 0$, which leads to $(p_v - p_u) \geq 0$, which is the same as Eq. (8.12).

If one wants to describe the capacity of a compressor group correctly, the detailed characteristic diagrams of the combination of different machines would have to be used: see Section 10.1.10. We consider a relaxation of all combined diagrams of a compressor group for all possible configurations $\Gamma_\mathscr{C}$ of the compressor group and therefore a superset of the possible combinations of $(p_u, p_v, q_\mathscr{C})$. The following simplified model might result in a combination of pressure increase and flow rate which cannot be reached by any of the admissible combinations in the station, but in practical applications on real-world networks it works quite well. We consider the following values, determined beforehand for each group separately using a detailed model:

▷ $\underline{q}_\mathscr{C} := \min_{\gamma \in \Gamma_\mathscr{C}} \underline{q}_\gamma$, $\overline{q}_\mathscr{C} := \max_{\gamma \in \Gamma_\mathscr{C}} \overline{q}_\gamma$,
▷ $\underline{\Delta p}_\mathscr{C} := \min_{\gamma \in \Gamma_\mathscr{C}} \underline{\Delta p}_\gamma$, $\overline{\Delta p}_\mathscr{C} := \max_{\gamma \in \Gamma_\mathscr{C}} \overline{\Delta p}_\gamma$,
▷ $\underline{\varepsilon}_\mathscr{C} := \min_{\gamma \in \Gamma_\mathscr{C}} \underline{\varepsilon}_\gamma$, $\overline{\varepsilon}_\mathscr{C} := \max_{\gamma \in \Gamma_\mathscr{C}} \overline{\varepsilon}_\gamma$.

Here γ refers to one "configuration of a compressor group." Using this minimal or maximal discharge, pressure increase, and pressure ratio, we can bound the capabilities of each compressor group. This leads to the following additional constraints on compressor groups:

3. **Minimal discharge**
   The compressor group can only be in *active* mode if the flow rate is sufficiently high. This limit is induced by the surgeline of the compressors and approximated here with the following constraint:
   $$(p_v - p_u) \cdot (\underline{q}_\mathscr{C} - q_\mathscr{C}) \leq 0. \tag{8.13}$$

   ▷ If the flow rate is too small, i.e., $q_\mathscr{C} < \underline{q}_\mathscr{C}$, we have $\underline{q}_\mathscr{C} - q_\mathscr{C} > 0$ and therefore $p_v - p_u \leq 0$. Together with Eq. (8.12) this implies $p_v = p_u$, i.e., the compressor group is in bypass.
   ▷ In the other case it holds that $\underline{q}_\mathscr{C} - q_\mathscr{C} \leq 0$, and inequality Eq. (8.13) becomes identical to Eq. (8.12).

4. **Maximal discharge**
   There is an upper limit to the flow rate the station could process in *active* mode:
   $$(p_v - p_u) \cdot (q_{\mathcal{G}} - \overline{q}_{\mathcal{G}}) \leq 0. \tag{8.14}$$

   ▷ In case the flow rate through the group is greater than the limit i.e., $q_{\mathcal{G}} > \overline{q}_{\mathcal{G}}$, the constraint together with Eq. (8.12) forces the switch to *bypass* mode.
   ▷ If $q_{\mathcal{G}} - \overline{q}_{\mathcal{G}} \leq 0$, Eq. (8.14) again reduces to Eq. (8.12).

5. **Minimal pressure increase**
   The compressor group cannot accomplish an arbitrarily small increase of pressure. Either both pressures are equal or the increase is greater than the lower bound:
   $$(p_v - p_u) \cdot (\underline{\Delta p}_{\mathcal{G}} - (p_v - p_u)) \leq 0.$$

6. **Maximal pressure increase**
   The compressor group cannot accomplish an arbitrarily big increase of pressure. Either both pressures are equal or the increase is below the upper bound:
   $$(p_v - p_u) \cdot ((p_v - p_u) - \overline{\Delta p}_{\mathcal{G}}) \leq 0.$$

7. **Minimal pressure ratio**
   The ratio between p_v and p_u is bounded from below:
   $$(p_v - p_u) \cdot \left(\underline{\varepsilon}_{\mathcal{G}} - \frac{p_v}{p_u} \right) \leq 0.$$

   ▷ If the pressures of gas at the end nodes are different, we have $\underline{\varepsilon}_{\mathcal{G}} - \frac{p_v}{p_u} \leq 0$ and therefore $\underline{\varepsilon}_{\mathcal{G}} \leq \frac{p_v}{p_u}$.

8. **Maximal pressure ratio**
   The ratio between p_v and p_u is bounded from above:
   $$(p_v - p_u) \cdot \left(\frac{p_v}{p_u} - \overline{\varepsilon}_{\mathcal{G}} \right) \leq 0. \tag{8.15}$$

   ▷ If the pressures of gas at the end nodes are different, the constraint forces $\frac{p_v}{p_u} - \overline{\varepsilon}_{\mathcal{G}} \leq 0$ and therefore $\frac{p_v}{p_u} \leq \overline{\varepsilon}_{\mathcal{G}}$.

8.2.2 ▪ Control valves

A control valve (CV) is used to decrease the pressure of the gas. This can be necessary for supporting pieces of the network with lower pressure or when gas should flow together from different pipelines. Control valves are capable of continuously reducing the pressure of an arbitrary amount of gas. In the following q_{CV} is a variable for the gas flow over the control valve, u is the tail of it, and v is the head.

1. **Pressure decrease**
   It always holds that the pressure of gas that flows through a control valve can either be decreased or stays the same. It is not possible that the pressure of the gas increases:
   $$(p_v - p_u) \cdot |q_{CV}| \leq 0. \tag{8.16}$$

2. **Reverse *bypass* mode**
   Some control valves allow for a flow in reverse direction when they are in *bypass* mode. In that case the pressure of the gas must not change:
   $$(p_v - p_u) \cdot q_{CV} \leq 0. \tag{8.17}$$

   ▷ If there is no gas flowing through the control valve, the pressures are decoupled by Eq. (8.17) and Eq. (8.16).
   ▷ If the gas flows backward through the control valve, it holds that $q_{CV} < 0$ and therefore $p_v - p_u \geq 0$. According to Eq. (8.16) it holds that $p_v = p_u$.
   ▷ If the gas flows forward through the control valve, it holds that $q_{CV} > 0$ and the constraint yields $p_v \leq p_u$.

8.2.3 ▪ Control valves without remote access

For control valves without remote access a target pressure p_{set} is fixed at the head of the control valve, which cannot be changed actively by the model. If the pressure at the tail of the control valve without remote access is higher than the target pressure and the pressure at its head is not forced to a pressure higher than this target by the flows through other arcs incident to the head, the pressure at the head will be equal to the preset pressure and gas flows through the element. If the pressure at the head node cannot be reduced to the preset pressure, the control valve closes automatically. If the pressure at the tail node is less than the preset pressure, the control valve opens fully to *bypass* mode and the pressures at both ends are equal. There exists the following dependency between the variables p_v, p_u, q and the fixed set pressure p_{set} as introduced in Section 2.3.4:

▷ Control valve without remote access is *closed*:
   ○ p_u, p_v independent,
   ○ $p_v \geq p_{set}$.

▷ Control valve without remote access is *bypassed*:
   ○ $p_u = p_v$,
   ○ $p_u \leq p_{set}$.

▷ Control valve without remote access is *active* (forward flow):
   ○ $p_u \geq p_v$,
   ○ $p_v = p_{set}$.

Because we model a nonlinear program (NLP), we cannot use binary variables to maintain these dependencies. We model them with the help of the maximum function:

1. **Pressure decrease**
   It always holds that the pressure of gas that flows through a control valve without remote access can either be decreased or remains the same. If the pressure of the gas at the outlet is higher than the inlet pressure, the control valve closes:
   $$(p_v - p_u) \cdot |q_{CV}| \leq 0.$$

2. **Reverse *bypass* mode**
   A reverse flow is possible for a control valve without remote access, when they are in *bypass* mode:
   $$(p_v - p_u) \cdot q_{CV} \leq 0.$$

3. Conditions resulting from pressure set

$$q_{CV} \cdot \max\{0, p_v - p_{set}\} = 0, \tag{8.18}$$

$$(p_u - p_v) \cdot \max\{q_{CV}, p_u - p_v\} \cdot \max\{0, p_{set} - p_v\} = 0. \tag{8.19}$$

▷ In case $p_v > p_{set}$, Eq. (8.19) is fulfilled automatically and (8.18) ensures $q_{CV} = 0$.
▷ In case $p_v = p_{set}$, Eqs. (8.18) and (8.19) are fulfilled.
▷ In case $p_v < p_{set}$, Eq. (8.18) is fulfilled and imposes no condition. Equation (8.19) implies in this case that either $p_u = p_v$ holds or $p_u < p_v$ together with $q_{CV} = 0$.

8.2.4 ▪ Resistors

There are two different types of a resistor (R) modeled (see Section 2.3.2). The first type reduces the pressure of the gas by a constant value (ξ_R). This decrease is independent from the amount of gas and the pressure. The second type reduces the pressure according to a nonlinear equation dependent on the discharge and the ingoing pressure. Thus, different constraints have to be used in order to model a resistor according to its type. In the following, q_R is a variable for the gas flow over the resistor, u is the tail of it, and v is the head. The nonlinear pressure loss equation is a reformulation of Eq. (2.30) with $\max\{p_u, p_v\}$ representing the ingoing pressure.

1. Constant pressure loss

$$(p_u - p_v) - \xi_R \cdot \mathrm{sgn}\, q_R = 0.$$

2. Nonlinear pressure loss

$$(p_u - p_v) - \frac{8.0 \cdot \rho_0 \cdot p_0 \cdot \zeta_R \cdot T_m \cdot q_R^2 \cdot z(\max\{p_u, p_v\}, T_m)}{\pi^2 \cdot z_0 \cdot T_0 \cdot D_W^4 \cdot 10^5 \cdot 3.6^2 \cdot \max\{p_u, p_v\}} \cdot \mathrm{sgn}\, q_R = 0. \tag{8.20}$$

8.2.5 ▪ Flow restriction

Some compressor groups, control valves, resistors, or pipes have restrictions on their flow rate. The reason for this can be technical or contractual. The restrictions can be a prescribed flow direction or limits on the amount of discharge:

$$\underline{q} \leq q \leq \overline{q}.$$

These inequalities are simple bounds for the variables q_N, while q_B is substituted by Eq. (8.6). This results in two-sided inequality constraints.

8.3 ▪ Objective function

As we try to check feasibility, we do not have a traditional objective function. Instead we try to support convergence of the feasibility problem or to approximate a cost function with the objective function.

Minimize flow around cycles Theoretically it is possible that a big amount of gas flows through a cycle; see Figure 8.3 for an example. In a compressor group, the pressure can be increased and a following control valve can decrease the pressure if both are within the cycle. In reality this should be avoided, because it causes a cost at the compressor groups

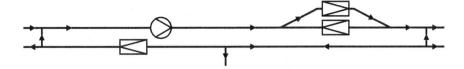

Figure 8.3. *Example for gas flowing in a cycle.*

and the temperature of the gas increases. With a corresponding term in the objective function we penalize flow in cycles and try to minimize it:

$$\sum_{n \in N} q_n^2.$$

Each fundamental cycle can be characterized using its cotree arc $n \in N$. We minimize the flow over these arcs, and since both flow directions are possible, we use the square of the variables.

Minimization of pressure increase in compressor groups We do not compute direct costs that are the result of pressure increase in compressor groups, but we try to reduce pressure increase that causes costs. Because the reduction of the flow around cycles outweighs the minimization of the pressure increase in compressor groups we choose a small factor $\omega < 1$:

$$\sum_{(u,v) \in A_{cg}} \omega \cdot (p_v - p_u)^2. \tag{8.21}$$

8.4 ▪ Summary of the model

In this section we summarize the model whose details where derived and motivated in the previous sections.

1. **Variables**
    (a) pressure in the root node: $\underline{p}_r \leq p_r \leq \overline{p}_r$.
    (b) pressure variables on nodes u, incident to a resistor, compressor group, or a control valve: $\underline{p}_u \leq p_u \leq \overline{p}_u$.
    (c) flow variables on cycle arcs: $\underline{q}_N \leq q_N \leq \overline{q}_N$.

   For the sake of simplicity of notation we use the abbreviation

   $$D := (\mathscr{A}_B^T)^{-1} \Big[\Delta_B + \alpha_B \Big| \mathscr{A}_B^{-1} q^{\text{nom}} - \mathscr{A}_B^{-1} \mathscr{A}_N q_N \Big| (\mathscr{A}_B^{-1} q^{\text{nom}} - \mathscr{A}_B^{-1} \mathscr{A}_N q_N) \Big].$$

2. **Constraints**
    (a) Pressure constraints for each node $u \in V$:

   $$\underline{p}_u^2 \leq p_r^2 + D_u \leq \overline{p}_u^2 \text{ if } u \neq \text{root nor incident to an active element,}$$
   $$p_r^2 + D_u - p_u^2 = 0 \text{ if } u = \text{root or incident to an active element.} \tag{8.22}$$

    (b) Fundamental cycle equations for all arcs in N:

   $$\mathscr{A}_N^T D - \alpha_N |q_N| q_N - \Delta_N = 0.$$

(c) Compressor group constraints for all arcs $a = (u,v) \in A_{cg}$ with bounds explained in Section 8.2.1:

  i. $p_v - p_u \geq 0$.
  ii. $(p_v - p_u) \cdot q_a \geq 0$.
  iii. $(p_v - p_u) \cdot (q_a - \overline{q}_a) \leq 0$.
  iv. $(p_v - p_u) \cdot (q_a - \underline{q}_a) \leq 0$.
  v. $(p_v - p_u) \cdot ((p_v - p_u) - \overline{\Delta p}_a) \leq 0$.
  vi. $(p_v - p_u) \cdot (\underline{\Delta p}_a - (p_v - p_u)) \leq 0$.
  vii. $(p_v - p_u) \cdot (\underline{\varepsilon}_a - \frac{p_v}{p_u}) \leq 0$.
  viii. $(p_v - p_u) \cdot (\frac{p_v}{p_u} - \overline{\varepsilon}_a) \leq 0$.

(d) Control valve constraints for all arcs $a = (u,v) \in A_{cv}$ with remote access:

  i. $(p_v - p_u) \cdot |q_a| \leq 0$,
  ii. $(p_v - p_u) \cdot q_a \leq 0$,

  and without remote access:

  i. $(p_v - p_u) \cdot |q_a| \leq 0$,
  ii. $(p_v - p_u) \cdot q_a \leq 0$,
  iii. $q_a \cdot \max\{0, p_v - p_{set}\} = 0$,
  iv. $(p_u - p_v) \cdot \max\{q_a, p_u - p_v\} \cdot \max\{0, p_{set} - p_v\} = 0$.

(f) Resistor constraints for all resistors $a = (u,v) \in A_{\text{lin-rs}}$ with constant pressure reduction:

$$(p_u - p_v) - \xi_a \cdot \operatorname{sgn} q_{ra} = 0.$$

(g) Resistor constraints for all resistors $a = (u,v) \in A_{\text{nl-rs}}$ with nonlinear pressure reduction:

$$(p_u - p_v) - \frac{8.0 \cdot \rho_0 \cdot p_0 \cdot \zeta_a \cdot T_m \cdot q_a^2 \cdot z(\max\{p_u, p_v\}, T_m)}{\pi^2 \cdot z_0 \cdot T_0 \cdot D_a^4 \cdot 10^5 \cdot 3.6^2 \cdot \max\{p_u, p_v\}} \cdot \operatorname{sgn} q_a = 0.$$

(g) Flow bounds for all arcs a with flow rate restriction
if a is an element of the cotree i.e., $a \in N$:

$$\underline{q}_a \leq q_a \leq \overline{q}_a;$$

if a is an element of the tree i.e., $a \in B$:

$$\underline{q}_a \leq (\mathcal{A}_B^{-1} q^{\text{nom}} - \mathcal{A}_B^{-1} \mathcal{A}_N q_N)_a \leq \overline{q}_a.$$

3. **Objective function**

$$\min \sum_{n \in N} q_n^2 + \sum_{(u,v) \in A_{cg}} \omega \cdot (p_v - p_u)^2.$$

8.5 ▪ Heuristics to fix binary decisions

Up to now we assumed that there are given switching states of all elements, declaring subsets of the compressor groups, control valves, and valves as *closed*. These *closed* elements are not within the graph $\tilde{G}$ used in Section 8.1. We use two different techniques to

8.5. Heuristics to fix binary decisions

fix these binary decisions. First we use a transshipment heuristic, and then we try some sets of given switching states for all active elements. These sets are constructed by using expert knowledge about the network and by collecting sets of switching states from the transshipment heuristic in cases where in the past these led to a feasible solution for some nominations for the current network.

Note that the a priori chosen states are not completely fixed in the described nonlinear program (NLP) model, since it is possible that a control valve assumed as open gets closed in a feasible solution due to constraints (8.16) and (8.17).

8.5.1 ▪ Transshipment heuristic

The transshipment heuristic solves a linear network flow problem on a condensed graph and analyzes the overall flow situation over the elements to be decided for the given nomination. The solution of the transshipment problem is translated into switching states for all active elements based on expert knowledge or analysis of the network and subnetwork operation modes under consideration.

8.5.1.1 ▪ The transshipment graph

We construct a directed transshipment graph $\tilde{G} = (\tilde{V}, \tilde{A})$ with costs $\tilde{c} \in \mathbb{R}^{\tilde{A}}$ and flow requirements $d_T \in \mathbb{R}^{\tilde{V}}$. In the first step, copy the original graph to an auxiliary graph $\hat{G} = (\hat{V}, \hat{A})$. Start with $\tilde{V} = \tilde{A} = \tilde{A}_{Station} = \tilde{A}_{Sub} = \emptyset$, and perform the following steps:

1. There are regions in the network where switching states are interdependent or have to be chosen from a set of prescribed combinations, the subnetwork operation modes (Section 2.4.3), only. In this section we call such region a station subnetwork. These station subnetworks are defined a priori via their boundary nodes. For each station subnetwork add one artificial node to $\tilde{G}$, representing the station subnetwork. Moreover, we add all its boundary nodes to $\tilde{G}$. The boundary nodes are assigned in the corresponding component of d_T their original in- or outflow requirements q_u^{nom}. Assign the sum of the flow requirements of the station subnetwork's internal nodes to the component of d_T for the new station node. Add arcs to $\tilde{A}_{Station} \subseteq \tilde{A}$ between the station node and its boundary nodes in both directions. All these arcs are assigned the same small artificial cost coefficient. Delete all internal nodes of the station subnetwork from $\hat{V}$ and its arcs from $\hat{A}$.

2. For all arcs a from A_{cg}, A_{va}, A_{cv} remaining in $\hat{A}$, add their endpoints with the original q_u^{nom} to $\tilde{V}$ and copy the arc to $\tilde{A}$. For each of these arcs a with $\underline{q}_a < 0$ additionally add a corresponding backward arc to $\tilde{A}$. Assign artificial cost coefficients. Delete arc a from $\hat{A}$.

3. For each connected component in $\hat{G}$, add one artificial node to $\tilde{V}$. These connected components being aggregated to one node are purely passive subnetworks of the original network. A similar network reduction step was done by Ríos-Mercado et al. (2002). We call the node representing it in $\tilde{V}$ a subnetwork node. Assign the flow balance of the component's internal nodes to the component of the vector d_T, corresponding to this subnetwork node. The boundary nodes of a component are boundary nodes of station subnetworks or end points of active arcs outside of a station subnetwork, which were added to $\tilde{V}$ in the previous steps. Add arcs in both directions to $\tilde{A}_{Sub} \subseteq \tilde{A}$, connecting the subnetwork node and the component's boundary nodes; assign the sum of the friction coefficients α_a within the component as the cost coefficient to these arcs.

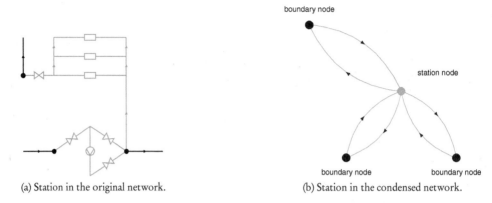

(a) Station in the original network. (b) Station in the condensed network.

Figure 8.4. *Aggregating the network* I.

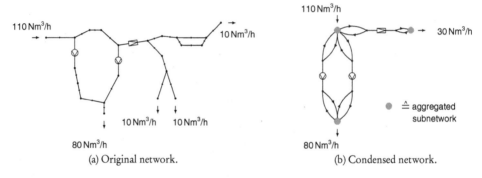

(a) Original network. (b) Condensed network.

Figure 8.5. *Aggregating the network* II.

The resulting graph $\tilde{G}$ represents the flow situation and guides the usage of arcs in the transshipment by cost coefficients, such that flow through pressure increasing and pressure decoupling network elements is favored.

As an illustration for the idea of station subnetwork aggregation we use the network in Figure 8.4(a) with one station subnetwork. All gray elements are inside the station subnetwork. The black nodes are the boundary nodes that connect the station subnetwork with the remaining network. The aggregated station subnetwork in $\tilde{G}$ is given in Figure 8.4(b). In this example, we assumed that gas can enter and leave the station subnetwork through each of its boundary nodes.

To illustrate subnetwork aggregation, we use the small network shown in Figure 8.5(a) without a predefined station subnetwork. We assume that no reverse flow is allowed for the compressor groups and the control valve. Thus we do not create arcs for the backward flow for these elements in the aggregated network $\tilde{G}$ shown in Figure 8.5(b). The aggregated subnetwork nodes are connected in both directions with the elements they are connected to in the original network, too. The exit amounts of $10\,\text{Nm}^3/\text{h}$ for each of the nodes being on the right-hand side of the control valve in Figure 8.5(a) are assigned to the aggregated subnetwork node in $\tilde{G}$, which results in an exit amount $30\,\text{Nm}^3/\text{h}$ as shown in Figure 8.5(b).

8.5.1.2 • Transshipment model

A standard linear transshipment problem is solved on $\tilde{G}$ with the aggregated supply/demand d_T.

Variables

The only variables are flow variables f_a for each arc $a \in \tilde{A}$. The arcs that do not exist in the original network have nonlimiting bounds

$$0 \le f_a \le \overline{f}_a := \infty, \qquad a \in \tilde{A}_{Station} \cup \tilde{A}_{Sub}.$$

Flow bounds for arcs added in step 2, which correspond to arcs in the original network, represent the original flow bounds:

$$0 \le f_a \le \overline{f}_a := \begin{cases} \overline{q}_a, & a \in A_{cg} \cup A_{rs} \cup A_{va} \text{ if } a \text{ is a forward arc,} \\ -\underline{q}_a, & a \in A_{cg} \cup A_{rs} \cup A_{va} \text{ if } a \text{ is a backward arc,} \end{cases}$$

where $\underline{q}_a$ denotes the lower bound for the flow over the arc in the original network, $\overline{q}_a$ the upper bound.

Constraints

The only constraints are the flow balance equations for each node in $\tilde{V}$ and the box constraints for the flow variables.

Objective function

For each arc a fictitious cost coefficient is assigned:

1. Cost coefficients for valves, control valves, and compressor groups are chosen a priori. Backward flow on control valves and compressor groups is more expensive than forward flow, because in the case of forward flow over these elements the pressure of the gas could be adjusted actively. Thus the pressures at the end nodes are decoupled to the extent of the capabilities of the element. On the other hand a reverse flow is possible in *bypass* mode only, forcing equality of the pressures at the end nodes, as is the case for valves. The coefficients are chosen such that the use of pressure decoupling elements is cheap and the use of (pressure increasing) compressors is more rewarding than that of (pressure decreasing) control valves.
2. Costs for the added connecting arcs between station nodes and their boundary nodes are chosen to be low in both directions.
3. Costs for arcs connecting aggregated subnetworks with active elements or boundary nodes of station subnetworks are chosen as the sum over all pressure loss coefficients of the pipes inside the aggregated passive subnetwork. To compute the coefficients without knowledge about flows we a priori fix the flow in Eq. (2.15) to the same constant (1000 Nm³/h) on all pipes. This sum overestimates the pressure loss of the gas flowing through the subnetwork.

The constant cost coefficients that we used in our implementation are

$$c_a = \begin{cases} 10^{-4} & \text{if the arc is a valve,} \\ 10^{-5} & \text{if it is a forward arc of a control valve,} \\ 10^{2} & \text{if it is a backward arc of a control valve,} \\ 10^{-11} & \text{if it is a forward arc of a compressor group,} \\ 10^{1} & \text{if it is a backward arc of a compressor group,} \\ 10^{-10} & \text{if the arc is a connecting arc of a station node} \\ & \text{with its boundary nodes.} \end{cases}$$

Model

$$\min \sum_{a \in \tilde{A}} c_a \cdot f_a \text{ s.t. } \mathscr{A}_T f = d_T, 0 \leq f \leq \overline{f}, \tag{8.23}$$

where

$\mathscr{A}_T \mathrel{\hat=}$ node-arc-incidence matrix of the condensed network,
$d_T \mathrel{\hat=}$ aggregated supplies/demands in the condensed network.

This standard linear transshipment problem can in practice be solved quickly with any linear program (LP) solver.

8.5.1.3 ▪ Interpretation of the solution of the transshipment problem

The solution of (8.23) consists of a vector with flow variables f. The values of these variables are interpreted in order to derive promising switching states for all active elements in the original network. These states are used as an input for solving the original problem stated more precisely in Section 8.4. We choose a tolerance τ ($= 10^{-4}$ in the computations) and distinguish between different kinds of arcs:

1. Connecting arcs of subnetworks do not correspond to any active element in the original network. That's why they are not used to decide switching states of the active elements.
2. For forward and backward arcs in $\tilde{A}$ corresponding to control valves, valves, and compressor groups, we check if the flow is above the tolerance, i.e.,

$$f_a > \tau, a \in A_{\text{va}} \cup A_{\text{rs}} \cup A_{\text{cg}}.$$

The switching state for the corresponding arcs is chosen as *open* in the reduced NLP. If, however, both corresponding flows are below the tolerance, these arcs are fixed as *closed*. In Figure 8.6(a) a solution of the network flow problem is shown. All arcs with flow $\leq \tau$ are deleted from the network to illustrate the interpretation of the solution of the network flow problem. In Figure 8.6(b) the interpretation as switching states in the original network is demonstrated. The light gray colored compressor group arc symbolizes that the arc is fixed as *closed*. Dark gray colored arcs are fixed to *open*.
3. Switching states for active elements within station subnetworks are chosen by a priori defined rules for each station subnetwork. The flow situation through the station subnetwork in the transshipment problem solution is represented by the flow variables on the arcs connecting the station node with its boundary nodes. Based on the direction and size of these flows a suitable combination of switching states is chosen from the subnetwork operation modes by the rules.

As a simple example for such rules we look at the part of a network with a station subnetwork in Figure 8.4(a). The black nodes are boundary nodes. In Figure 8.4(b) the condensed network for the station subnetwork is shown. According to the solution of the network flow problem we have a closer look at the flow values of the arcs going out of or into the station subnetwork. In Figures 8.7 and 8.8 possible solutions of the network flow problem are shown with their interpretation as switching states. All arcs between boundary nodes and the station node with flow $f < \tau$ are deleted to illustrate the solution of the network flow problem. Again light gray colored elements are fixed to *closed* and dark gray ones are chosen as *open*. Rules taking also the amount of flow into account are used for station subnetworks with more choices from the subnetwork operation modes like the one depicted in Figure 2.10. In this example for the choice from the subnetwork operation modes besides the direction of flows it is important how many of the compressor groups should be used in which direction.

8.5. Heuristics to fix binary decisions

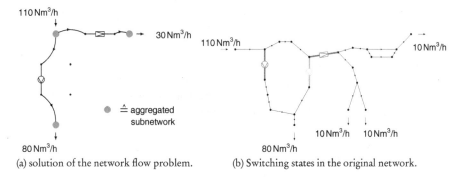

Figure 8.6. *Interpretation of the solution in connected components.*

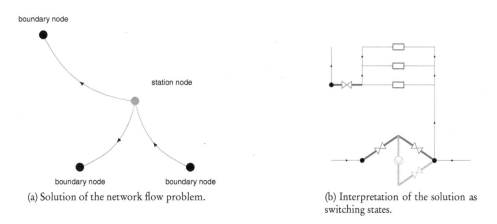

Figure 8.7. *Interpretation of the solution in a station subnetwork I.*

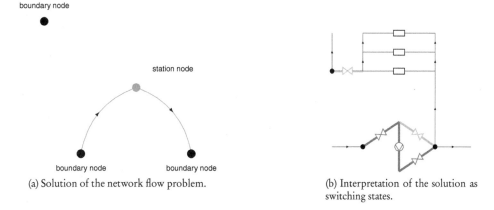

Figure 8.8. *Interpretation of the solution in a station subnetwork II.*

8.5.2 • Fixing switching states using sets of given binary decisions

To answer advanced questions like the validation of a booking, the reduced NLP has to be solved frequently for different nominations on the same underlying network. In addition to that, often there are nominations being comparable to each other, i.e., basically requiring the same flow situation in the network. These comparable nominations can often be solved with the same binary decisions. That's why as a second heuristic we test switching states that led to feasible solutions in other nominations. We iterate over the given sets of switching states until either a feasible solution is found or all sets have been tested.

This method can be compared to what actually is done manually by the network planners working for the network operator. For a set of nominations reflecting the existing contract situation and requests for additional contracts, an operation mode for the network elements is sought which results in no limit violations when checked with simulation software as described in Chapter 4. In addition to decisions for the switching states of all active elements, a simulation tool requires the fixing of the downstream pressure or the flow for all control valves and compressor groups. Therefore, expert knowledge is available and can be used very well in this heuristic.

8.6 ▪ Conclusion

The presented heuristic basically can be separated into two parts. In the first part the state of all active network elements is fixed. In the second part we solve the remaining nonlinear and nonconvex gas transportation problem. Compared to the ValNLP, which is described in detail in Chapter 10.4, we simplify gas physics. Simplifications are done within the context of pressure loss over pipes. In this heuristic differences in geodetic height are neglected and we use a fairly rough approximation for the mean pressure in the pipe within the computation of the mean real gas factor $p_\mathrm{m} = 0.5 \cdot (\max(\underline{p}_u, \underline{p}_v) + \min(\overline{p}_u, \overline{p}_v))$. The ValNLP uses another approximation for the mean pressure, namely Eq. (10.24).

Simplifications also are done within the context of compressor groups. For this heuristic these are modeled approximately as detailed in Section 8.2.1. In contrast to the ValNLP we totally neglect drives and all constraints that are induced by them. The detailed drive model that is used for the ValNLP can be found in Section 10.1.10. Here we only model each compressor group in its entirety. We do not go into the detail of different machines and the possibility to combine them within the group. This is a rough approximation compared to the ValNLP.

Chapter 9
An MPEC based heuristic

Martin Schmidt, Marc C. Steinbach, Bernhard M. Willert

Abstract *In this chapter we discuss the problem of validation of nominations as a nonsmooth and nonconvex mixed-integer nonlinear feasibility problem. For this problem we present a primal heuristic that is based on reformulation techniques that smooth the appearing nonsmooth aspects and that reformulate discrete aspects with complementarity constraints and problem specific relaxations. The resulting mathematical program with equilibrium constraints (MPEC) model can be regularized by standard techniques leading to a nonlinear program (NLP) type model. Solutions to the latter can finally be used as approximative solutions to the underlying feasibility problem.*

In this chapter, we review the problem of validation of nominations as a *nonsmooth mixed-integer nonlinear feasibility problem* and develop a problem-specific primal heuristic. In contrast to other approaches like the ones described in Chapter 6 or 7, we explicitly distinguish between

1. discontinuous aspects including discrete decisions leading to mixed-integer formulations using binary variables and
2. nonsmooth, but continuous aspects.

Since our aim is to employ standard solvers for (smooth) nonlinear optimization, we *reformulate* both the integer and the nonsmooth aspects. It will turn out that most of the integer parts of the validation of nominations model can be exactly reformulated by using complementarity constraints leading to a mathematical program with equilibrium constraints (MPEC). This model type is a generalization of nonlinear programs (NLPs) in which so-called *equilibrium* or *complementarity constraints* of the form

$$\phi_i(x)\psi_i(x) = 0, \quad \phi_i(x) \geq 0, \quad \psi_i(x) \geq 0, \quad i = 1,\ldots,k, \tag{9.1}$$

appear, where $\phi_i, \psi_i : \mathbb{R}^n \to \mathbb{R}$ are smooth functions. See Luo, Pang, and Ralph (1996) for an overview of MPECs. The remaining nonsmooth parts are *smoothed* by applying both standard and problem-specific smoothing techniques. The resulting smooth MPEC has the well-known drawback that it lacks standard constraint qualifications (CQs) like LICQ or MFCQ (see Ye and Zhu (1995)). Since these CQs are a major assumption for the convergence theory of almost every NLP solver, there is no convergence theory when

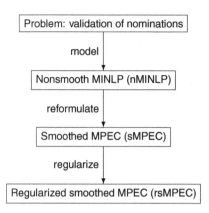

Figure 9.1. *Relationship of models studied in this chapter.*

applying standard NLP solvers directly to mathematical program with equilibrium constraintss (MPECs). Because we want to use standard NLP solvers for our primal heuristic, we finally have to *regularize* the MPEC to get an NLP satisfying standard CQs. The description and analysis of some standard regularization schemes for MPECs can be found in Scholtes (2001), DeMiguel et al. (2005), and Hu and Ralph (2004). Here we use a penalization scheme that regularizes the MPEC by moving the complementarity constraints (9.1) to the objective function. The details will be discussed later. Summarizing, we get a hierarchy of model formulations. An overview is given in Figure 9.1.

One central goal of our heuristic approach is that it should produce feasible solutions of the underlying nonsmooth mixed-integer nonlinear program (nMINLP) for real-world instances in a short time. Thus, it is not reasonable to incorporate all gas physics and engineering aspects. On the other hand, we have to incorporate as many physics and engineering details as necessary such that the consecutive NLP validation (see Chapter 10) has a high rate of positive validations. Thus, we performed numerical experiments that led us to the following modeling decisions:

1. We neglect all composition-specific gas parameters and the corresponding mixing model. The gas parameters appearing in the constraints of our model are mean values in dependence of the network data and the concrete nomination.
2. We only consider the isothermal case, i.e., the gas temperature is assumed to be constant. Thus, we can neglect all constraints concerning heat dynamics.
3. We disregard small pressure losses in control valve stations and compressor groups, which are caused by inner station piping, flow through measuring systems, or filters; see Chapter 2. Thus, we neglect up- and downstream resistors (see Chapter 10) that are used to model these potential pressure losses. In contrast, we account for resistors located outside of control valve stations and compressor groups.
4. We incorporate the highly nonlinear and nonconvex models of the operating ranges of compressor machines (see Section 2.3.5) without any further simplifications.

The chapter is organized as follows. In Section 9.1 we give a formulation of the problem of validation of nominations as an nMINLP. In addition, we directly develop the MPEC based reformulations of the mixed-integer parts and describe the smoothing techniques applied to nonsmooth aspects. The section ends with a complete description of both nMINLP as well as its smoothed and MPEC based reformulation sMPEC. Afterwards, Section 9.2 describes the used regularization scheme for MPECs that enables us to apply standard NLP solvers for solving the final problem formulation rsMPEC. Since the resulting regularized MPEC is extremely hard to solve for current solvers, we split up the solution process into two stages that are described in Section 9.3.

9.1 ▪ Model

In this section, we present a nonsmooth mixed-integer nonlinear model of every component of the gas transport networks under consideration (see Chapter 2). Based on this model, we develop reformulations using smoothing techniques and complementarity constraints that are used to finally build up the NLP based primal heuristic. The reader interested in a more detailed description of the used reformulation techniques can find additional information in Schmidt (2013) and Schmidt, Steinbach, and Willert (2013). A related model of natural gas transport using complementarity constraints was recently published by Baumrucker and Biegler (2010). These authors use MPEC techniques for handling nonsmooth model aspects like flow reversals and flow state transitions. In contrast, we use problem-specific smoothing techniques for these aspects and apply MPEC techniques for modeling the discrete control of active elements.

For ease of notation, the nonsmooth mixed-integer nonlinear program (nMINLP) model is stated in standard NLP form,

$$\min_{x \in \mathscr{X}} \ f(x) \quad \text{s.t.} \quad c_{\mathscr{E}}(x) = 0, \quad c_{\mathscr{I}}(x) \geq 0,$$

and will be discussed componentwise in the following sections. The decision vector $x \in \mathscr{X} := \mathbb{R}^n \times \{0,1\}^m$ consists of real and binary variables. A subindex refers to the sub-vector of the corresponding network element or set of elements. For instance, x_u denotes the variables of the component model of node u, and $x_{A_{\text{pi}}}$ denotes the variables of the component models of all pipes. Single constraints $c: \mathscr{X} \to \mathbb{R}$ are subindexed with the corresponding vertex u or arc a and superindexed with an abbreviated name describing the semantics of the constraint. For instance, c_u^{flow} is the constraint modeling the flow balance at node u. Complete component models are written as vectors of constraints $c: \mathscr{X} \to \mathbb{R}^k$, with subindices indicating single elements or sets thereof, as for the variables. If necessary, a subindex $\mathscr{E}$ or $\mathscr{I}$ is used to distinguish between equality and inequality constraints. Finally, objective functions or portions thereof are denoted by f.

9.1.1 ▪ Nodes

Nodes $u \in V$ are modeled as elements without capacity, satisfying the mass balance equation (see (2.8))

$$0 = c_u^{\text{flow}}(x) = \sum_{a \in \delta^+(u)} q_a - \sum_{a \in \delta^-(u)} q_a - q_u^{\text{nom}}, \tag{9.2}$$

where q_a is the mass flow on arc a and $q^{\text{nom}} \in \mathbb{R}^V$ is the flow supply/demand vector:

$$q_u^{\text{nom}} \geq 0, \quad u \in V_+ \quad \text{(entries)},$$
$$q_u^{\text{nom}} = 0, \quad u \in V_0 \quad \text{(inner nodes)},$$
$$q_u^{\text{nom}} \leq 0, \quad u \in V_- \quad \text{(exits)}.$$

In addition, every node u has a gas pressure variable p_u with bounds $[\underline{p}_u, \overline{p}_u]$ that depend on technical and/or contractual data. The complete node model reads

$$0 = c_u(x) = c_u^{\text{flow}}(x), \quad x_u = p_u.$$

It does not require any reformulation.

9.1.2 • Pipes

A pipe $a = (u,v) \in A_{\text{pi}}$ generally requires a complicated PDE model describing the gas dynamics in terms of mass, momentum, and energy balances: the Euler equations for compressible fluids (see Feistauer (1993); Lurie (2008); Schmidt, Steinbach, and Willert (2014)). These equations and all other relevant basics of the fluid model can be found in Section 2.3.1. For the ease of understanding, we use the same notation as in Section 2.3.1. For the isothermal and stationary case considered here, the mass balance (continuity equation) yields constant mass flow q along the pipe, the energy equation is not needed, and we are left with a stationary variant of the *momentum equation*; see (2.13). The state quantities pressure p, density ρ and temperature T (constant in our case) are coupled by an *equation of state*; we use the thermodynamical standard equation

$$\rho = \rho(p, T) = \frac{p}{R_s z(p, T) T},$$

where R_s is the specific gas constant; see (2.20). Finally, we need empirical models for the *compressibility factor* $z(p, T)$, describing the deviation of real gas from ideal gas, and for the *friction coefficient* $\lambda(q)$. The latter will be discussed later; for the former we use the AGA formula (2.5). For details and alternative models see Chapter 2 or Schmidt, Steinbach, and Willert (2014).

The stationary momentum equation essentially yields the pressure loss along the pipe for which various approximation formulas exist; see Saleh (2002) or Schmidt, Steinbach, and Willert (2014). Here we use a quadratic equation of Weymouth type (see Katz (1959); Weymouth (1912); de Nevers (1970)) derived in Lurie (2008),

$$0 = c_a^{\text{p-loss}}(x) = p_v^2 - \left(p_u^2 - \Lambda_a q_a |q_a| \frac{e^{S_a} - 1}{S_a} \right) e^{-S_a}, \qquad (9.3)$$

where

$$\Lambda_a = \frac{L_a}{A_a^2 D_a} z_{a,m} T R_s \lambda_a, \quad S_a = \frac{2 L_a g s_a}{R_s z_{a,m} T}.$$

The used quantities are the length of the pipe L_a, the constant slope of the pipe s_a, its cross-sectional area A_a and its diameter D_a. The gravitational acceleration is denoted by g. See Section 2.3.1.2 for the details. Both coefficients Λ_a and S_a depend on p_u, p_v via $z_{a,m}$ and $p_{a,m}$: we use approximate mean values of the compressibility factor and pressure defined by

$$0 = c_a^{\text{p-mean}}(x) = p_{a,m} - \frac{2}{3}\left(p_u + p_v - \frac{p_u p_v}{p_u + p_v} \right),$$
$$0 = c_a^{\text{z-mean}}(x) = z_{a,m} - z(p_{a,m}, T).$$

Finally, we model the friction coefficient $\lambda(q)$ by the Hagen–Poiseuille formula for laminar flow (2.16),

$$\lambda^{\text{HP}}(q) = \frac{64}{Re(q)}, \quad q \leq q_{\text{crit}},$$

and by the implicit empirical model of Prandtl–Colebrook (2.17) for turbulent flow,

$$\frac{1}{\sqrt{\lambda^{\text{PC}}(q)}} = -2 \log_{10}\left(\frac{2.51}{Re(q)\sqrt{\lambda^{\text{PC}}(q)}} + \frac{k}{3.71 D} \right), \quad q > q_{\text{crit}}.$$

Here, k is the roughness of the inner pipe wall and Re is the Reynolds number. If we replace the variable λ_a in $c_a^{\text{p-loss}}$ (9.3) with the new variable λ_a^{HPPC} subject to the nonsmooth constraint

$$0 = c_a^{\text{HPPC}}(x) = \lambda_a^{\text{HPPC}} - \begin{cases} \lambda^{\text{HP}}(q_a), & q_a \leq q_{\text{crit}}, \\ \lambda^{\text{PC}}(q_a), & q_a > q_{\text{crit}}, \end{cases} \tag{9.4}$$

we end up with the nonsmooth pipe model

$$0 = c_a(x) = \begin{pmatrix} c_a^{\text{p-loss}}(x) \\ c_a^{\text{p-mean}}(x) \\ c_a^{\text{z-mean}}(x) \\ c_a^{\text{HPPC}}(x) \end{pmatrix}, \quad x_a = \begin{pmatrix} q_a \\ z_{a,m} \\ p_{a,m} \\ \lambda_a^{\text{HPPC}} \end{pmatrix}. \tag{9.5}$$

Pipe model reformulation: Smoothing The pipe model (9.5) is discontinuous at $q_a = q_{\text{crit}}$ due to c_a^{HPPC} (9.4) and second-order discontinuous at $q_a = 0$ due to the term $q_a|q_a|$ in $c_a^{\text{p-loss}}$ (9.3). We address both difficulties simultaneously by replacing $\Lambda_a q_a |q_a|$ in (9.3) by $\tilde{\Lambda}_a \phi_a$:

$$0 = c_a^{\text{p-loss-s}}(x) = p_v^2 - \left(p_u^2 - \tilde{\Lambda}_a \phi_a \frac{e^{S_a} - 1}{S_a} \right) e^{-S_a},$$

where $\tilde{\Lambda}_a := \Lambda_a / \lambda_a$ and ϕ_a approximates the term $\lambda_a^{\text{HPPC}} q_a |q_a|$,

$$0 = c_a^{\text{HPPC-s}}(x) = \phi_a - r_a \left(\sqrt{q_a^2 + e_a^2} + b_a + \frac{c_a}{\sqrt{q_a^2 + d_a^2}} \right) q_a.$$

This smoothing has been shown to provide an asymptotically correct second-order approximation of $\lambda_a^{\text{HPPC}} q_a |q_a|$ if $e_a, d_a > 0$ and

$$r_a = (2 \log_{10} \beta_a)^{-2}, \quad b_a = 2\delta_a, \quad c_a = (\ln \beta_a + 1)\delta_a^2 - \frac{e_a^2}{2},$$

with

$$\alpha_a = \frac{2.51 A_a \eta}{D_a}, \quad \beta_a = \frac{k_a}{3.71 D_a}, \quad \delta_a = \frac{2\alpha_a}{\beta_a \ln 10},$$

(see Burgschweiger, Gnädig, and Steinbach (2009); Schmidt, Steinbach, and Willert (2014)). Here, η is the dynamic viscosity of the gas which we assume to be a constant; see Section 2.3.1. In summary, we obtain the smooth pipe model

$$0 = c_a^{\text{smooth}}(x) = \begin{pmatrix} c_a^{\text{p-loss-s}}(x) \\ c_a^{\text{p-mean}}(x) \\ c_a^{\text{z-mean}}(x) \\ c_a^{\text{HPPC-s}}(x) \end{pmatrix}, \quad x_a^{\text{smooth}} = \begin{pmatrix} q_a \\ z_{a,m} \\ p_{a,m} \\ \phi_a \end{pmatrix}.$$

9.1.3 ▪ Resistors

A resistor $a = (u, v) \in A_{\text{rs}}$ is a fictitious network element modeling the approximate pressure loss across gadgets, partly closed valves, filters etc.; see also Section 2.3.2. The pressure loss has the same sign as the mass flow and is either assumed to be (piecewise) constant,

$$0 = c_a^{\text{p-loss-lin}}(x) = p_u - p_v - \zeta_a \operatorname{sgn}(q_a), \tag{9.6}$$

or (piecewise) quadratic according to the law of Darcy–Weisbach (see Lurie (2008); Finnemore and Franzini (2002)),

$$0 = c_a^{\text{p-loss-nl}}(x) = p_u - p_v - \frac{8\zeta_a}{\pi^2 D_a^4} \frac{q_a |q_a|}{\rho_{a,w}}. \tag{9.7}$$

Here $\zeta_a \geq 0$ is the resistance coefficient, and $\rho_{a,w}$ is the inflow gas density, which is determined via

$$0 = c_a^{\text{dens-in}}(x) = \rho_{a,w} - \rho(p_w, T) \quad \text{with} \quad w := \begin{cases} u, & q_a \geq 0, \\ v, & q_a < 0. \end{cases} \tag{9.8}$$

The compressibility factor z has to be evaluated at the inflow node as well,

$$0 = c_a^{\text{z-in}}(x) = z_{a,w} - z(p_w, T).$$

In summary, the piecewise constant resistor model ($a \in A_{\text{lin-rs}}$) reads

$$0 = c_a(x) = c_a^{\text{p-loss-lin}}(x), \quad x_a = q_a, \tag{9.9}$$

and the piecewise quadratic resistor model ($a \in A_{\text{nl-rs}}$) reads

$$0 = c_a(x) = \begin{pmatrix} c_a^{\text{p-loss-nl}}(x) \\ c_a^{\text{dens-in}}(x) \\ c_a^{\text{z-in}}(x) \end{pmatrix}, \quad x_a = \begin{pmatrix} q_a \\ z_{a,w} \\ \rho_{a,w} \end{pmatrix}. \tag{9.10}$$

Resistor model reformulation: Smoothing The presented resistor models (9.9) and (9.10) are nonsmooth, because of three reasons:

1. the discontinuous sgn function in (9.6),
2. the second-order discontinuous term $|q_a|q_a$ in (9.7), and
3. the direction dependence of the inflow gas density $\rho_{a,k}$ in (9.7).

In the piecewise constant resistor model, we smooth the sgn function via

$$\text{sgn}(x) = \frac{x}{|x|} = \frac{x}{\sqrt{x^2}} \approx \frac{x}{\sqrt{x^2 + \varepsilon_a}}, \tag{9.11}$$

with a smoothing parameter $\varepsilon_a > 0$. For $a \in A_{\text{lin-rs}}$ this yields

$$0 = c_a^{\text{smooth}}(x) = c_a^{\text{p-loss-lin-s}}(x),$$

with

$$0 = c_a^{\text{p-loss-lin-s}}(x) = p_u - p_v - \zeta_a \frac{q_a}{\sqrt{q_a^2 + \varepsilon_a}}.$$

The same approximation of the absolute value function as in (9.11) is applied to the piecewise quadratic resistor model (9.7):

$$0 = c_a^{\text{p-loss-nl-s}}(x) = p_u - p_v - \frac{8\zeta_a}{\pi^2 D_a^4} \frac{q_a \sqrt{q_a^2 + \varepsilon_a}}{\rho_{a,m}}.$$

Finally, the direction dependence of the inflow gas density $\rho_{a,k}$ is approximatively addressed by using the mean density

$$0 = c_a^{\text{dens-mean}}(x) = \rho_{a,m} - \frac{1}{2}(\rho_{a,\text{in}} + \rho_{a,\text{out}}).$$

9.1. Model

As a consequence, we need to evaluate the equation of state and the compressibility factor at both nodes u and v,

$$0 = c_a^{\text{dens-in}}(x) = \rho_{a,\text{in}} - \rho(p_u, T), \qquad 0 = c_a^{\text{dens-out}}(x) = \rho_{a,\text{out}} - \rho(p_v, T),$$
$$0 = c_a^{\text{z-in}}(x) = z_{a,\text{in}} - z(p_u, T), \qquad 0 = c_a^{\text{z-out}}(x) = z_{a,\text{out}} - z(p_v, T).$$

This yields for $a \in A_{\text{nl-rs}}$ the smoothed model

$$0 = c_a^{\text{smooth}}(x) = \begin{pmatrix} c_a^{\text{p-loss-nl-s}}(x) \\ c_a^{\text{dens-in}}(x) \\ c_a^{\text{dens-out}}(x) \\ c_a^{\text{dens-mean}}(x) \\ c_a^{\text{z-in}}(x) \\ c_a^{\text{z-out}}(x) \end{pmatrix}, \qquad x_a^{\text{smooth}} = \begin{pmatrix} q_a \\ \rho_{a,\text{in}} \\ \rho_{a,\text{out}} \\ \rho_{a,\text{m}} \\ z_{a,\text{in}} \\ z_{a,\text{out}} \end{pmatrix}.$$

9.1.4 ▪ Short cuts

A short cut $a = (u, v) \in A_{\text{sc}}$ is a fictitious network element involving only a simple pressure equality:

$$0 = c_a^{\text{p-coupl}}(x) = p_u - p_v.$$

Thus, we have the model

$$0 = c_a(x) = c_a^{\text{p-coupl}}(x), \qquad x_a = q_a.$$

No reformulation is required.

9.1.5 ▪ Valves

A valve $a = (u, v) \in A_{\text{va}}$ has two discrete states: *open* and *closed*. Across open valves, the pressures are identical and the flow is arbitrary within its technical bounds; see (2.33)–(2.35),

$$p_v = p_u, \qquad q_a \in [\underline{q}_a, \overline{q}_a]. \tag{9.12}$$

Closed valves block the gas flow and the pressures are arbitrary within their bounds; see (2.36),

$$q_a = 0, \qquad p_u \in [\underline{p}_u, \overline{p}_u], \qquad p_v \in [\underline{p}_v, \overline{p}_v]. \tag{9.13}$$

This behavior can be modeled with one binary variable $s_a \in \{0, 1\}$ together with big-M constraints:

$$0 \le c_a^{\text{flow-lb}}(x) = q_a - s_a \underline{q}_a,$$
$$0 \le c_a^{\text{flow-ub}}(x) = -q_a + s_a \overline{q}_a,$$
$$0 \le c_a^{\text{p-coupl-1}}(x) = M_{a,1}(1 - s_a) - p_v + p_u,$$
$$0 \le c_a^{\text{p-coupl-2}}(x) = M_{a,2}(1 - s_a) - p_u + p_v.$$

Here and in what follows, the big M's are chosen sufficiently large in order to deactivate the constraint in case of $s_a = 0$. In the concrete case above, the smallest possible values are $M_{a,1} = \overline{p}_v - \underline{p}_u$ and $M_{a,2} = \overline{p}_u - \underline{p}_v$. The resulting valve model reads

$$0 \le c_a(x) = \begin{pmatrix} c_a^{\text{flow-lb}}(x) \\ c_a^{\text{flow-ub}}(x) \\ c_a^{\text{p-coupl-1}}(x) \\ c_a^{\text{p-coupl-2}}(x) \end{pmatrix}, \qquad x_a = \begin{pmatrix} q_a \\ s_a \end{pmatrix}.$$

Valve model reformulation: Complementarity constraints It is easily seen that the switching between *open* and *closed* valve states can be formulated equivalently with the complementarity constraint

$$0 = c_a^{\text{state}}(x) = (p_u - p_v)q_a.$$

Note that this is only possible because the two states are not *disjoint*: there exist triples (q_a, p_u, p_v) that satisfy both (9.12) and (9.13). The complete reformulation reads

$$0 = c_a^{\text{mpec}}(x) = c_a^{\text{state}}(x), \quad x_a^{\text{mpec}} = q_a.$$

It offers two advantages: no binary variables are required and the number of constraints reduces from four to one. The drawback is the loss of model regularity in terms of constraint qualifications.

9.1.6 • Control valves

A control valve $a = (u, v) \in A_{\text{cv}}$ is used to decrease the gas pressure in a technically prescribed direction which is given implicitly by the direction of the arc a. It possesses three discrete states: *active*, *bypass*, and *closed*. An *active* control valve reduces the inflow pressure by a certain amount (see (2.37)),

$$p_v = p_u - \Delta_a, \quad \Delta_a \in [\underline{\Delta}_a, \overline{\Delta}_a], \quad q_a \in [\underline{q}_a, \overline{q}_a] \cap \mathbb{R}_{\geq 0}.$$

A *closed* control valve acts like a closed regular valve, leading to the simple state model (9.13). A control valve in *bypass* mode acts like an open regular valve, with arbitrary flow direction and without decreasing the pressure, see (9.12). Our complete mixed-integer linear model is based on the variable vector

$$x_a = \begin{pmatrix} q_a \\ s_{a,1} \\ s_{a,2} \end{pmatrix},$$

where $s_{a,1}, s_{a,2} \in \{0, 1\}$ have the following interpretation:

$$s_{a,1} = \begin{cases} 0, & a \text{ is closed,} \\ 1, & a \text{ is open,} \end{cases} \quad s_{a,2} = \begin{cases} 0, & a \text{ is inactive,} \\ 1, & a \text{ is active.} \end{cases}$$

The three states given above are represented as the following combinations:

$$(s_{a,1}, s_{a,2}) = (0, 0): a \text{ is closed,}$$
$$(s_{a,1}, s_{a,2}) = (1, 0): a \text{ is in bypass mode,}$$
$$(s_{a,1}, s_{a,2}) = (1, 1): a \text{ is active.}$$

In terms of the constraints

$$0 \leq c_a^{\text{flow-lb-open}}(x) = q_a - s_{a,1}\underline{q}_a,$$
$$0 \leq c_a^{\text{flow-ub-open}}(x) = -q_a + s_{a,1}\overline{q}_a,$$
$$0 \leq c_a^{\text{flow-lb-active}}(x) = q_a - (1 - s_{a,2})\underline{q}_a,$$
$$0 \leq c_a^{\text{p-coupl-1}}(x) = M_{a,1}(1 - s_{a,1}) + \overline{\Delta}_a s_{a,2} - (p_u - p_v),$$
$$0 \leq c_a^{\text{p-coupl-2}}(x) = M_{a,2}(1 - s_{a,1}) - \underline{\Delta}_a s_{a,2} - (p_v - p_u),$$
$$0 \leq c_a^{\text{consistent-states}}(x) = s_{a,1} - s_{a,2},$$

the resulting mixed-integer model then becomes

$$0 \leq c_a(x) = \begin{pmatrix} c_a^{\text{flow-lb-open}}(x) \\ c_a^{\text{flow-ub-open}}(x) \\ c_a^{\text{flow-lb-active}}(x) \\ c_a^{\text{p-coupl-1}}(x) \\ c_a^{\text{p-coupl-2}}(x) \\ c_a^{\text{consistent-states}}(x) \end{pmatrix}. \tag{9.14}$$

Control valve model reformulation: Complementarity constraints As for valves, the reformulation here is based on a complementarity constraint:

$$0 = c_a^{\text{state}}(x) = (p_v - p_u + \Delta_a) q_a.$$

In addition, the restriction to nonnegative flows in the active state is modeled as

$$0 \leq c_a^{\text{active-flow}}(x) = \Delta_a q_a.$$

Note that this model is equivalent to (9.14) only if $\underline{\Delta}_a = 0$. However, this appears to be a moderate restriction in practice: it holds in all cases we have encountered. The complete MPEC type control valve model now reads:

$$0 = c_{a,\mathscr{E}}^{\text{mpec}}(x) = c_a^{\text{state}}(x),$$
$$0 \leq c_{a,\mathscr{I}}^{\text{mpec}}(x) = c_a^{\text{active-flow}}(x),$$
$$x_a = \begin{pmatrix} q_a \\ \Delta_a \end{pmatrix}.$$

9.1.7 • Compressor groups

A compressor group $a = (u, v) \in A_{\text{cg}}$ usually has several compressor machines, powered by corresponding drives, to propel the gas by increasing its pressure (see Chapter 2). To serve a sufficiently broad range of operating conditions, i.e., flow-pressure combinations (q_a, p_u, p_v), every group has a suitable set $\mathscr{K}_a$ of technically possible *configurations*: serial combinations of parallel arrangements of compressor machines.

We introduce a triple $(q_{a,k}, p_{a,\text{in},k}, p_{a,\text{out},k})$ for every configuration $k \in \mathscr{K}_a$ and extend it to a variable vector $x_{a,k}$ to model the operation of configuration k with complete compressor machine and drive models. (In our smooth model formulation we will have $(q_{a,k}, p_{a,\text{in},k}) = (q_a, p_u)$, but $p_{a,\text{out},k} \neq p_v$ in general.) Now let $\mathscr{F}_{a,k}$ denote the set of *feasible* vectors $x_{a,k}$ in the sense that the gas pressure $p_{a,\text{in},k}$ can be increased to $p_{a,\text{out},k}$ at mass flow $q_{a,k}$ in configuration k. Moreover, choose a set of smooth constraints $c_{a,k,\mathscr{E}}^{\text{op-range}}, c_{a,k,\mathscr{I}}^{\text{op-range}}$ representing $\mathscr{F}_{a,k}$:

$$c_{a,k,\mathscr{E}}^{\text{op-range}}(x_{a,k}) = 0, \quad c_{a,k,\mathscr{I}}^{\text{op-range}}(x_{a,k}) \geq 0 \iff x_{a,k} \in \mathscr{F}_{a,k}.$$

Details of these models are irrelevant here, but can be found in Schmidt, Steinbach, and Willert (2014) or Chapter 10.

Since only one configuration k can be active, we model the selection of a configuration with the special-ordered-set constraint

$$0 = c_a^{\text{sos}}(x) = \sum_{k \in \mathscr{K}_a} s_{a,k} - 1, \quad s_{a,k} \in \{0, 1\},$$

in combination with suitable big-M indicator constraints,

$$0 \leq c_{a,k}^{\text{ind-1}}(x) = M_{a,k,1}\left(1 - s_{a,k}\right) + c_{a,k,\mathscr{E}}^{\text{op-range}}(x_{a,k}),$$
$$0 \leq c_{a,k}^{\text{ind-2}}(x) = M_{a,k,2}\left(1 - s_{a,k}\right) - c_{a,k,\mathscr{E}}^{\text{op-range}}(x_{a,k}),$$
$$0 \leq c_{a,k}^{\text{ind-3}}(x) = M_{a,k,3}\left(1 - s_{a,k}\right) + c_{a,k,\mathscr{I}}^{\text{op-range}}(x_{a,k}).$$

Here, $M_{a,k,i}, i = 1,2,3$, are vectors of big M's that are determined by estimations of the components of $c_{a,k,\mathscr{E}}^{\text{op-range}}$ and $c_{a,k,\mathscr{I}}^{\text{op-range}}$. Then the mixed-integer nonlinear model of the compressor group has variables

$$x_a = \begin{pmatrix} (x_{a,k})_{k \in \mathscr{K}_a} \\ (s_{a,k})_{k \in \mathscr{K}_a} \end{pmatrix}$$

and reads

$$0 = c_{a,\mathscr{E}}(x) = c_a^{\text{sos}}(x),$$
$$0 \leq c_{a,\mathscr{I}}(x) = \left(c_{a,k}^{\text{ind-}i}(x)\right)_{k \in \mathscr{K}_a}^{i=1,2,3}.$$

9.1.7.1 ▪ Compressor group model reformulation: Step 1—convex combination

Our key idea for finding a *feasible configuration* is based on a convex combination of all configurations with the following (fictitious) interpretation:

1. All configurations are simultaneously active in our model. As with the mixed-integer model, this allows conclusions on the feasibility of configurations.
2. All configurations have identical inflow gas pressures and serve the complete gas flow,

$$0 = c_{a,k}^{\text{p-in-coupl}}(x) = p_u - p_{a,\text{in},k},$$
$$0 = c_{a,k}^{\text{flow-distr}}(x) = q_a - q_{a,k}.$$

3. The pressure increase $p_v - p_u$ of the entire group is a convex combination of pressure increases of the configurations with weights $\gamma_{a,k} \in [0,1]$:

$$0 = c_a^{\text{press-inc}}(x) = p_v - p_u - \sum_{k \in \mathscr{K}_a} \gamma_{a,k}\left(p_{a,\text{out},k} - p_{a,\text{in},k}\right),$$
$$0 = c_a^{\text{conv-combi}}(x) = \sum_{k \in \mathscr{K}_a} \gamma_{a,k} - 1.$$

We remark that the constraints $c_{a,k,\mathscr{E}}^{\text{op-range}}$ and $c_{a,k,\mathscr{I}}^{\text{op-range}}$ ensure that $p_{a,\text{out},k} \geq p_{a,\text{in},k}$ holds. It might be possible in practice that more than one configuration can serve as a feasible active configuration for a given operating condition (q_a, p_u, p_v). For these cases, a selection method is described in Section 9.3.2. On the other hand, a given operating condition may be infeasible for some of the configurations. However, the corresponding weights can be set to zero so that the convex combination model is a feasible relaxation if (q_a, p_u, p_v) is feasible for at least one configuration.

9.1.7.2 ▪ Compressor group model reformulation: Step 2—relaxation

As the configurations of a group are designed to handle different operating conditions, their "overlap" is typically small: most of the concrete conditions (q_a, p_u, p_v) can only be

9.1. Model

handled by a few configurations. As a consequence, the above model will often be infeasible. The main reason for this is that every configuration has to compress the complete gas flow q_a, which is not always possible. Thus, we relax every set $\mathscr{F}_{a,k}$ to $\tilde{\mathscr{F}}_{a,k}$ with a standard slack formulation: we have $(x_{a,k}, \sigma_{a,k}) \in \tilde{\mathscr{F}}_{a,k}$ if there exist slack variables $\sigma_{a,k} \geq 0$ such that

$$\tilde{c}^{\text{op-range}}_{a,k,\mathscr{E}}(x_{a,k}, \sigma_{a,k}) = 0 \quad \text{and} \quad \tilde{c}^{\text{op-range}}_{a,k,\mathscr{I}}(x_{a,k}, \sigma_{a,k}) \geq 0.$$

Here the relaxed equality constraints are defined as

$$\tilde{c}^{\text{op-range}}_{a,k,\mathscr{E}}(x_{a,k}, \sigma_{a,k}) = c^{\text{op-range}}_{a,k,\mathscr{E}}(x_{a,k}) + \sigma^+_{a,k,\mathscr{E}} - \sigma^-_{a,k,\mathscr{E}},$$

the relaxed inequality constraints is given by

$$\tilde{c}^{\text{op-range}}_{a,k,\mathscr{I}}(x_{a,k}, \sigma_{a,k}) = c^{\text{op-range}}_{a,k,\mathscr{I}}(x_{a,k}) + \sigma^+_{a,k,\mathscr{I}},$$

and the complete slack vector is $\sigma_{a,k} = (\sigma^+_{a,k,\mathscr{E}}, \sigma^-_{a,k,\mathscr{E}}, \sigma^+_{a,k,\mathscr{I}})$. With $\sigma_a = (\sigma_{a,k})_{k \in \mathscr{K}_a}$, the relaxed convex combination model now reads

$$0 = c^{\text{rel-conv}}_{a,\mathscr{E}}(x, \sigma_a) = \begin{pmatrix} \left(c^{\text{flow-distr}}_{a,k}(x)\right)_{k \in \mathscr{K}_a} \\ c^{\text{press-inc}}_{a}(x) \\ c^{\text{conv-combi}}_{a}(x) \\ \left(c^{\text{p-in-coupl}}_{a,k}(x)\right)_{k \in \mathscr{K}_a} \\ \left(\tilde{c}^{\text{op-range}}_{a,k,\mathscr{E}}(x_{a,k}, \sigma_{a,k})\right)_{k \in \mathscr{K}_a} \end{pmatrix}, \tag{9.15a}$$

$$0 \leq c^{\text{rel-conv}}_{a,\mathscr{I}}(x, \sigma_a) = \left(\left(\tilde{c}^{\text{op-range}}_{a,k,\mathscr{I}}(x_{a,k}, \sigma_{a,k})\right)_{k \in \mathscr{K}_a} \right), \tag{9.15b}$$

$$x^{\text{rel-conv}}_a = \begin{pmatrix} q_a \\ (x_{a,k})_{k \in \mathscr{K}_a} \\ (\gamma_{a,k})_{k \in \mathscr{K}_a} \end{pmatrix}. \tag{9.15c}$$

We remark that the convex combination approach leads to the fact that (9.15) does not contain binary variables.

As usual, we try to enforce feasibility by minimizing the slacks $\sigma = (\sigma_a)_{a \in A_{\text{cg}}}$ in a suitable norm. This is also the objective of the entire smoothed MPEC:

$$f_{\text{sMPEC}}(\sigma) = \|\sigma\|.$$

9.1.8 ▪ Model summary

In the previous sections we have described components of gas transport networks, both as nonsmooth mixed-integer nonlinear models and, if necessary, as smooth MPEC reformulations. Now we collect the component models to obtain complete nMINLP and sMPEC formulations.

9.1.8.1 ▪ The complete nonsmooth mixed-integer nonlinear model

The complete feasibility problem in nMINLP form reads

$$\exists ?\, x : \quad c_{\mathscr{E}, \text{nMINLP}}(x) = 0, \quad c_{\mathscr{I}, \text{nMINLP}}(x) \geq 0,$$

where

$$c_{\mathcal{E},\text{nMINLP}}(x) = \begin{pmatrix} c_V(x) \\ c_{A_{\text{pi}}}(x) \\ c_{A_{\text{lin-rs}}}(x) \\ c_{A_{\text{nl-rs}}}(x) \\ c_{A_{\text{sc}}}(x) \\ c_{A_{\text{cg}},\mathcal{E}}(x) \end{pmatrix}, \quad c_{\mathcal{I},\text{nMINLP}}(x) = \begin{pmatrix} c_{A_{\text{va}}}(x) \\ c_{A_{\text{cv}}}(x) \\ c_{A_{\text{cg}},\mathcal{I}}(x) \end{pmatrix}$$

and

$$x = (x_V, x_{A_{\text{pi}}}, x_{A_{\text{lin-rs}}}, x_{A_{\text{nl-rs}}}, x_{A_{\text{sc}}}, x_{A_{\text{va}}}, x_{A_{\text{cv}}}, x_{A_{\text{cg}}})^T.$$

Here, nonsmooth aspects arise in all passive elements except for short cuts: $c_{A_{\text{pi}}}(x)$, $c_{A_{\text{lin-rs}}}(x)$, $c_{A_{\text{nl-rs}}}(x)$. Discrete decisions ("genuine" binary variables) arise in the active elements: $c_{A_{\text{cg}},\mathcal{E}}(x)$, $c_{A_{\text{cg}},\mathcal{I}}(x)$, $c_{A_{\text{va}}}(x)$, $c_{A_{\text{cv}}}(x)$. The node and short cut models $c_V(x)$, $c_{A_{\text{sc}}}(x)$ are smooth and will be kept in original form.

9.1.8.2 ▪ The complete smoothed MPEC model

Combining all smoothed and complementarity constrained component models, we obtain the following smooth MPEC model:

$$\min_{x,\sigma} \; f_{\text{sMPEC}}(\sigma) \quad \text{s.t.} \quad c_{\mathcal{E},\text{sMPEC}}(x,\sigma) = 0, \quad c_{\mathcal{I},\text{sMPEC}}(x,\sigma) \geq 0, \tag{9.16}$$

where

$$c_{\mathcal{E},\text{sMPEC}}(x,\sigma) = \begin{pmatrix} c_V(x) \\ c_{A_{\text{pi}}}^{\text{smooth}}(x) \\ c_{A_{\text{lin-rs}}}^{\text{smooth}}(x) \\ c_{A_{\text{nl-rs}}}^{\text{smooth}}(x) \\ c_{A_{\text{sc}}}(x) \\ c_{A_{\text{va}}}^{\text{mpec}}(x) \\ c_{A_{\text{cv}},\mathcal{E}}^{\text{mpec}}(x) \\ c_{A_{\text{cg}},\mathcal{E}}^{\text{rel-conv}}(x,\sigma) \end{pmatrix}, \quad c_{\mathcal{I},\text{sMPEC}}(x,\sigma) = \begin{pmatrix} c_{A_{\text{cv}},\mathcal{I}}^{\text{mpec}}(x) \\ c_{A_{\text{cg}},\mathcal{I}}^{\text{rel-conv}}(x,\sigma) \end{pmatrix}, \tag{9.17}$$

and

$$x = \begin{pmatrix} x_V \\ x_{A_{\text{pi}}}^{\text{smooth}} \\ x_{A_{\text{lin-rs}}} \\ x_{A_{\text{nl-rs}}}^{\text{smooth}} \\ x_{A_{\text{sc}}} \\ x_{A_{\text{va}}}^{\text{mpec}} \\ x_{A_{\text{cv}}}^{\text{mpec}} \\ x_{A_{\text{cg}}}^{\text{rel-conv}} \end{pmatrix}, \quad \sigma = (\sigma_a)_{a \in A_{\text{cg}}}. \tag{9.18}$$

9.2 • MPEC regularization

It is well known that MPEC models like (9.16) do not satisfy standard constraint qualifications (CQs), such as the *linear independence constraint qualification* (LICQ) or the *Mangasarian–Fromowitz constraint qualification* (MFCQ). As standard NLP solvers rely on such conditions, they cannot be applied directly to (9.16) without losing their theoretical convergence properties. In the literature one finds various regularization, smoothing, or penalization techniques to address this difficulty (see Scholtes (2001); DeMiguel et al. (2005); Hu and Ralph (2004)). Provided that the MPEC satisfies certain generalized CQs, they yield regular NLP formulations that depend on a (regularization, smoothing, or penalization) parameter $\tau > 0$. A sequence of parameters $\tau_\nu \to 0$ then yields a sequence of NLPs whose solutions converge to a solution of the given MPEC.

The standard penalization scheme turned out to be a very successful approach on our real-world instances. This scheme moves the complementarity constraints into the objective as a penalty term. More formally, given a general MPEC model

$$\begin{aligned}
\min_{x} \quad & f(x) \\
\text{s.t.} \quad & c_{\mathcal{E}}(x) = 0, \quad c_{\mathcal{I}}(x) \geq 0, \\
& \phi_i(x)\psi_i(x) = 0, \quad i = 1,\ldots,k, \\
& \phi_i(x), \psi_i(x) \geq 0, \quad i = 1,\ldots,k,
\end{aligned}$$

and a suitable penalty function π, the sequence of penalized NLPs is defined as

$$\begin{aligned}
\min_{x} \quad & f(x) + \frac{1}{\tau_\nu} \sum_{i=1}^{k} \pi(\phi_i(x), \psi_i(x)) \\
\text{s.t.} \quad & c_{\mathcal{E}}(x) = 0, \quad c_{\mathcal{I}}(x) \geq 0, \\
& \phi_i(x), \psi_i(x) \geq 0, \quad i = 1,\ldots,k.
\end{aligned}$$

A typical choice for the penalty function is $\pi(\phi_i(x), \psi_i(x)) = \phi_i(x)\psi_i(x)$.

For our purpose, solving an entire sequence of parameterized NLPs has two major drawbacks:

1. The sequence τ_ν is usually generated by a continuation technique and many parameterized NLPs may have to be solved until a given tolerance is reached. This contradicts the central goal of a primal heuristic, namely that it yields a feasible solution *quickly*.
2. The approach has the theoretical property that it only works if *every* parameterized NLP is solved successfully. However, our smoothed MPEC formulation is hard to solve because of potentially unstable smoothings and the large number of complementarity constraints. Thus, having to solve a large number of NLPs increases the risk of a failure.

To avoid these drawbacks, we solve only a *single* parameterized NLP where the parameter $\tau > 0$ is fixed at the value 10^{-6}, which proves to be a good value in our numerical experiments. The regular NLP reformulation of (9.16), rsMPEC, then reads

$$\begin{aligned}
\min_{x,\sigma} \quad & f_{\text{rsMPEC}}(x,\sigma) := f_{\text{sMPEC}}(\sigma) + f_{\text{penalty}}(x) \\
\text{s.t.} \quad & c_{\mathcal{E},\text{rsMPEC}}(x,\sigma) = 0, \quad c_{\mathcal{I},\text{rsMPEC}}(x,\sigma) \geq 0,
\end{aligned} \quad (9.19)$$

where $c_{\mathcal{E},\text{rsMPEC}}$ is constructed from $c_{\mathcal{E},\text{sMPEC}}$ by moving all complementarity constraints to the penalty term and keeping $c_{\mathcal{I},\text{rsMPEC}} = c_{\mathcal{I},\text{sMPEC}}$. The specific penalty term in our formulation is

$$f_{\text{penalty}}(x) = \frac{1}{\tau}\left(\sum_{a\in A_{\text{va}}} c_a^{\text{state}}(x)^2 + \sum_{a\in A_{\text{cv}}} c_a^{\text{state}}(x)^2\right),$$

where the squares are introduced to account for complementarity constraints that may become negative, such as $\tilde{p}_a q_a$ for valves ($\tilde{p}_a = p_u - p_v$) or control valves ($\tilde{p}_a = p_v - p_u + \Delta_a$).

9.3 • Solution technique: A two-stage approach

In the previous sections, we have described how to transform the nonsmooth MINLP model into the regularized smooth MPEC model (9.19). This simplification comes at a price: the rsMPEC model, although in standard NLP form, is very hard to solve for most of our real-world instances. There are several reasons. First, the smoothing applied to nonsmooth constraints of pipes and resistors may lead to ill-conditioning and numerical instabilities. Second, rsMPEC involves a large number of penalized complementarity constraints in real-world instances, which may create additional numerical difficulties in the solution process. Finally, the convex combination model for compressor group configurations is highly nonlinear and nonconvex: it contains every compressor machine with full physical and technical details.

To overcome these difficulties we use a *two-stage approach* for solving (9.19). The goal of the first stage is to determine the major network flow situation including the discrete state of active elements and the directions of flow through resistors. In the second stage, these discrete decisions and flow directions are fixed, and the goal is to find feasible configurations for all active compressor groups.

9.3.1 • Solving stage 1

In the first stage we try to determine a feasible flow situation in the network for the given nomination situation in terms of

1. the discrete states of valves, control valves, and compressor groups;
2. the directions of flow through resistors.

For this purpose we simplify the compressor group model substantially: rather than modeling configurations and operating ranges of compressor machines, we simply assume that every group can deliver a certain pressure increase $\Delta_a \in [0, \overline{\Delta}_a]$ that is independent of the flow. Here, $\overline{\Delta}_a$ is obtained by the preprocessing described in Section 5.4. The resulting MPEC formulation is similar to the MPEC model of control valves (Section 9.1.6):

$$0 = c_a^{\text{state}}(x) = \tilde{p}_a q_a, \quad 0 \leq c_a^{\text{active-flow}}(x) = \Delta_a q_a,$$

with the pressure coupling variable

$$\tilde{p}_a := p_v - p_u - \Delta_a, \quad \Delta_a \in [0, \overline{\Delta}_a].$$

Thus, we have to replace $c_{a,\mathcal{E}}^{\text{rel-conv}}$, $c_{a,\mathcal{I}}^{\text{rel-conv}}$, $x_a^{\text{rel-conv}}$ in (9.15) by

$$0 = c_{a,\mathcal{E}}^{\text{simple}}(x) = c_a^{\text{state}}(x),$$
$$0 \leq c_{a,\mathcal{I}}^{\text{simple}}(x) = c_a^{\text{active-flow}}(x),$$

9.3. Solution technique: A two-stage approach

Table 9.1. *Fixing discrete states of valves.*

$\tilde{p}_a$	q_a	State	Binary variable
≈ 0	≈ 0	arbitrary	$s_a \in \{0,1\}$
≈ 0	$> \varepsilon_q$	open	$s_a = 1$
$> \varepsilon_p$	≈ 0	closed	$s_a = 0$
$> \varepsilon_p$	$> \varepsilon_q$	infeasible	—

Table 9.2. *Fixing discrete states of control valves and compressor groups.*

$\tilde{p}_a$	q_a	Δ_a	State	Binary variables	
≈ 0	≈ 0	≈ 0	arbitrary	$s_{a,1} \in \{0,1\}$	$s_{a,2} \in \{0,1\}$
≈ 0	≈ 0	$> \varepsilon_p$	closed	$s_{a,1} = 0$	$s_{a,2} = 0$
≈ 0	$> \varepsilon_q$	≈ 0	bypass or active	$s_{a,1} = 1$	$s_{a,2} \in \{0,1\}$
≈ 0	$> \varepsilon_q$	$> \varepsilon_p$	active	$s_{a,1} = 1$	$s_{a,2} = 1$
$> \varepsilon_p$	≈ 0	≈ 0	closed	$s_{a,1} = 0$	$s_{a,2} = 0$
$> \varepsilon_p$	≈ 0	$> \varepsilon_p$	closed	$s_{a,1} = 0$	$s_{a,2} = 0$
$> \varepsilon_p$	$> \varepsilon_q$	≈ 0	infeasible	—	
$> \varepsilon_p$	$> \varepsilon_q$	$> \varepsilon_p$	infeasible	—	

and

$$x_a^{\text{simple}} = \begin{pmatrix} q_a \\ \Delta_a \end{pmatrix}.$$

Our numerical experience shows that the first-stage solutions often have large flows in active elements that may lead to an infeasible second-stage model. Therefore we extend the first-stage objective to penalize flow in active elements,

$$f_{\text{rsMPEC-1}}(x) := f_{\text{penalty}}(x) + \sum_{a \in A_{\text{va}} \cup A_{\text{cv}} \cup A_{\text{cg}}} \omega_a q_a^2,$$

where $\omega_a \geq 0$ are instance-specific scaling factors. The objective term $f_{\text{sMPEC}}(x)$ of (9.19) is absent here, since we do not consider any configurations. Finally, the first-stage constraints and variables are obtained from (9.17) and (9.18) by replacing $c_{A_{\text{cg}},\mathscr{E}}^{\text{rel-conv}}, c_{A_{\text{cg}},\mathscr{I}}^{\text{rel-conv}}, x_{A_{\text{cg}}}^{\text{rel-conv}}$ with $c_{A_{\text{cg}},\mathscr{E}}^{\text{simple}}, c_{A_{\text{cg}},\mathscr{I}}^{\text{simple}}, x_{A_{\text{cg}}}^{\text{simple}}$.

9.3.2 ▪ Solving stage 2

The first-stage solution is used to fix discrete decisions and resistor flow directions in the second-stage model. Because of the penalization scheme used to regularize the complementarity constraints in the first-stage model (see Section 9.2), we cannot expect that the complementarity constraints hold exactly; this is only guaranteed in the limit $\tau \to 0$. Therefore, we choose pressure and flow tolerances $\varepsilon_p, \varepsilon_q > 0$ to decide whether $\tilde{p}_a$ or q_a are sufficiently small to be regarded as zero. This determines the discrete states of active elements according to Table 9.1 and Table 9.2. Finally, the resistor flow directions are fixed according to the sign of the flow variable.

The effects of fixing discrete states and resistor flow directions in Stage 2 are twofold. First, all nonsmooth aspects resulting from dependencies on flow directions are resolved.

Second, all discrete decisions of the underlying nonsmooth mixed-integer nonlinear program (MINLP) are fixed, except for the compressor group configurations. Thus, a smooth model can be formulated:

$$\min_{x,\sigma} \quad f_{\text{rsMPEC-2}}(\sigma) := f_{\text{sMPEC}}(\sigma) \tag{9.20}$$

$$\text{s.t.} \quad c_V(x) = 0, \qquad\qquad c_{A_{\text{pi}}}^{\text{smooth}}(x) = 0,$$

$$c_{A_{\text{sc}}}(x) = 0, \qquad\qquad c_{A_{\text{lin-rs}}}^{\text{fix-flow}}(x) = 0,$$

$$c_{A_{\text{nl-rs}}}^{\text{fix-flow}}(x) = 0, \qquad\qquad c_{A_{\text{va}}}^{\text{fix-state}}(x) = 0,$$

$$c_{A_{\text{cv}}}^{\text{fix-state}}(x) = 0, \qquad\qquad c_{A_{\text{cg}},\mathscr{E}}^{\text{rel-conv}}(x,\sigma) = 0,$$

$$c_{A_{\text{cg}},\mathscr{I}}^{\text{rel-conv}}(x,\sigma) \geq 0.$$

Here, the constraints selected by fixed discrete decisions read

$$0 = c_a^{\text{fix-state}}(x) = \begin{cases} p_u - p_v & \text{if } a \text{ is open,} \\ q_a & \text{if } a \text{ is closed,} \end{cases}$$

for $a \in A_{\text{va}}$, and

$$0 = c_a^{\text{fix-state}}(x) = \begin{cases} p_u - p_v - \Delta_a & \text{if } a \text{ is active,} \\ p_u - p_v & \text{if } a \text{ is in bypass mode,} \\ q_a & \text{if } a \text{ is closed,} \end{cases}$$

for $a \in A_{\text{cv}}$. Finally, given resistor flows q_a^* of the first-stage solution, the discontinuous sgn term of the piecewise constant resistor model becomes constant,

$$0 = c_a^{\text{fix-flow}}(x) = p_u - p_v - \zeta_a \operatorname{sgn}(q_a^*), \quad a \in A_{\text{lin-rs}},$$

the nonsmooth term $q_a|q_a|$ in the piecewise quadratic model becomes $\operatorname{sgn}(q_a^*)q_a^2$ in $c_{A_{\text{nl-rs}}}^{\text{fix-flow}}$, and the unknown node index k in (9.8) is resolved as

$$k = \begin{cases} u & \text{if } q_a^* \geq 0, \\ v & \text{if } q_a^* < 0. \end{cases}$$

Selecting compressor group configurations After solving (9.20) it remains to decide on the configurations of active compressor groups. This is done in a heuristic way based on the convex coefficients $\gamma_{a,k}$ and the slack variables $\sigma_{a,k}$. To this end, let $\mathscr{K}_{a,\mathscr{F}}$ denote the set of configurations $k \in \mathscr{K}_a$ with vanishing slack variables,

$$\mathscr{K}_{a,\mathscr{F}} := \left\{ k \in \mathscr{K}_a : \left\| \sigma_{a,k} \right\| = 0 \right\}.$$

The decision is now made by the following case distinction:
1. If $|\mathscr{K}_{a,\mathscr{F}}| = 1$, choose the unique configuration $k \in \mathscr{K}_{a,\mathscr{F}}$.
2. If $|\mathscr{K}_{a,\mathscr{F}}| > 1$, choose $\operatorname{argmax}_{k \in \mathscr{K}_{a,\mathscr{F}}} \{\gamma_{a,k}\}$.
3. If $|\mathscr{K}_{a,\mathscr{F}}| = 0$, choose $\operatorname{argmin}_{k \in \mathscr{K}_a} \{\|\sigma_{a,k}\|\}$.

9.3.3 ▪ Conclusion

In this chapter, we presented reformulation techniques for the nonsmooth mixed-integer nonlinear problem of validation of nominations. These reformulation techniques cover the usage of complementarity constraints and a problem-specific convex combination approach for modeling the discrete parts of the model. In addition, all nonsmooth aspects are smoothed by standard or problem-specific approaches. The result is a smooth MPEC model, which is finally regularized by standard techniques from the literature in order to achieve an NLP that satisfies standard constraint qualifications. Thus, the final model can be solved by NLP solvers and the solutions can be used as approximative solution candidates for the NLP validation.

A drawback of this approach is that the reformulation of discrete aspects is not directly applicable to other discrete parts like subnetwork operation modes (see Section 2.4.3). As a consequence, this aspect is not regarded at all.

The main advantage of this approach is that all nonlinear aspects of the problem of validation of nominations can be included without significantly increasing the hardness of the problem. Thus, besides the complementarity constraints, the convex combination approach for compressor group configurations and all applied smoothings, the rsMPEC model contains the same model formulations as the ValNLP model.

Chapter 10
The precise NLP model

Martin Schmidt, Marc C. Steinbach, Bernhard M. Willert

Abstract *In this chapter we describe a highly detailed nonlinear program (NLP) of gas networks for the case of fixed discrete decisions. By including nonlinear physics and a detailed description of the technical network devices, a level of accuracy is reached that is comparable to current commercial simulation software. Our NLP model is used to validate the solutions of the previously described models. A successful validation provides a solution of the underlying mixed-integer nonlinear program (MINLP).*

Short-term and mid-term planning problems in gas networks, such as the validation of nominations, involve gas dynamics in combination with complex technical devices. The laws of thermodynamics introduce partial differential equations (PDEs) and further nonlinear aspects to the problem. In addition, the switching between working modes of active (i.e., controllable) network elements is modeled by discrete decisions. A full consideration of all aspects leads to a *nonsmooth discrete-continuous control problem*, which after discretization of PDEs becomes a *nonsmooth mixed-integer nonlinear optimization problem* (nonsmooth MINLP).

The preceding Chapters 6 through 9 present four approaches that simplify the nonlinear aspects of the MINLP—amongst others by approximating the description of the gas flow—in order to determine suitable discrete decisions for the problem of validation of nominations. These previously discussed approaches are referred to as *decision approaches*, and any solution of a decision approach is considered a *solution candidate* for the original MINLP.

In contrast, this chapter presents a nonlinear optimization problem (NLP) based on a highly detailed model of stationary gas physics and technical devices. In this context we suppose that discrete decisions are already given, typically as result of one of the decision approaches, so that all discrete aspects can be removed from the MINLP. In fact, the main purpose of the NLP model is the *validation* of solution candidates obtained from the decision approaches: after fixing the discrete decisions, an initial iterate for the NLP model is generated from the continuous variables of the solution candidate (see Section 5.6). If the NLP solver converges to a (usually different) optimum, we have a feasible solution of the original MINLP that is locally optimal with respect to the fixed discrete decisions, and the given solution candidate has been confirmed to be a *valid* approximation to this solution. If the validating NLP is *not* solved to optimality, no immediate conclusions can be drawn. For these cases, we introduce relaxed NLP models. Their impact is analyzed in detail in Section 10.3 and Chapter 11.

One important point remains to be mentioned: after discretizing the PDE constraints and fixing discrete decisions, we still have a nonsmooth optimization problem, but standard NLP theory, algorithms and solvers rely on smoothness (C^2), specifically to guarantee stability of solutions. Therefore we will apply reformulations and smoothing techniques to arrive at a standard NLP model,

$$\min_{x \in \mathbb{R}^n} \quad f(x)$$
$$\text{s.t.} \quad c_{\mathcal{E}}(x) = 0,$$
$$c_{\mathcal{I}}(x) \geq 0, \qquad (10.1)$$
$$x \in [\underline{x}, \overline{x}],$$

where $f: \mathbb{R}^n \to \mathbb{R}$ is the objective function and $c_{\mathcal{E}}: \mathbb{R}^n \to \mathbb{R}^m$, $c_{\mathcal{I}}: \mathbb{R}^n \to \mathbb{R}^k$ denote equality and inequality constraints, respectively, with $f, c_{\mathcal{E}}, c_{\mathcal{I}} \in C^2$.

The central part of this chapter is Section 10.1: here we formulate suitable component models for the relevant aspects of gas physics and for the individual element types of gas networks. We also discuss the discretization of PDE constraints and the direct approximation of their solutions, and we develop smoothing techniques where applicable. The component models are then combined to a standard NLP (10.1) in Section 10.4. Typical objective functions are presented in Section 10.2 and in Section 10.3 we consider relaxations of the NLP that we use to obtain further information in cases where no feasible solution is found.

10.1 ▪ Component models

The gas network is modeled as a directed graph $G = (V, A)$ with node set V and arc set A. The node set is partitioned into entries V_+, exits V_-, and inner nodes V_0. Gas is supplied to the network at entries and discharged at exits. At inner nodes, no gas is supplied or discharged.

The arc set is partitioned into pipes A_{pi}, resistors A_{rs}, short cuts A_{sc}, valves A_{va}, control valve stations A_{cv}, and compressor groups A_{cg}. Every arc $a = (u, v)$ connects a node u (called *tail*) to a different node v (called *head*). The notations $\delta^-(u)$ and $\delta^+(u)$ refer to the respective sets of incoming and outgoing arcs of u. The set of all incident arcs is $\delta(u) = \delta^-(u) \cup \delta^+(u)$.

Every network element has associated vectors of constraints and (continuous) variables. Throughout this chapter, constraints are denoted by c and vectors of variables by x, subscripted with indices or index sets; e.g., $c_{\mathcal{E}} = (c_i)_{i \in \mathcal{E}}$ is the vector of all equality constraints and $x_{A_{\text{pi}}} = (x_a)_{a \in A_{\text{pi}}}$ is the vector of all pipe variables.

Some gas quantities vary along arcs, so that their values at the head and tail may differ. If x is such a quantity, e.g., gas temperature, then the value on $a = (u, v)$ at u is written as $x_{a:u}$ and the value on a at v is denoted by $x_{a:v}$. These quantities are often mixed at nodes and thus are discontinuous across nodes, so that $x_{a:u} \neq x_{b:u}$ if $a \neq b \in \delta(u)$.

Finally, certain constraints depend on the *direction* of flow. To this end, we define flows from tail to head as positive and flows from head to tail as negative, and for every

10.1. Component models

node u we define respective sets of *inflow arcs* and *outflow arcs*:

$$\mathscr{I}(u) = \{a \in \delta^-(u) \mid q_a \geq 0\} \cup \{a \in \delta^+(u) \mid q_a \leq 0\},$$
$$\mathscr{O}(u) = \{a \in \delta^-(u) \mid q_a < 0\} \cup \{a \in \delta^+(u) \mid q_a > 0\}.$$

Now let x be a quantity that varies along arcs. The respective values of x at the *inflow node* and *outflow node* of $a = (u, v)$ are then written as

$$x_{a,\text{in}} = \begin{cases} x_{a:u}, & q_a \geq 0, \\ x_{a:v}, & q_a < 0, \end{cases} \qquad x_{a,\text{out}} = \begin{cases} x_{a:v}, & q_a \geq 0, \\ x_{a:u}, & q_a < 0. \end{cases} \tag{10.2}$$

Note that constraints that depend on $x_{a,\text{in}}$ or $x_{a,\text{out}}$ will typically be nonsmooth at $q_a = 0$.

We remark that most of the aspects discussed in this section are based on Chapter 2, where additional information can be found. Nevertheless, many formulas of Chapter 2 are restated here in order to fix the notation required to formulate the concrete NLP models.

10.1.1 ▪ Common model aspects

Here we present models of four basic physical phenomena that are relevant to several types of network elements. These phenomena depend on the gas composition, which is characterized by a *gas quality parameter vector X* with seven components (see Section 2.2): molar mass m, calorific value H_c, pseudocritical pressure p_c and temperature T_c, and coefficients of the molar isobaric heat capacity, $\tilde{A}, \tilde{B}, \tilde{C}$,

$$X = (m, H_c, p_c, T_c, \tilde{A}, \tilde{B}, \tilde{C}). \tag{10.3}$$

The first common aspect is the deviation of real (natural) gas from ideal gas, which is measured by the *compressibility factor z*; for ideal gas $z = 1$. Various empirical compressibility models have been developed; see Section 2.3.1. In our NLP model we use either the equation of the American Gas Association (AGA) (2.5), which is known to be sufficiently accurate for pressures up to 70 bar,

$$z^{\text{aga}}(p, T, p_c, T_c) = 1 + 0.257 p_r - 0.533 \frac{p_r}{T_r}, \quad \text{where} \quad p_r = \frac{p}{p_c}, \quad T_r = \frac{T}{T_c}, \tag{10.4}$$

or Papay's equation (2.4), which is appropriate for pressures up to 150 bar,

$$z^{\text{papay}}(p, T, p_c, T_c) = 1 - 3.52 p_r e^{-2.26 T_r} + 0.274 (p_r)^2 e^{-1.878 T_r}. \tag{10.5}$$

The constraint resulting from the chosen compressibility model is formulated as

$$0 = c^{\text{compr}}(z, p, T, p_c, T_c) = z - z^{\text{aga}}(p, T, p_c, T_c) \tag{10.6a}$$

or

$$0 = c^{\text{compr}}(z, p, T, p_c, T_c) = z - z^{\text{papay}}(p, T, p_c, T_c), \tag{10.6b}$$

where the concrete choice is up to the modeler. Several constraints described in the next sections make use of the partial derivatives

$$\frac{\partial z}{\partial p}(p, T, p_c, T_c) \quad \text{and} \quad \frac{\partial z}{\partial T}(p, T, p_c, T_c)$$

of the chosen compressibility model. We choose to substitute them directly into the constraints in which they appear, instead of representing them by auxiliary variables and additional constraints.

The second aspect is the *specific isobaric heat capacity* c_p, or equivalently the *molar isobaric heat capacity* $\tilde{c}_p = mc_p$; see Section 2.2. These quantities express the energy that is required to increase the temperature of one kilogram (or mol) of gas by one Kelvin at constant pressure. We use the molar heat capacity of real gas, which is expressed by the molar heat capacity of ideal gas $\tilde{c}_p^0$ and a correction term for real gas $\Delta \tilde{c}_p$:

$$
\begin{aligned}
0 &= c^{\text{mhc-real}}(m, c_p, \tilde{c}_p^0, \Delta \tilde{c}_p) = mc_p - (\tilde{c}_p^0 + \Delta \tilde{c}_p), \\
0 &= c^{\text{mhc-ideal}}(\tilde{c}_p^0, T, \tilde{A}, \tilde{B}, \tilde{C}) = \tilde{c}_p^0 - (\tilde{A} + \tilde{B}T + \tilde{C}T^2), \\
0 &= c^{\text{mhc-corr}}(\Delta \tilde{c}_p, p, T, p_c, T_c) \\
&= \Delta \tilde{c}_p + R \int_0^p \frac{1}{\tilde{p}} \left(2T \frac{\partial z}{\partial T}(\tilde{p}, T, p_c, T_c) + T^2 \frac{\partial^2 z}{\partial T^2}(\tilde{p}, T, p_c, T_c) \right) d\tilde{p}.
\end{aligned}
$$

Here, $\tilde{c}_p^0$ is modeled by a least-squares fit with parameters $\tilde{A}, \tilde{B}, \tilde{C}$. The partial derivatives of the compressibility factor are substituted in-place, hence $c^{\text{mhc-corr}}$ does not have z as an explicit argument. If the AGA formula (10.4) is chosen for the compressibility factor, the correction term for real gas vanishes, and the model of heat capacity reduces to the model for ideal gas. If Papay's equation is chosen instead, the analytical solution of the integral is used directly in $c^{\text{mhc-corr}}$, i.e.,

$$
\begin{aligned}
& c^{\text{mhc-corr}}(\Delta \tilde{c}_p, p, T, p_c, T_c) \\
&= \Delta \tilde{c}_p + R \left(\left(\gamma \delta + \frac{1}{2} \gamma \delta^2 T_r \right) p_r^2 T_r e^{\delta T_r} - (2\alpha\beta + \alpha\beta^2 T_r) p_r T_r e^{\beta T_r} \right)
\end{aligned}
$$

with constants

$$\alpha = 3.52, \quad \beta = -2.26, \quad \gamma = 0.274, \quad \delta = -1.878.$$

The full heat capacity model is given by the constraint vector

$$
0 = c^{\text{heat-cap}}(p, T, X, x^{\text{heat-cap}}) = \begin{pmatrix} c^{\text{mhc-real}}(m, c_p, \tilde{c}_p^0, \Delta \tilde{c}_p) \\ c^{\text{mhc-ideal}}(\tilde{c}_p^0, T, \tilde{A}, \tilde{B}, \tilde{C}) \\ c^{\text{mhc-corr}}(\Delta \tilde{c}_p, p, T, p_c, T_c) \end{pmatrix} \tag{10.7}
$$

with associated variables

$$x^{\text{heat-cap}} = (c_p, \tilde{c}_p^0, \Delta \tilde{c}_p).$$

Another common aspect is the Joule–Thomson effect (see Section 2.2 for physical details), which describes the temperature change that results from any pressure change according to (2.6) and (2.7),

$$T_{\text{out}} - T_{\text{in}} = \int_{p_{\text{in}}}^{p_{\text{out}}} \mu_{\text{JT}}(p, T, m, c_p, p_c, T_c) \, dp,$$

where

$$\mu_{\text{JT}}(p, T, m, c_p, p_c, T_c) = \frac{T}{p} \frac{R}{mc_p} \left(T \frac{\partial z}{\partial T}(p, T, p_c, T_c) \right).$$

Since the calculation of gas temperatures involves substantial inaccuracies, because exact environmental temperatures are usually not known for mid-term planning, a simple two-point finite difference approximation is usually sufficiently accurate, yielding the constraint

$$0 = c^{\mathrm{jt}}(p_{\mathrm{in}}, p_{\mathrm{out}}, T_{\mathrm{in}}, T_{\mathrm{out}}, X, \mu_{\mathrm{JT}}, c_{p,\mathrm{out}})$$
$$= \begin{pmatrix} \mu_{\mathrm{JT}} - \dfrac{T_{\mathrm{out}}}{p_{\mathrm{out}}} \dfrac{R}{mc_{p,\mathrm{out}}} \left(T_{\mathrm{out}} \dfrac{\partial z}{\partial T}(p_{\mathrm{out}}, T_{\mathrm{out}}, p_{\mathrm{c}}, T_{\mathrm{c}}) \right) \\ T_{\mathrm{out}} - T_{\mathrm{in}} - (p_{\mathrm{out}} - p_{\mathrm{in}}) \mu_{\mathrm{JT}} \end{pmatrix}. \quad (10.8)$$

The last common phenomenon of gas physics is the interrelation of density ρ, pressure p, and temperature T. These gas quantities are coupled by the *thermodynamical standard equation of state for real gases* (2.20):

$$0 = c^{\mathrm{eos}}(p, T, \rho, m, z) = \rho z R T - p m. \quad (10.9)$$

10.1.2 ▪ Nodes

Every node $u \in V$ has a pressure variable p_u, a temperature variable T_u, and a vector of mixed gas parameters X_u (10.3). The incident arcs define relations between the pressures and temperatures of the connected nodes.

Nodes are assumed to have zero volume and hence satisfy a mass balance equation of Kirchhoff type:

$$0 = c_u^{\mathrm{flow}}(q_{\delta(u)}) = \sum_{a \in \delta^+(u)} q_a - \sum_{a \in \delta^-(u)} q_a - q_u^{\mathrm{nom}}. \quad (10.10)$$

Here, $q_{\delta(u)}$ contains the mass flows along the incident arcs, and q_u^{nom} denotes the externally supplied or discharged mass flow at u, which satisfies

$$q_u^{\mathrm{nom}} \geq 0, \quad u \in V_+,$$
$$q_u^{\mathrm{nom}} = 0, \quad u \in V_0,$$
$$q_u^{\mathrm{nom}} \leq 0, \quad u \in V_-.$$

These flows are typically fixed to values that are determined by the considered nomination. Thus, they are regarded as constants in this context. However, variables representing exchanged flows can easily be added to the model, if the supplied and discharged flows are not known a priori.

Gas flows entering node u are assumed to mix perfectly. The components of X mix according to the distribution of molar inflows $\hat{q} = q/m$, yielding a convex combination X_u for the outflow composition. The resulting constraint reads (see (2.9))

$$0 = c_u^{\mathrm{mix}}(q_{\mathscr{I}(u)}, X_{\mathscr{I}(u)}, X_u)$$
$$= X_u \left(\hat{q}_u^{\mathrm{nom}} + \sum_{a \in \mathscr{I}(u)} \hat{q}_a \right) - \left(\hat{q}_u^{\mathrm{nom}} X_u^{\mathrm{nom}} + \sum_{a \in \mathscr{I}(u)} \hat{q}_a X_a \right). \quad (10.11)$$

Here and in what follows, X_a is the vector of gas parameters of the gas in arc a and X_u represents the mixed gas of inflow arcs at node u. Both $\hat{q}_u^{\mathrm{nom}}$ and X_u^{nom} are given constants at entries $u \in V_+$ and are set to zero at other nodes $u \in V_- \cup V_0$. The mixed quality parameters are propagated to all outgoing arcs,

$$0 = c_{u,a}^{\mathrm{prop}}(X_u, X_a) = X_u - X_a \quad \text{for all } a \in \mathcal{O}(u).$$

The exact mixing equation for gas temperatures slightly differs from (10.11): it can be derived from the conservation of energy and involves the molar isobaric heat capacity $\tilde{c}_p$, yielding a convex combination with weights determined by the distribution of $\tilde{c}_p \hat{q} = c_p q$ (see (2.10)):

$$T_u = \frac{\tilde{c}_{p,u}^{\text{nom}} \hat{q}_u^{\text{nom}} T_u^{\text{nom}} + \sum_{a \in \mathscr{I}(u)} \tilde{c}_{p,a:u} \hat{q}_a T_{a:u}}{\tilde{c}_{p,u}^{\text{nom}} \hat{q}_u^{\text{nom}} + \sum_{a \in \mathscr{I}(u)} \tilde{c}_{p,a:u} \hat{q}_a}, \tag{10.12}$$

where T_u^{nom} is the (constant) temperature of the supplied gas at entries $u \in V_+$ and zero for all other nodes $u \in V_- \cup V_0$. Since (10.12) results in a quite complicated model, we approximate the equations of temperature mixing and propagation by assuming identical heat capacities, so that the corresponding factors in (10.12) cancel each other. The investigation of the quality of this approximation remains a topic of further research. The resulting constraints are:

$$\begin{aligned} 0 &= c_u^{\text{mix-temp}}(q_{\mathscr{I}(u)}, m_{\mathscr{I}(u)}, (T_{a:u})_{a \in \mathscr{I}(u)}, T_u) \\ &= T_u\Big(\hat{q}_u^{\text{nom}} + \sum_{a \in \mathscr{I}(u)} \hat{q}_a\Big) - \Big(\hat{q}_u^{\text{nom}} T_u^{\text{nom}} + \sum_{a \in \mathscr{I}(u)} \hat{q}_a T_{a:u}\Big), \end{aligned} \tag{10.13}$$

$$0 = c_{u,a}^{\text{prop-temp}}(T_u, T_{a:u}) = T_u - T_{a:u} \quad \text{for all } a \in \mathscr{O}(u). \tag{10.14}$$

Note that all mixing constraints are discontinuous since $\mathscr{I}(u)$ and $\mathscr{O}(u)$ depend on the flow directions. To obtain a smooth model, we fix all flow directions according to the candidate solution of the decision approach (by setting 0 as lower or upper bound on q_a). Letting x_a^{base} denote the common variables of all arc types (see the immediately following Section 10.1.3), the full set of constraints of node u becomes

$$0 = c_u(x_u, x_{\delta(u)}^{\text{base}}) = \begin{pmatrix} c_u^{\text{flow}}(q_{\delta(u)}) \\ c_u^{\text{mix}}(q_{\mathscr{I}(u)}, X_{\mathscr{I}(u)}, X_u) \\ (c_{u,a}^{\text{prop}}(X_u, X_a))_{a \in \mathscr{O}(u)} \\ c_u^{\text{mix-temp}}(q_{\mathscr{I}(u)}, m_{\mathscr{I}(u)}, (T_{a:u})_{a \in \mathscr{I}(u)}, T_u) \\ (c_{u,a}^{\text{prop-temp}}(T_u, T_{a:u}))_{a \in \mathscr{O}(u)} \end{pmatrix}$$

and the vector of variables at node u reads

$$x_u = (p_u, T_u, X_u).$$

10.1.3 ▪ Arcs

Every arc $a = (u,v) \in A$ has a mass flow variable q_a, variables for the gas temperatures at tail and head, $T_{a:u}$ and $T_{a:v}$, respectively, and a quality parameter vector X_a, yielding a common basic variable vector of all arc models:

$$x_a^{\text{base}} = (q_a, T_{a:u}, T_{a:v}, X_a).$$

The specific constraints of every arc type will be described separately below; in some cases they involve additional variables.

10.1.4 ▪ Pipes

Pipes outnumber all other elements of a gas network. They are the most essential elements and the only ones with a nonnegligible length, which transport the gas over large

10.1. Component models

distances. Due to friction at the inner wall and due to gravity (if the pipe is inclined), pressure and temperature change when gas flows through a pipe. In addition, there is heat exchange with the surrounding. As already explained in Section 2.3.1, this leads to highly complex gas dynamics described by a system of hyperbolic, partial differential equations (PDEs).

We consider a cylindrical pipe with diameter D, cross-sectional area $A = D^2\pi/4$, and slope $s \in [-1,+1]$ (the tangent of the inclination angle), so that the transient gas dynamics can be expressed by the one-dimensional PDE system of Euler equations (2.12)–(2.14) as derived in Feistauer (1993) and Lurie (2008):

$$\frac{\partial \rho}{\partial t} + \frac{1}{A}\frac{\partial q}{\partial x} = 0, \tag{10.15a}$$

$$\frac{1}{A}\frac{\partial q}{\partial t} + \frac{\partial p}{\partial x} + \frac{1}{A}\frac{\partial (qv)}{\partial x} + g\rho s + \lambda(q)\frac{|v|v}{2D}\rho = 0, \tag{10.15b}$$

$$A\rho c_p \left(\frac{\partial T}{\partial t} + v\frac{\partial T}{\partial x}\right) - A\left(1 + \frac{T}{z}\frac{\partial z}{\partial T}\right)\frac{\partial p}{\partial t} \\ - Av\frac{T}{z}\frac{\partial z}{\partial T}\frac{\partial p}{\partial x} + A\rho v g s + \pi D c_{\mathrm{HT}}(T - T_{\mathrm{soil}}) = 0, \tag{10.15c}$$

Here x and t denote the spatial coordinate and time coordinate, respectively, and g, c_{HT}, and T_{soil} stand for the gravitational acceleration, the heat transfer coefficient, and the soil temperature, respectively. The friction factor $\lambda(q)$ in (10.15b) models frictional forces at the inner pipe walls; see Section 2.3.1. Given a specific pipe $a \in A_{\mathrm{pi}}$, the gas velocity v is related to the mass flow q, density ρ, and the cross-sectional pipe area A_a via (2.11) yielding the constraint

$$0 = c_a^{\mathrm{vel\text{-}flow}}(q,v,\rho) = A_a \rho v - q. \tag{10.16}$$

Since we consider the stationary case, all partial derivatives with respect to time vanish, and with (10.16) the above PDE system reduces to a semi-implicit system of ordinary differential equations (ODEs) for q, p, and T; ρ and p are related by (10.9). Applying some minor transformations (e.g., multiplying with ρ and using the chain-rule) yields

$$\frac{\partial q}{\partial x} = 0, \tag{10.17a}$$

$$\rho\frac{\partial p}{\partial x} - \frac{q^2}{A^2}\frac{\partial \rho}{\partial x}\frac{1}{\rho} + g\rho^2 s + \lambda(q)\frac{|q|q}{2A^2 D} = 0, \tag{10.17b}$$

$$qc_p\frac{\partial T}{\partial x} - \frac{qT}{\rho z}\frac{\partial z}{\partial T}\frac{\partial p}{\partial x} + qgs + \pi D c_{\mathrm{HT}}(T - T_{\mathrm{soil}}) = 0. \tag{10.17c}$$

The continuity equation (10.17a) readily implies that q is constant along the pipe, which justifies the use of a single mass flow variable q_a for every pipe a. The momentum equation (10.17b) models the pressure gradient in terms of impact pressure, gravitational forces and frictional forces, and the energy equation (10.17c) models the temperature gradient in terms of pressure changes, gravitational effects, and heat exchange with the surrounding soil. Figure 10.1 illustrates possible profiles of pressure and temperature along a pipe and the influence of the pipe diameter.

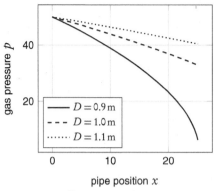 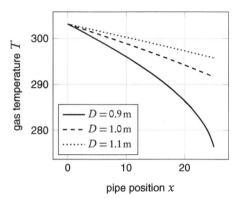

(a) Pressure profile; pressure in bar vs. pipe position in km.

(b) Temperature profile; temperature in K vs. pipe position in km.

Figure 10.1. *Pressure and temperature profiles of three horizontal pipes with different diameters ($L = 25$ km, $k = 0.06$ mm, $q = 500$ kg/s). (Source: Schmidt, Steinbach, and Willert (2014).)*

10.1.4.1 ▪ Smooth approximation of the friction term

Depending on which model of $\lambda(q)$ is chosen, the friction term $\lambda(q)|q|q$ in the momentum equation (10.17b) will be nonsmooth: the product $|q|q$ creates a second-order discontinuity at $q = 0$ if $\lambda(0) \neq 0$, and the piecewise reference model of $\lambda(q)$ defined by the formulas of Prandtl–Colebrook (2.17) and Hagen–Poiseuille (2.16) has a jump discontinuity at the transition between laminar and turbulent flow; see Chapter 2 for details.

To avoid discontinuities, we replace the entire friction term $\lambda(q)|q|q$ by a globally smooth approximation $\phi(q)$. This approximation has originally been developed for water networks by Burgschweiger, Gnädig, and Steinbach (2009) and Burgschweiger, Gnädig, and Steinbach (2009), and applies as well to gas networks (see Schmidt, Steinbach, and Willert (2014)). The corresponding constraint for a specific pipe $a \in A_{\mathrm{pi}}$ reads

$$0 = c_a^{\mathrm{friction}}(\phi_a, q_a) = \phi_a - \tilde{\lambda}_a \left(\sqrt{q_a^2 + e_a^2} + b_a + \frac{c_a}{\sqrt{q_a^2 + d_a^2}} \right) q_a. \tag{10.18}$$

As illustrated in Figure 10.2, our approximation (10.18) is asymptotically correct for $|q| \to \infty$ if the following parameters are used (see Burgschweiger, Gnädig, and Steinbach (2009)):

$$\tilde{\lambda}_a = (2 \log_{10} \beta_a)^{-2}, \qquad b_a = 2\delta_a, \qquad c_a = (\ln \beta_a + 1)\delta_a^2 - \frac{e_a^2}{2},$$

$$\alpha_a = \frac{2.51 A_a \eta}{D_a}, \qquad \beta_a = \frac{k_a}{3.71 D_a}, \qquad \delta_a = \frac{2\alpha_a}{\beta_a \ln 10}.$$

Here, η and k_a denote the dynamic viscosity of gas and the roughness of the pipe (see Section 2.3.1). The two smoothing parameters $d_a, e_a > 0$ remain to be chosen by the modeler (see Schmidt, Steinbach, and Willert (2014); Burgschweiger, Gnädig, and Steinbach (2009)).

10.1. Component models

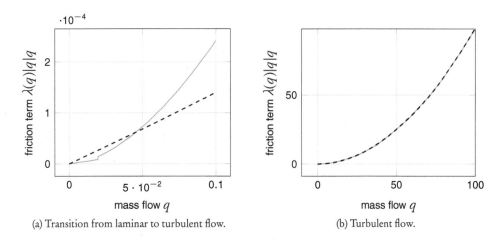

Figure 10.2. *Friction term $\lambda(q)|q|q$ according to Hagen–Poiseuille and Prandtl–Colebrook (——) and smooth approximation $\phi(q)$ with $d = e = 2.2$ (- - -) vs. mass flow q in kg/s. (Source: Schmidt, Steinbach, and Willert (2014).)*

10.1.4.2 ▪ ODE discretization

After obtaining a constant mass flow q_a from the continuity equation and smoothing the friction term, stationary gas dynamics are expressed by a spatial system of ODEs: a smooth version of the momentum equation (10.17b) and the unmodified smooth energy equation (10.17c). This system of two equations for the three variables (ρ, p, T) is completed by the equation of state (10.9); see the explanation for the general case in Section 2.3.1.

The standard procedure for converting ODE constraints to finite-dimensional NLP constraints is a discretization. Depending on the chosen grid, this results in a highly accurate model with a large number of nonlinear constraints. For simplicity, both in terms of implementation and presentation, we use an a priori discretization rather than an adaptive one: given a pipe $a \in A_{\text{pi}}$, we choose some fixed grid, $0 = x_{a,0} < \cdots < x_{a,d} = L_a$. The continuous variables in (10.17b) and (10.17c) are then evaluated at the spatial grid points. To this end, we define the following abbreviations for $k = 0, \ldots, d$:

$$
\begin{aligned}
& p_{a,k} = p(x_{a,k}), & & z_{a,k} = z(p_{a,k}, T_{a,k}, p_{c,a}, T_{c,a}) & & \text{(see (10.6))}, \\
& T_{a,k} = T(x_{a,k}), & & c_{p,a,k} = c_p(p_{a,k}, T_{a,k}, X_a) & & \text{(see (10.7))}, \\
& \rho_{a,k} = \rho(x_{a,k}), & & z_{T,a,k} = \frac{\partial z}{\partial T}(p_{a,k}, T_{a,k}, p_{c,a}, T_{c,a}).
\end{aligned}
$$

Note that pressures $p_{a,0}$ and $p_{a,d}$ are identified with the pressure variables at the tail and head, p_u and p_v, respectively. In analogy, the temperature variables T_0 and T_d are identified with $T_{a:u}$ and $T_{a:v}$. Also for simplicity, we illustrate the ODE discretization with a two-point finite difference approach (backwards in space). Denoting step sizes by $\Delta x_{a,k} = x_{a,k} - x_{a,k-1}, k = 1, \ldots, d$, this yields the discrete pressure gradient

$$
p'(x_{a,k}) \approx \frac{p(x_{a,k}) - p(x_{a,k-1})}{\Delta x_{a,k}} =: \frac{\Delta p_{a,k}}{\Delta x_{a,k}}, \quad k = 1, \ldots, d,
$$

and similar gradient approximations for temperature and density. The discretized ODE constraints can then be written (for $k = 1, \ldots, d$) as

$$0 = c_a^{\text{mom-discr}}(q_a, p_{a,k}, p_{a,k-1}, \rho_{a,k}, \rho_{a,k-1}, \phi_a)$$
$$= \rho_{a,k} \frac{\Delta p_{a,k}}{\Delta x_{a,k}} - \frac{q_a^2}{A_a^2} \frac{\Delta \rho_{a,k}}{\Delta x_{a,k}} \frac{1}{\rho_{a,k}} + g\rho_{a,k}^2 s_a + \frac{\phi_a}{2A_a^2 D_a}, \tag{10.19a}$$

$$0 = c_a^{\text{ener-discr}}(q_a, p_{a,k}, p_{a,k-1}, T_{a,k}, T_{a,k-1}, \rho_{a,k}, z_{a,k}, c_{p,a,k}, p_{c,a}, T_{c,a})$$
$$= q_a c_{p,a,k} \frac{\Delta T_{a,k}}{\Delta x_{a,k}} - \frac{q_a T_{a,k}}{\rho_{a,k} z_{a,k}} z_{T,a,k} \frac{\Delta p_{a,k}}{\Delta x_{a,k}} + q_a g s_a$$
$$+ \pi D_a c_{\text{HT},a}(T_{a,k} - T_{\text{soil},a}). \tag{10.19b}$$

The partial derivative $z_{T,a,k}$ is substituted directly in the constraint instead of introducing an additional variable and an according constraint. Next, we have to add the constraints for the equation of state, compressibility, and heat capacity in every grid point, obtaining (for $k = 1, \ldots, d$)

$$0 = c_{a,k}^{\text{dyn}}(x_a^{\text{base}}, x_{a,k}^{\text{dyn}}, x_{a,k-1}^{\text{dyn}}, \phi_a, x_u, x_v)$$
$$= \begin{pmatrix} c_a^{\text{mom-discr}}(q_a, p_{a,k}, p_{a,k-1}, \rho_{a,k}, \rho_{a,k-1}, \phi_a) \\ c_a^{\text{ener-discr}}(q_a, p_{a,k}, p_{a,k-1}, T_{a,k}, T_{a,k-1}, \rho_{a,k}, z_{a,k}, c_{p,a,k}, p_{c,a}, T_{c,a}) \\ c^{\text{eos}}(p_{a,k}, T_{a,k}, \rho_{a,k}, m_a, z_{a,k}) \\ c^{\text{compr}}(z_{a,k}, p_{a,k}, T_{a,k}, p_{c,a}, T_{c,a}) \\ c^{\text{heat-cap}}(p_{a,k}, T_{a,k}, X_a, x_{a,k}^{\text{heat-cap}}) \end{pmatrix}, \tag{10.20}$$

where the required additional dynamic variables are defined as

$$x_{a,k}^{\text{dyn}} = (p_{a,k}, T_{a,k}, \rho_{a,k}, z_{a,k}, x_{a,k}^{\text{heat-cap}}), \qquad k = 1, \ldots, d-1,$$
$$x_{a,k}^{\text{dyn}} = (\rho_{a,k}, z_{a,k}, x_{a,k}^{\text{heat-cap}}), \qquad k \in \{0, d\}.$$

Together with the smooth approximation of the friction term, the complete model of the discretized dynamic system finally reads

$$0 = c_a^{\text{dyn}}(x_a^{\text{base}}, x_a^{\text{dyn}}, x_u, x_v) = \begin{pmatrix} c_a^{\text{friction}}(\phi_a, q_a) \\ (c_{a,k}^{\text{dyn}}(x_a^{\text{base}}, x_{a,k}^{\text{dyn}}, x_{a,k-1}^{\text{dyn}}, \phi_a, x_u, x_v))_{k=1}^d \end{pmatrix},$$
$$x_a^{\text{dyn}} = (\phi_a, (x_{a,k}^{\text{dyn}})_{k=0}^d).$$

Finally, we remark that one may also apply suitable higher order discretizations instead of the presented two-point finite differences scheme.

10.1.4.3 • ODE approximation

A second possible approach for converting ODE constraints to finite-dimensional NLP constraints is the direct approximation of ODE solutions, which leads to fewer nonlinear equations with a reduced degree of accuracy as compared to a discretization.

In fact, the decision approaches of the preceding chapters drop the energy equation (10.17c) and use approximate solutions of the momentum equation (10.17b). This is an approach with a long tradition in gas engineering, so that suitable approximations are well known and well tested. Specifically, by further neglecting the impact pressure term and assuming a mean compressibility factor $z_{m,a} = z(p_{m,a}, T_{m,a}, p_c, T_c)$ (that can be assumed by using a mean pressure $p_{m,a}$ and a mean temperature $T_{m,a}$), solutions of the momentum equation (10.17b) can be approximated by the equation stated in Lemma 2.1,

$$0 = c_a^{\text{mom-approx}}(p_u, p_v, \phi_a, T_{m,a}, z_{m,a}, m_a)$$
$$= p_v^2 - \left(p_u^2 - \Lambda_a \phi_a \frac{e^{S_a} - 1}{S_a}\right) e^{-S_a}, \tag{10.21}$$

where the roughly quadratic dependence on q_a is hidden in the friction variable, $\phi_a \approx \lambda(q_a)|q_a|q_a$, and Λ_a, S_a are defined in terms of the above mean values:

$$\Lambda_a = \Lambda_a(T_{m,a}, z_{m,a}, m_a) = \frac{L_a}{A_a^2 D_a} \frac{z_{m,a} T_{m,a} R}{m_a},$$
$$S_a = S_a(T_{m,a}, z_{m,a}, m_a) = 2 g L_a s_a \frac{m_a}{z_{m,a} T_{m,a} R}.$$

More details on this approximation can be found in Chapter 2; suitable choices for the approximate mean values $p_{m,a}, T_{m,a}$ will be discussed below.

An approximation replaces the energy equation (10.17c) when an isothermal approach is not desired. It is derived by assuming a constant value for the specific heat capacity c_p and applying a two point finite difference. In addition, mean values for pressure $p_{m,a}$ and $T_{m,a}$ are used to obtain mean values for the compressibility factor $z_{m,a}$ and the density $\rho_{m,a}$, yielding (see Schmidt, Steinbach, and Willert (2014))

$$0 = c_a^{\text{ener-approx}}(q_a, p_{a,\text{in}}, p_{a,\text{out}}, T_{a,\text{in}}, T_{a,\text{out}}, \rho_{m,a}, p_{m,a}, T_{m,a}, z_{m,a}, p_{c,a}, T_{c,a})$$
$$= q_a \left(T_{a,\text{out}} - T_{a,\text{in}} + \frac{g s_a L_a}{c_p}\right) - \frac{z_{T,m,a}}{c_p \rho_{m,a} z_{m,a}} T_{a,\text{out}} q_a (p_{a,\text{out}} - p_{a,\text{in}})$$
$$+ \frac{\pi D_a c_{\text{HT},a} L_a}{c_p} (T_{a,\text{out}} - T_{\text{soil},a}), \tag{10.22}$$

where

$$z_{T,m,a} = \frac{\partial z}{\partial T}(p_{m,a}, T_{m,a}, p_{c,a}, T_{c,a})$$

is substituted directly. Recall the definition of direction-dependent variables: for positive flow we have

$$p_{a,\text{in}} = p_u, \qquad p_{a,\text{out}} = p_v, \qquad T_{a,\text{in}} = T_{a:u}, \qquad T_{a,\text{out}} = T_{a:v},$$

and in case of (strictly) negative flow the definitions read

$$p_{a,\text{in}} = p_v, \qquad p_{a,\text{out}} = p_u, \qquad T_{a,\text{in}} = T_{a:v}, \qquad T_{a,\text{out}} = T_{a:u}.$$

For $\rho_{m,a}$ we need an additional constraint for the equation of state (10.9). Alternatively, we can work with a further simplified model where $\rho_{m,a}$ is replaced by a constant mean value.

Figure 10.3 compares the temperature change according to the ODE (10.19b) and its approximation (10.22). On the left, the temperature profile along a pipe is illustrated and on the right the temperature profile for varying mass flow is presented.

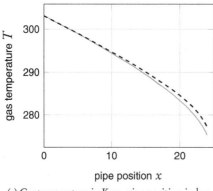
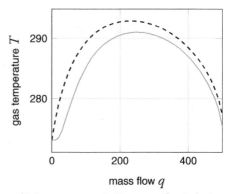

(a) Gas temperature in K vs. pipe position in km. (b) Gas temperature in K vs. mass flow in kg/s.

Figure 10.3. *Gas temperature according to ODE discretization with 40 discretization steps (——) and approximation (- - -) (L = 24 km, D = 1 m, k = 0.1 mm, q = 500 kg/s). (Source: Schmidt, Steinbach, and Willert (2014).)*

The approximating constraints (10.21) and (10.22) make both use of approximate mean pressures $p_{m,a}$ and temperatures $T_{m,a}$. Several possibilities exist to define these values and to incorporate them into an NLP. A simple choice defines the values as constant averages of given variable bounds in

$$p_{m,a} = \frac{1}{2}\big(\max(\underline{p}_u, \underline{p}_v) + \min(\overline{p}_u, \overline{p}_v)\big), \qquad (10.23\text{a})$$

$$T_{m,a} = \frac{1}{2}\big(\max(\underline{T}_{a:u}, \underline{T}_{a:v}) + \min(\overline{T}_{a:u}, \overline{T}_{a:v})\big). \qquad (10.23\text{b})$$

In case of globally identical temperature bounds, (10.23b) can be simplified to

$$T_{m,a} = \frac{1}{2}(\underline{T} + \overline{T}).$$

A more sophisticated choice adds a constraint to include the mean values as variables that depend on pressures and temperatures at the pipe; see Menon (2005) and Eq. (2.28) in Section 2.3.1.2:

$$0 = c_a^{\text{mean}}(p_{m,a}, T_{m,a}, p_u, p_v, T_{a:u}, T_{a:v}) = \begin{pmatrix} p_{m,a} - \frac{2}{3}\big(p_u + p_v - \frac{p_u p_v}{p_u + p_v}\big) \\ T_{m,a} - \frac{2}{3}\big(T_{a:u} + T_{a:v} - \frac{T_{a:u} T_{a:v}}{T_{a:u} + T_{a:v}}\big) \end{pmatrix}. \qquad (10.24)$$

Since pressure and temperature profiles along a pipe have similar trends—see Figure 10.1—the same formula for the mean value is used. Alternatively, the arithmetic mean can be used for temperature. The full approximation model of a pipe is thus expressed by the constraints

$$0 = c_a^{\text{dyn}}(x_a^{\text{base}}, x_a^{\text{dyn}}, x_u, x_v)$$

$$= \begin{pmatrix} c_a^{\text{friction}}(\phi_a, q_a) \\ c_a^{\text{mom-approx}}(p_u, p_v, \phi_a, T_{m,a}, z_{m,a}, m_a) \\ c_a^{\text{ener-approx}}(q_a, p_{a,\text{in}}, p_{a,\text{out}}, T_{a,\text{in}}, T_{a,\text{out}}, \rho_{m,a}, p_{m,a}, T_{m,a}, z_{m,a}, p_{c,a}, T_{c,a}) \\ c^{\text{eos}}(p_{m,a}, T_{m,a}, \rho_{m,a}, m_a, z_{m,a}) \\ c^{\text{compr}}(z_{m,a}, p_{m,a}, T_{m,a}, p_{c,a}, T_{c,a}) \\ c_a^{\text{mean}}(p_{m,a}, T_{m,a}, p_u, p_v, T_{a:u}, T_{a:v}) \end{pmatrix} \qquad (10.25)$$

10.1. Component models

and variables
$$x_a^{\text{dyn}} = (\phi_a, p_{\text{m},a}, T_{\text{m},a}, z_{\text{m},a}, \rho_{\text{m},a}).$$

We remark that constraint (10.25) has the same name as constraint (10.20) since both can be used as alternatives in the complete NLP model. If (10.23) is chosen for the mean values, the model simplifies: the constraint c_a^{mean} is dropped, and $p_{\text{m},a}$ as well as $T_{\text{m},a}$ become constants.

10.1.4.4 ▪ Velocity constraint

The pressure change along a pipe induces changes of density and temperature, and also of gas velocity. Since high velocities can generate vibrations of the pipe (which in turn may lead to substantial noise emission or, worse, bursting of pipes), our model includes velocity limits for the gas flow. By (10.16), the velocity is related to density and mass flow, and (10.17) leads to a monotone change of pressure and temperature (with respect to the spatial coordinate). This results in a monotone change of density along the pipe due to the equation of state, and therefore the gas velocity is also monotone. It is thus sufficient to control the velocity at the end points of pipe $a = (u, v)$. The required additional variables, $x_a^{\text{vel}} = (v_{a:u}, v_{a:v}, \rho_{a:u}, \rho_{a:v}, z_{a:u}, z_{a:v})$, are determined by the constraints

$$0 = c_a^{\text{vel-1}}(x_a^{\text{base}}, x_a^{\text{vel}}, x_u, x_v) = \begin{pmatrix} c_a^{\text{vel-flow}}(q_a, v_{a:u}, \rho_{a:u}) \\ c_a^{\text{vel-flow}}(q_a, v_{a:v}, \rho_{a:v}) \\ c^{\text{eos}}(p_u, T_{a:u}, \rho_{a:u}, m_a, z_{a:u}) \\ c^{\text{eos}}(p_v, T_{a:v}, \rho_{a:v}, m_a, z_{a:v}) \\ c^{\text{compr}}(z_{a:u}, p_u, T_{a:u}, p_{c,a}, T_{c,a}) \\ c^{\text{compr}}(z_{a:v}, p_v, T_{a:v}, p_{c,a}, T_{c,a}) \end{pmatrix}. \quad (10.26)$$

If an ODE discretization scheme is used, the densities $\rho_{a:u}$ and $\rho_{a:v}$ coincide with the densities at the grid endpoints $\rho_{a,0}$ and $\rho_{a,d}$. This also holds for the compressibility factors, hence the two equations of state and the two constraints for the compressibility factors are already part of the model, and the velocity constraints reduce to

$$0 = c_a^{\text{vel-2}}(x_a^{\text{base}}, x_a^{\text{vel}}, x_a^{\text{dyn}}) = \begin{pmatrix} c_a^{\text{vel-flow}}(q_a, v_{a:u}, \rho_{a,0}) \\ c_a^{\text{vel-flow}}(q_a, v_{a:v}, \rho_{a,d}) \end{pmatrix}, \qquad x_a^{\text{vel}} = (v_{a:u}, v_{a:v}).$$

Any velocity limits are thus represented by bounds on the velocity variables x_a^{vel}.

10.1.4.5 ▪ Complete pipe model

Combining the constraints c_a^{dyn} and c_a^{vel}, we obtain the complete pipe model

$$0 = c_a(x_a, x_u, x_v) = \begin{pmatrix} c_a^{\text{dyn}}(x_a^{\text{base}}, x_a^{\text{dyn}}, x_u, x_v) \\ c_a^{\text{vel-1}}(x_a^{\text{base}}, x_a^{\text{vel}}, x_u, x_v) \end{pmatrix}$$

or

$$0 = c_a(x_a, x_u, x_v) = \begin{pmatrix} c_a^{\text{dyn}}(x_a^{\text{base}}, x_a^{\text{dyn}}, x_u, x_v) \\ c_a^{\text{vel-2}}(x_a^{\text{base}}, x_a^{\text{vel}}, x_a^{\text{dyn}}) \end{pmatrix},$$

with the variable vector $x_a = (x_a^{\text{base}}, x_a^{\text{dyn}}, x_a^{\text{vel}})$, depending on the required constraints for the gas velocity.

10.1.4.6 ▪ Comparison of model choices

As we have seen, several physical aspects allow for multiple modeling choices. The possible combinations lead to a broad variety of variants of the pipe constraints c_a, featuring different advantages and disadvantages.

A modestly accurate model combines the approximation (10.25) with the smooth friction model (10.18), the AGA compressibility model (10.4), and constant values of mean pressure and temperature (10.23), but assumes isothermal gas flow and thus drops the approximation of the energy equation from the model. Of all the presented alternative models for the relevant physical phenomena, this combination features the largest degree of similarity to the decision approaches described in Chapters 6–9, and a candidate solution generated by any decision approach will typically be a good initial point for NLP solution (see Section 5.6). The drawback is that this combination is in fact one of the most inaccurate choices of pipe models. Nevertheless, our experiences show that it proves to be reasonably accurate for several practical situations.

A very accurate pipe model combines a discretization of the Euler equations (10.20) with Papay's compressibility model (10.5). This leads to larger discrepancies with the models on which the decision approaches are based and tends to reduce the quality of candidate solutions as initial points for NLP solution. Moreover, it also tends to increase the likelihood that a feasible flow situation for the NLP will require different discrete decisions, and hence the probability that candidate solutions of the decision approaches will be rejected.

For a more detailed elaboration of these aspects we refer to Chapter 11.

10.1.5 ▪ Resistors

A resistor $a = (u, v) \in A_{\mathrm{rs}}$ is a fictitious network element that is used to account for pressure losses from components without an exact description in our model, such as measurement devices, narrow bends of pipes, filters, or internal station piping; see Section 2.3.2. The simpler of the two empirical models, which assumes a constant pressure loss $\xi_a > 0$ in the direction of flow, has a jump discontinuity in our context; see (2.32). The more elaborate resistor model of Darcy–Weisbach type formulates the pressure drop with a fictitious diameter D_a and a resistance coefficient $\zeta_a > 0$; see Section 2.3.2:

$$p_u - p_v = \frac{8\zeta_a}{\pi^2 D_a^4} \frac{q_a |q_a|}{\rho_{a,\mathrm{in}}}. \tag{10.27}$$

Here, (10.27) is obtained from (2.30) by replacing the velocity by the mass flow. Rather than a jump discontinuity, this model has only a second-order discontinuity at $q_a = 0$. Both nonsmooth aspects are handled by fixing the direction of flow according to the given candidate solution. In summary, the pressure loss of a resistor is given by the constraint

$$0 = \begin{cases} c_a^{\mathrm{p\text{-}loss}}(q_a, p_u, p_v) = p_u - p_v - \mathrm{sgn}(q_a)\xi_a & \text{or} \\ c_a^{\mathrm{p\text{-}loss}}(q_a, p_u, p_v, \rho_{a,\mathrm{in}}) = p_u - p_v - \dfrac{8\zeta_a}{\pi^2 D_a^4} \dfrac{q_a|q_a|}{\rho_{a,\mathrm{in}}}. \end{cases}$$

The inflow density $\rho_{a,\mathrm{in}}$ depends on the flow direction and correlates to the pressure, temperature, and compressibility factor either at the tail or at the head of the arc; see (10.2).

10.1. Component models

As always, the pressure loss induces a corresponding temperature decrease due to the Joule–Thomson effect (10.8). Thus, the full resistor model of a linear resistor reads

$$0 = c_a(x_u, x_v, x_a) = \begin{pmatrix} c_a^{\text{p-loss}}(q_a, p_u, p_v) \\ c_a^{\text{heat-cap}}(p_v, T_{a:v}, X_a, x_{a,\text{out}}^{\text{heat-cap}}) \\ c^{\text{jt}}(p_u, p_v, T_{a:u}, T_{a:v}, X_a, \mu_{\text{JT},a}, c_{p,a,\text{out}}) \end{pmatrix}, \quad (10.28)$$

$$x_a = (x_a^{\text{base}}, x_{a,\text{out}}^{\text{heat-cap}}, \mu_{\text{JT},a}).$$

The nonlinear model of a resistor is

$$0 = c_a(x_u, x_v, x_a) = \begin{pmatrix} c_a^{\text{p-loss}}(q_a, p_u, p_v, \rho_{a,\text{in}}) \\ c_a^{\text{heat-cap}}(p_v, T_{a:v}, X_a, x_{a,\text{out}}^{\text{heat-cap}}) \\ c^{\text{jt}}(p_u, p_v, T_{a:u}, T_{a:v}, X_a, \mu_{\text{JT},a}, c_{p,a,\text{out}}) \\ c^{\text{eos}}(p_{a,\text{in}}, T_{a,\text{in}}, \rho_{a,\text{in}}, m_a, z_{a,\text{in}}) \\ c^{\text{compr}}(z_{a,\text{in}}, p_{a,\text{in}}, T_{a,\text{in}}, p_{c,a}, T_{c,a}) \end{pmatrix}, \quad (10.29)$$

$$x_a = (x_a^{\text{base}}, x_{a,\text{out}}^{\text{heat-cap}}, \mu_{\text{JT},a}, \rho_{a,\text{in}}, z_{a,\text{in}}).$$

If the gas composition is uniform and the gas temperature is approximated by a mean value, i.e., no mixing equations are required, the pressure loss at resistors is the only model aspect left that depends on the flow direction. If a smooth approximation is applied, flow directions do not have to be fixed at any arc. See Section 9.1.3 for suitable nonlinear smoothings.

10.1.6 ▪ Valves

A valve $a = (u, v) \in A_{\text{va}}$ is an active element which can be *open* or *closed*. Valves are used to route the gas flow through the network, to decouple subnetworks, or to shut down sections of the network for maintenance. The discrete valve status is always fixed in our NLP context. An open valve has no impact on the flow (see Section 2.3.3) yielding identical inflow and outflow pressure and temperature,

$$0 = c_a(x_u, x_v, x_a) = \begin{pmatrix} p_u - p_v \\ T_{a,\text{in}} - T_{a,\text{out}} \end{pmatrix}, \quad x_a = x_a^{\text{base}}. \quad (10.30)$$

A closed valve simply acts like an absent arc: there is no flow through the valve, and the pressure and temperature of the connected nodes are completely decoupled,

$$0 = c_a(x_u, x_v, x_a) = q_a, \quad x_a = x_a^{\text{base}}. \quad (10.31)$$

10.1.7 ▪ Short cuts

Short cuts $a = (u, v) \in A_{\text{sc}}$ are again fictitious network elements introduced exclusively for modeling purposes, such as splitting a single physical exit with several customers into several virtual exits with just one customer each. A short cut does not impair the flow in any way, having the same model as an open valve; see (10.30).

10.1.8 ▪ Control valve stations with remote access

A control valve station with remote access $a = (u, v) \in A_{\text{cv}}$ is used to reduce the pressure in a controlled way. This is necessary in cases where customers or downstream networks

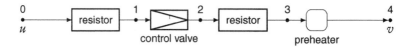

Figure 10.4. *Active control valve station (schematic overview). (Source: Schmidt, Steinbach, and Willert (2014).)*

require lower pressure levels than the usual pressure in transport pipelines. The remote access offers the operator a direct control of the pressure decrease. Note that control valve stations represent subnetworks that consist of several basic elements; see Section 2.3.4 and Section 2.4.1. The concrete layout depends on the considered network model. In our case, we use Figure 10.4 (but see also Figure 2.7, which models the bypass explicitly). In any case, control valve stations have a single inlet and a single outlet, so we consider entire stations as arcs in the NLP model.

Control valve stations possess three discrete operation modes: *active*, *bypass*, and *closed*. A *closed* control valve station acts like a closed valve. It blocks the gas flow and decouples the adjacent nodes; see (10.31). A control valve station in *bypass* mode lets the gas flow through the station in arbitrary direction and has no impact on pressure and temperature. It is modeled like an open valve (10.30).

The remainder of this section describes an *active* control valve station, which reduces the inflow pressure by a controllable amount, $\Delta_a \in [\underline{\Delta}_a, \overline{\Delta}_a]$. The pressure reduction works always in the positive direction of flow, i.e., from u to v; negative flow can only occur in bypass mode.

Additional devices like station piping or measurement devices generate further (uncontrolled) pressure losses that are accounted for with inlet and outlet resistors.

Because of the Joule–Thomson effect (10.8), the pressure reduction is always accompanied by a temperature decrease. Large pressure reductions may cause excessive temperature losses, which can lead to the undesirable process of gas hydrate formation. To prevent this, control valve stations contain a *gas preheater* that keeps the gas temperature above a given threshold value $\underline{T}_{a:v}$.

The control valve station is modeled as a subgraph; see Figure 10.4. When labeling the control valve itself as subarc $(1,2)$, the inlet resistor is subarc $(0,1)$, the outlet resistor is subarc $(2,3)$, and the preheater is subarc $(3,4)$. A complete active control valve station thus consists of four subarcs, where subnode 0 is identified with u and subnode 4 with v. Associated with every inner subnode are additional pressure, temperature, and heat capacity variables: $x_{a,i} = (p_{a,i}, T_{a,i}, x_{a,i}^{\text{heat-cap}})$, $i = 1, 2, 3$.

If the inlet resistor $(0,1) = (u,1)$ and outlet resistor $(2,3)$ are linear resistors, they are modeled based on (10.28),

$$0 = c_{a,(i,i+1)}(x_{a,i}, x_{a,i+1}, x_a^{\text{base}}, x_{a,(i,i+1)})$$
$$= \begin{pmatrix} c_{a,(i,i+1)}^{\text{p-loss}}(q_a, p_{a,i}, p_{a,i+1}) \\ c_{a,(i,i+1)}^{\text{heat-cap}}(p_{a,i+1}, T_{a,i+1}, X_a, x_{a,i+1}^{\text{heat-cap}}) \\ c_{a,(i,i+1)}^{\text{jt}}(p_{a,i}, p_{a,i+1}, T_{a,i}, T_{a,i+1}, X_a, \mu_{\text{JT},a,(i,i+1)}, c_{p,a,i+1}) \end{pmatrix}, \quad i = 0, 2, \quad (10.32)$$

with technical adjustments: The variable vector x_a of the model (10.28) is split into x_a^{base} and additional local variables $x_{a,(i,i+1)} = \mu_{\text{JT},a,(i,i+1)}$, $i = 0, 2$, since the variables $x_{a,(i,i+1)}$ are specific for the inlet or outlet resistor, but mass flow and gas composition are associated with the station arc. In case of nonlinear resistors, the inlet and outlet resistors are modeled based on (10.29) with similar adjustments.

10.1. Component models

The control valve $(1,2)$ is modeled by a simple linear constraint,

$$0 = c_{a,(1,2)}^{\text{p-decr}}(p_{a,1}, p_{a,2}, \Delta_a) = p_{a,1} - p_{a,2} - \Delta_a.$$

Completed by the temperature decrease due to the Joule–Thomson effect, the control valve model reads

$$0 = c_{a,(1,2)}(x_{a,1}, x_{a,2}, x_a^{\text{base}}, x_{a,(1,2)})$$

$$= \begin{pmatrix} c_{a,(1,2)}^{\text{p-decr}}(p_{a,1}, p_{a,2}, \Delta_a) \\ c_{a,(1,2)}^{\text{heat-cap}}(p_{a,2}, T_{a,2}, X_a, x_{a,2}^{\text{heat-cap}}) \\ c_{a,(1,2)}^{\text{jt}}(p_{a,1}, p_{a,2}, T_{a,1}, T_{a,2}, X_a, \mu_{\text{JT},a,(1,2)}, c_{p,a,2}) \end{pmatrix},$$

$$x_{a,(1,2)} = (\Delta_a, \mu_{\text{JT},a,(1,2)}).$$

The gas preheater is actually a feedback controller that measures the gas temperature at the station outlet to keep it above the threshold temperature $\underline{T}_{a:v}$ by preheating the gas at the station inlet, i.e., before pressure reduction. Since none of the decision approaches described in Chapters 6 through 9 consider a preheater, the activity status is not known a priori. Instead of modeling the feedback control, we obtain a simple model by considering a heating of the outlet gas,

$$T_{a:v} = \max(\underline{T}_{a:v}, T_{a,3}). \tag{10.33}$$

Since the maximum function is nonsmooth, we actually use the following smooth approximation of (10.33),

$$0 = c_{a,(3,4)}(p_{a,3}, p_v, T_{a,3}, T_{a:v}) = \begin{pmatrix} \frac{p_{a,3} - p_v}{T_{a:v} - \underline{T}_{a:v} - \frac{1}{2}(\sqrt{\Delta T_a^2 + \varepsilon} + \Delta T_a)} \end{pmatrix},$$

where $\Delta T_a = T_{a,3} - \underline{T}_{a:v}$ and $\varepsilon > 0$ is a suitable smoothing parameter. If the resistors of the station are linear, this model is logically equivalent to a feedback control model, since the temperature does not influence any other quantity. However, the pressure loss at nonlinear resistors depends on the inflow density at the resistor, which in turn depends on the inflow temperature. A higher inflow temperature results in a smaller density and thus a larger pressure loss at the resistor. The larger pressure loss at the resistor in a feedback control model is compensated by the control valve in the simple model, as long as the maximum pressure reduction of the control valve is not reached.

The complete model of an active control valve station with remote access then reads

$$0 = c_a(x_u, x_v, x_a) = \begin{pmatrix} c_{a,(0,1)}(x_u, x_{a,1}, x_a^{\text{base}}, x_{a,(0,1)}) \\ c_{a,(1,2)}(x_{a,1}, x_{a,2}, x_a^{\text{base}}, x_{a,(1,2)}) \\ c_{a,(2,3)}(x_{a,2}, x_{a,3}, x_a^{\text{base}}, x_{a,(2,3)}) \\ c_{a,(3,4)}(x_{a,3}, x_v, x_a^{\text{base}}) \end{pmatrix}, \tag{10.34}$$

$$x_a = (x_a^{\text{base}}, x_{a,1}, x_{a,2}, x_{a,3}, x_{a,(0,1)}, x_{a,(1,2)}, x_{a,(2,3)}).$$

10.1.9 ▪ Control valve stations without remote access

Some control valve stations lack the remote access featured by the previously described control valves. Instead of controlling the pressure loss directly, a threshold pressure value p^{set} is preset in this case. Since changing this preset pressure requires manual adjustments on-site, it is handled as a constant parameter in our model.

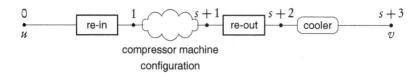

Figure 10.5. *Active compressor group (schematic overview). (Source: Schmidt, Steinbach, and Willert (2014).)*

The state of a control valve station without remote access $a = (u, v)$ depends on the pressures inside the station in relation to the preset pressure. The control valve station is active if the inflow pressure of the control valve lies above the preset pressure. If the pressure level in the downstream network requires a higher pressure at the head v than the threshold value, the control valve closes automatically. If the control valve does not need to reduce the pressure, i.e., the pressure at the head is smaller than the preset pressure threshold and the inflow and outflow pressures of the control valve are equal, the control valve is in bypass mode.

The model of a control valve station without remote access uses the same station graph as in Section 10.1.8. Note that, in contrast to control valves with remote access, the station resistors and the gas preheater are not circumvented in bypass, but cause a pressure reduction and temperature change.

Based on the given discrete decisions, a closed control valve is modeled by setting

$$\underline{p}_v = p_{a,1}, \quad q_a = 0.$$

In case of an active control valve station, the pressure loss at the control valve $\Delta_a = p_1 - p_2$ and the flow q_a must be nonnegative. Furthermore, the pressure at the head p_u is fixed to p_a^{set}. The pressure reduction results in a temperature change due to the Joule–Thomson effect.

The bypass mode is modeled by setting

$$\Delta_a = 0, \quad \overline{p}_v = p_a^{\text{set}}.$$

The resistors and the gas preheater are modeled as described in Section 10.1.8.

10.1.10 • Compressor groups

While control valve stations reduce the gas pressure, compressor groups $a = (u, v) \in A_{\text{cg}}$ increase the gas pressure to compensate for the pressure loss caused by friction. Like control valve stations, compressor groups represent subnetworks with a single inlet and a single outlet. They also operate in the three discrete modes *active*, *bypass*, and *closed*, where the closed mode is again modeled like a closed valve, (10.31), and the bypass mode is again modeled like an open valve, (10.30). Here, the operation mode is always fixed according to a solution candidate from a decision approach.

In the remainder of this section, we concentrate on the active mode, where the direction of flow must be nonnegative, $q_a \geq 0$. The compressor group is modeled as a subgraph consisting of the following elements: several compressor machines that can be operated in a number of arrangements called *configurations*, an inlet resistor and an outlet resistor, and a *gas cooler* that reduces the outlet gas temperature if necessary. A compressor configuration is a serial connection of s subgroups (*stages*) of parallel compressor machines, to be described in more detail below. As illustrated in Figure 10.5, these elements are modeled as subarcs $(i, i+1)$, $i = 0, \ldots, s+2$, where subnode 0 is identified with u and subnode

10.1. Component models

$s+3$ with v. Associated with every inner subnode is a variable vector, $x_{a,i} = (p_{a,i}, T_{a,i})$, $i = 1,\ldots,s+2$.

The models of all other elements described so far represent all physical or technical bounds by variable bounds, e.g., limits of the gas velocity are represented by bounds of the velocity variable at pipes. In contrast, the model of an active compressor group consists of equality constraints $c_{\mathcal{E},a}$ and nontrivial inequality constraints $c_{\mathcal{G},a}$. The inequality constraints will be needed to describe operating ranges of individual compressor machines. (The closed mode and bypass mode do not require inequality constraints.)

In the following we describe the subelements of an active compressor group. Subarc $(0,1)$ represents the inlet resistor,

$$0 = c_{a,(0,1)}(x_{a,0}, x_{a,1}, x_a^{\text{base}}, x_{a,(0,1)}),$$
$$x_{a,(0,1)} = (x_{a,1}^{\text{heat-cap}}, \mu_{\text{JT},a,(0,1)}),$$

and subarc $(s+1, s+2)$ represents the outlet resistor

$$0 = c_{a,(s+1,s+2)}(x_{a,s+1}, x_{a,s+2}, x_a^{\text{base}}, x_{a,(s+1,s+2)}),$$
$$x_{a,(s+1,s+2)} = (x_{a,s+2}^{\text{heat-cap}}, \mu_{\text{JT},a,(s+1,s+2)}).$$

Both resistors are modeled identically to the resistors in a control valve station; see (10.32). The gas cooler represented by the final subarc $(s+2, s+3)$ is modeled like the gas preheater in a control valve station, except that it keeps the outlet gas temperature below the threshold value $\overline{T}_{a:v}$ rather than above it:

$$T_{a:v} = \min(T_{a,s+2}, \overline{T}_{a:v}).$$

Again we approximate the nonsmooth equation with a smooth constraint,

$$0 = c_{a,(s+2,s+3)}(p_{a,s+2}, p_v, T_{a,s+2}, T_{a:v}) = \left(T_{a:v} - \overline{T}_{a:v} + \tfrac{1}{2}\left(\sqrt{\Delta T_a^2 + \varepsilon} - \Delta T_a\right)\right),\frac{p_{a,s+2} - p_v}{},$$

where $\Delta T_a = T_{a,s+2} - \overline{T}_{a:v}$, and $\varepsilon > 0$ is a suitable smoothing parameter.

10.1.10.1 ▪ Configurations

The configurations of a compressor group form a discrete set of possible choices. The particular choice to be considered in the NLP model is determined by the candidate solution of a decision approach. Let us denote the subarcs associated with the given configuration by $(l, l+1)$, $l = 1,\ldots,s_a$, where s_a is the number of serial stages of the configuration. Each stage l may contain m_l compressor machines. For example, the schematic compressor station in Figure 5.1(c) consists of two stages ($s_a = 2$) and two compressor machines each ($m_1 = m_2 = 2$). Finally, individual compressor machines at stage l of compressor group a will be referred to by $a(l, k)$, $k = 1,\ldots,m_l$, and the mass flow through machine $a(l, k)$ will be denoted as $q_{a(l,k)}$. The total mass flow through the compressor group is distributed over the parallel machines at every stage,

$$0 = c_{a,l}^{\text{flow-dist}}(q_a, (q_{a(l,k)})_{k=1}^{m_l}) = q_a - \sum_{k=1}^{m_l} q_{a(l,k)}, \qquad l = 1,\ldots,s_a.$$

Thus, the flows through the individual parallel machines do not have to be equal. In general, the inlet values of p and T at the parallel machines at stage $l+1$ are identical and determined by the outlet values of stage l. The outlet pressures of parallel machines are also identical; we denote their common value by $p_{a,l+1}$. In contrast, the outlet temperatures $T^{\text{out}}_{a(l,k)}$ are usually different, since individual compressor machines may work at different operating points. The outlet temperatures then mix according to (10.13). Since all gas streams have identical composition, the molar masses cancel each other and (10.13) simplifies to

$$0 = c^{\text{mix-temp}}((q_{a(l,k)})_{k=1}^{m_l}, T_{a,l+1}, (T^{\text{out}}_{a(l,k)})_{k=1}^{m_l})$$
$$= T_{a,l+1} \sum_{k=1}^{m_l} q_{a(l,k)} - \sum_{k=1}^{m_l} q_{a(l,k)} T^{\text{out}}_{a(l,k)}$$

for all stages $l = 1, \ldots, s_a$.

10.1.10.2 ▪ Compressor machines

The actual pressure increase at a compressor group is realized by compressor machines. Essentially two kinds of machines are commonly used: turbo compressors and piston compressors. These two types are based on different mechanical principles, leading to different physical and technical properties and to different applications (see Section 2.3.5.1 for details). We first describe basic features and principles that both compressor types have in common.

The energy that is required to compress a certain mass of gas is expressed by the specific change in adiabatic enthalpy H_{ad}, which depends primarily on the compression ratio $p_{\text{out}}/p_{\text{in}}$. For machine $a(l,k)$, it is modeled by the constraint

$$0 = c^{\text{ad-ent}}(H_{\text{ad},a(l,k)}, p_{a,l}, T_{a,l}, p_{a,l+1}, z_{a,l}, m_a, \varkappa_{a(l,k)})$$
$$= H_{\text{ad},a(l,k)} - \frac{z_{a,l} T_{a,l} R}{m_a r_{a(l,k)}} \left(\left(\frac{p_{a,l+1}}{p_{a,l}} \right)^{r_{a(l,k)}} - 1 \right), \quad r_{a(l,k)} = \frac{\varkappa_{a(l,k)} - 1}{\varkappa_{a(l,k)}}. \quad (10.35)$$

The compression is assumed to be *isentropic*, more precisely it is adiabatic and reversible, and $\varkappa_{a(l,k)}$ denotes the *isentropic exponent* (see following subsection).

The power P that is required to increase the pressure depends on the mass flow q through the machine, the specific change in adiabatic enthalpy H_{ad}, and the adiabatic efficiency η_{ad}:

$$0 = c^{\text{power}}(P_{a(l,k)}, q_{a(l,k)}, H_{\text{ad},a(l,k)}, \eta_{\text{ad},a(l,k)}) = P_{a(l,k)} - \frac{q_{a(l,k)} H_{\text{ad},a(l,k)}}{\eta_{\text{ad},a(l,k)}}. \quad (10.36)$$

Isentropic exponent The value $\varkappa_{a(l,k)}$ used above depends on the gas state during the entire compression process. Several approximate models of this *isentropic exponent* exist. In our most detailed model choice, $\varkappa_{a(l,k)}$ is an arithmetic mean value,

$$c^{\text{isen-exp-mean}}(\varkappa_{a(l,k)}, \varkappa^{\text{in}}_{a(l,k)}, \varkappa^{\text{out}}_{a(l,k)}) = \varkappa_{a(l,k)} - \frac{1}{2}(\varkappa^{\text{in}}_{a(l,k)} + \varkappa^{\text{out}}_{a(l,k)}),$$

where $\varkappa^{\text{in}}_{a(l,k)}$ and $\varkappa^{\text{out}}_{a(l,k)}$ are the respective isentropic exponents at the compressor inlet and outlet, respectively. The latter are defined by (see Doering, Schedwill, and Dehli (2012)):

$$0 = c^{\text{isen-exp-def}}(\varkappa, p, T, z, c_p, m, p_c, T_c)$$
$$= \varkappa - \frac{m c_p z}{m c_p Z_p(z, p, T, p_c, T_c) - R Z_T(z, p, T, p_c, T_c)^2}, \quad (10.37)$$

10.1. Component models

where we use the abbreviations

$$Z_p(z, p, T, p_c, T_c) = z - p \frac{\partial z}{\partial p}(p, T, p_c, T_c),$$

$$Z_T(z, p, T, p_c, T_c) = z + T \frac{\partial z}{\partial T}(p, T, p_c, T_c).$$

The complete model of the isentropic exponent thus becomes

$$0 = c^{\text{isen-exp}}(p_{a,l}, T_{a,l}, z_{a,l}, p_{a,l+1}, T^{\text{out}}_{a(l,k)}, z^{\text{out}}_{a(l,k)}, x^{\text{base}}_a, x^{\text{isen-exp}}_{a(l,k)})$$

$$= \begin{pmatrix} c^{\text{isen-exp-mean}}(x_{a(l,k)}, x^{\text{in}}_{a(l,k)}, x^{\text{out}}_{a(l,k)}) \\ c^{\text{isen-exp-def}}(x^{\text{in}}_{a(l,k)}, p_{a,l}, T_{a,l}, z_{a,l}, c_{p,a(l,k),\text{in}}, m_a, p_{c,a}, T_{c,a}) \\ c^{\text{heat-cap}}(p_{a,l}, T_{a,l}, X_a, x^{\text{heat-cap}}_{a(l,k),\text{in}}) \\ c^{\text{isen-exp-def}}(x^{\text{out}}_{a(l,k)}, p_{a,l+1}, T^{\text{out}}_{a(l,k)}, z^{\text{out}}_{a(l,k)}, c_{p,a(l,k),\text{out}}, m_a, p_{c,a}, T_{c,a}) \\ c^{\text{heat-cap}}(p_{a,l+1}, T^{\text{out}}_{a(l,k)}, X_a, x^{\text{heat-cap}}_{a(l,k),\text{out}}) \end{pmatrix} \quad (10.38)$$

with additional variables

$$x^{\text{isen-exp}}_{a(l,k)} = (x_{a(l,k)}, x^{\text{in}}_{a(l,k)}, x^{\text{out}}_{a(l,k)}, x^{\text{heat-cap}}_{a(l,k),\text{in}}, x^{\text{heat-cap}}_{a(l,k),\text{out}}).$$

Here, $z_{a,l}$ and $z^{\text{out}}_{a(l,k)}$ represent the respective compressibility factors at the compressor inlet and outlet, which we need to compute $x^{\text{in}}_{a(l,k)}$ and $x^{\text{out}}_{a(l,k)}$. The values $c_{p,a(l,k),\text{in}}$ and $c_{p,a(l,k),\text{out}}$ are the corresponding specific isobaric heat capacities; these are part of the vectors $x^{\text{heat-cap}}_{a(l,k),\text{in}}$ and $x^{\text{heat-cap}}_{a(l,k),\text{out}}$, respectively.

A suitable simplified model for x is obtained if we replace (10.37) by a linear function of the temperature (see LIWACOM (2004)),

$$0 = c^{\text{isen-exp}}(T_{a,l}, T^{\text{out}}_{a(l,k)}, x^{\text{isen-exp}}_{a(l,k)})$$

$$= \begin{pmatrix} x_{a(l,k)} - 1.296 + 5.8824 \times 10^{-4}(T_{m,a(l,k)} - T_0) \\ T_{m,a(l,k)} - \frac{1}{2}(T_{a,l} + T^{\text{out}}_{a(l,k)}) \end{pmatrix}, \quad (10.39)$$

$$x^{\text{isen-exp}}_{a(l,k)} = (x_{a(l,k)}, T_{m,a(l,k)}).$$

The coarsest model choice for x is a constant value, such as $x_{a(l,k)} = 1.296$, which is obtained from (10.39) with $T_{m,a(l,k)} = T_0$. This model is either formulated by the constraint

$$0 = c^{\text{isen-exp}}(x_{a(l,k)}) = x_{a(l,k)} - 1.296, \quad (10.40)$$

or by replacing the variable x with the constant value in all other constraints.

Temperature increase Due to the Joule–Thomson effect, the pressure increase at a compressor machine $a(l,k)$ causes a corresponding temperature increase for which several empirical models exist (see LIWACOM (2004)): the *isentropic equation model*, the *standard model*, and the *RG1991 model*. All these models can be interpreted as special cases of a fixed-point iteration that is initialized with the temperature increase of an ideal gas (see Schmidt, Steinbach, and Willert (2014)):

$$T^{\text{out},i+1}_{a(l,k)} = T^{\text{out,ideal}}_{a(l,k)} \frac{z(p_{a,l}, T_{a,l})}{z(p_{a,l+1}, T^{\text{out},i}_{a(l,k)})}, \quad T^{\text{out},0}_{a(l,k)} = T^{\text{out,ideal}}_{a(l,k)}, \quad i = 0, 1, 2, \ldots.$$

The models mentioned above differ in the number of iterations and in the choice of the initial iterate $T^{\text{out,ideal}}_{a(l,k)}$. Here we consider only the standard model which uses a single fixed-point iteration, i.e., $T^{\text{out}}_{a(l,k)} = T^{\text{out},1}_{a(l,k)}$, and defines

$$T^{\text{out,ideal}}_{a(l,k)} = T_{a,l}\left(\frac{p_{a,l+1}}{p_{a,l}}\right)^{(\varkappa_{a(l,k)}-1)/(\varkappa_{a(l,k)}\eta_{\text{ad},a(l,k)})}.$$

This results in the constraints

$$0 = c^{\text{temp-inc}}(T^{\text{out,ideal}}_{a(l,k)}, T_{a,l}, p_{a,l}, p_{a,l+1}, \varkappa_{a(l,k)}, \eta_{\text{ad},a(l,k)},$$
$$z^{\text{out,temp-ideal}}_{a(l,k)}, p_{c,a}, T_{c,a}, T^{\text{out}}_{a(l,k)}, z_{a,l})$$
$$= \begin{pmatrix} T^{\text{out,ideal}}_{a(l,k)} - T_{a,l}\left(\frac{p_{a,l+1}}{p_{a,l}}\right)^{(\varkappa_{a(l,k)}-1)/(\varkappa_{a(l,k)}\eta_{\text{ad},a(l,k)})} \\ c^{\text{compr}}(z^{\text{out,temp-ideal}}_{a(l,k)}, p_{a,l+1}, T^{\text{out,ideal}}_{a(l,k)}, p_{c,a}, T_{c,a}) \\ T^{\text{out}}_{a(l,k)} - T^{\text{out,ideal}}_{a(l,k)} z_{a,l}/z^{\text{out,temp-ideal}}_{a(l,k)} \end{pmatrix}. \quad (10.41)$$

Common machine model We are now ready to state the common part of the models for turbo compressors and piston compressors. Both types of compressor machines involve the volumetric flow in their specific descriptions, $Q = q/\rho$. In addition to the constraints described so far, we thus need a further constraint c^{eos} to determine the gas density at the machine inlet. Together with (10.35), (10.36), (10.38) or (10.39), and (10.41) this results in the constraint

$$0 = c^{\text{base}}_{a(l,k)}(x^{\text{base}}_a, x_{a,l}, x_{a,l+1}, x^{\text{base}}_{a(l,k)}, T^{\text{out,ideal}}_{a(l,k)}, z^{\text{out,temp-ideal}}_{a(l,k)})$$
$$= \begin{pmatrix} c^{\text{ad-ent}}(H_{\text{ad},a(l,k)}, p_{a,l}, T_{a,l}, p_{a,l+1}, z_{a,l}, m_a, \varkappa_{a(l,k)}) \\ c^{\text{power}}(P_{a(l,k)}, q_{a(l,k)}, H_{\text{ad},a(l,k)}, \eta_{\text{ad},a(l,k)}) \\ c^{\text{isen-exp}}(p_{a,l}, T_{a,l}, z_{a,l}, p_{a,l+1}, T^{\text{out}}_{a(l,k)}, z^{\text{out}}_{a(l,k)}, x^{\text{base}}_a, x^{\text{isen-exp}}_{a(l,k)}) \\ c^{\text{temp-inc}}(T^{\text{out,ideal}}_{a(l,k)}, T_{a,l}, p_{a,l}, p_{a,l+1}, \varkappa_{a(l,k)}, \eta_{\text{ad},a(l,k)}, \\ z^{\text{out,temp-ideal}}_{a(l,k)}, p_{c,a}, T_{c,a}, T^{\text{out}}_{a(l,k)}, z_{a,l}) \\ c^{\text{compr}}(z_{a,l}, p_{a,l}, T_{a,l}, p_{c,a}, T_{c,a}) \\ c^{\text{eos}}(p_{a,l}, T_{a,l}, \rho^{\text{in}}_{a(l,k)}, m_a, z_{a,l}) \end{pmatrix} \quad (10.42)$$

with variables

$$x^{\text{base}}_{a(l,k)} = (q_{a(l,k)}, P_{a(l,k)}, H_{\text{ad},a(l,k)}, \eta_{\text{ad},a(l,k)}, T^{\text{out}}_{a(l,k)}, z_{a,l}, z^{\text{out}}_{a(l,k)}, \rho^{\text{in}}_{a(l,k)}, x^{\text{isen-exp}}_{a(l,k)}).$$

The complete set of constraints of every compressor machine then consists of (10.42) and type-specific technical restrictions. Many of these additional restrictions are modeled by least-squares fits based on measurements. For the resulting quadratic and biquadratic polynomials we use the notation introduced in Chapter 2; see (2.45) and (2.46). The specific model components of turbo compressors and piston compressors will now be discussed in detail.

Turbo compressor Turbo compressors are designed for large throughput at moderate compression ratios. From a mathematical point of view they are the most complex

10.1. Component models

network elements besides pipes. More details on turbo compressors are given in Section 2.3.5.1.

Every turbo compressor has an *operating range* representing the feasible *working points*. A working point is a pair of volumetric flow rate and specific change in adiabatic enthalpy. The operating range is described by a *characteristic diagram* as in Figure 2.4(a). This characteristic diagram is defined by the isolines of speed (2.47) and isolines of adiabatic efficiency (2.48). In our NLP model, the isolines of compressor $a(l,k)$ are given by the respective constraints

$$0 = c^{\text{speed}}_{a(l,k)}(H_{\text{ad},a(l,k)}, q_{a(l,k)}, \rho^{\text{in}}_{a(l,k)}, n_{a(l,k)})$$
$$= H_{\text{ad},a(l,k)} - \chi\left(\frac{q_{a(l,k)}}{\rho^{\text{in}}_{a(l,k)}}, n_{a(l,k)}; A^{\text{speed}}_{a(l,k)}\right),$$

$$0 = c^{\text{eff}}_{a(l,k)}(\eta_{\text{ad},a(l,k)}, q_{a(l,k)}, \rho^{\text{in}}_{a(l,k)}, n_{a(l,k)})$$
$$= \eta_{\text{ad},a(l,k)} - \chi\left(\frac{q_{a(l,k)}}{\rho^{\text{in}}_{a(l,k)}}, n_{a(l,k)}; A^{\text{eff}}_{a(l,k)}\right),$$

where the volumetric flow rate $Q_{a(l,k)}$ is expressed as $q_{a(l,k)}/\rho^{\text{in}}_{a(l,k)}$ and χ is the biquadratic polynomial from (2.46).

The curved lower and upper boundaries of the operating range are defined by the isolines of the speed limits, $n_{a(l,k)} \in [\underline{n}_{a(l,k)}, \overline{n}_{a(l,k)}]$. To the left, the operating range of the compressor machine is bounded by the *surgeline*,

$$0 \leq c^{\text{surge}}_{a(l,k)}(q_{a(l,k)}, \rho^{\text{in}}_{a(l,k)}, H_{\text{ad},a(l,k)}) = \psi\left(\frac{q_{a(l,k)}}{\rho^{\text{in}}_{a(l,k)}}; \alpha^{\text{surge}}_{a(l,k)}\right) - H_{\text{ad},a(l,k)}.$$

To the right, the operating range is bounded by the *chokeline*,

$$0 \leq c^{\text{choke}}_{a(l,k)}(q_{a(l,k)}, \rho^{\text{in}}_{a(l,k)}, H_{\text{ad},a(l,k)}) = H_{\text{ad},a(l,k)} - \psi\left(\frac{q_{a(l,k)}}{\rho^{\text{in}}_{a(l,k)}}; \alpha^{\text{choke}}_{a(l,k)}\right).$$

For the quadratic polynomial ψ see (2.45). In summary, the complete model of a turbo compressor reads

$$0 = c_{\mathcal{E},a(l,k)}(x^{\text{base}}_a, x_{a,l}, x_{a,l+1}, x_{a(l,k)})$$
$$= \begin{pmatrix} c^{\text{base}}_{a(l,k)}(x^{\text{base}}_a, x_{a,l}, x_{a,l+1}, x^{\text{base}}_{a(l,k)}) \\ c^{\text{speed}}_{a(l,k)}(H_{\text{ad},a(l,k)}, q_{a(l,k)}, \rho^{\text{in}}_{a(l,k)}, n_{a(l,k)}) \\ c^{\text{eff}}_{a(l,k)}(\eta_{\text{ad},a(l,k)}, q_{a(l,k)}, \rho^{\text{in}}_{a(l,k)}, n_{a(l,k)}) \end{pmatrix},$$

$$0 \leq c_{\mathcal{I},a(l,k)}(x_{a,l}, x_{a,l+1}, x_{a(l,k)}) = \begin{pmatrix} c^{\text{surge}}_{a(l,k)}(q_{a(l,k)}, \rho^{\text{in}}_{a(l,k)}, H_{\text{ad},a(l,k)}) \\ c^{\text{choke}}_{a(l,k)}(q_{a(l,k)}, \rho^{\text{in}}_{a(l,k)}, H_{\text{ad},a(l,k)}) \end{pmatrix}.$$

The only additional variable besides $x^{\text{base}}_{a(l,k)}$ is the shaft speed $n_{a(l,k)}$,

$$x_{a(l,k)} = (x^{\text{base}}_{a(l,k)}, n_{a(l,k)}).$$

Piston compressor Piston compressors are designed to generate high compression ratios with moderate throughput. This type of compressor machines appears less frequently than turbo compressors; see again Section 2.3.5.1 for more details.

The characteristic diagram of a piston compressor is defined in the coordinates volumetric flow Q and shaft torque M; it has a simple box shape as illustrated in Figure 2.4(b). The volumetric flow through a piston compressor depends on the operating volume $V_{o,a(l,k)}$ (the volume of gas that is compressed during one cycle) and the speed of the crankshaft that drives the machine:

$$0 = c^{\text{vol}}_{a(l,k)}(q_{a(l,k)}, \rho^{\text{in}}_{a(l,k)}, n_{a(l,k)}) = \frac{q_{a(l,k)}}{\rho^{\text{in}}_{a(l,k)}} - V_{o,a(l,k)} n_{a(l,k)}.$$

Since the shaft speed is bounded by $n_{a(l,k)} \in [\underline{n}_{a(l,k)}, \overline{n}_{a(l,k)}]$, one obtains corresponding limits of the volumetric flow as left and right boundaries of the operating range. The shaft torque $M_{a(l,k)}$ is given by the constraint

$$0 = c^{\text{torque}}_{a(l,k)}(\rho^{\text{in}}_{a(l,k)}, M_{a(l,k)}, H_{\text{ad},a(l,k)}) = M_{a(l,k)} - \frac{V_{o,a(l,k)} H_{\text{ad},a(l,k)}}{2\pi \eta_{\text{ad},a(l,k)}} \rho^{\text{in}}_{a(l,k)},$$

where $\rho^{\text{in}}_{a(l,k)}$ denotes the gas density at the compressor inlet. For piston compressors, the adiabatic efficiency $\eta_{\text{ad},a(l,k)}$ is a constant parameter. Depending on the specific machine and the available technical data, the compression ability is limited in one of the following ways:

$$0 \leq c^{\text{limit}}(p_{a,l}, p_{a,l+1}, M_{a(l,k)}) = \begin{cases} \overline{\varepsilon} - p_{a,l+1}/p_{a,l}, \\ p_{a,l} - p_{a,l+1} + \overline{\Delta p}, \\ \overline{M}_{a(l,k)} - M_{a(l,k)}. \end{cases}$$

Here $\overline{\varepsilon}$ denotes an upper limit on the compression ratio, $\overline{\Delta p}$ an upper limit on the pressure increase, and $\overline{M}_{a(l,k)}$ an upper torque bound.

In summary, a piston compressor is modeled by the constraints

$$0 = c_{\mathscr{E},a(l,k)}(x_a^{\text{base}}, x_{a,l}, x_{a,l+1}, x_{a(l,k)}) = \begin{pmatrix} c^{\text{base}}_{a(l,k)}(x_a^{\text{base}}, x_{a,l}, x_{a,l+1}, x^{\text{base}}_{a(l,k)}) \\ c^{\text{torque}}_{a(l,k)}(\rho^{\text{in}}_{a(l,k)}, M_{a(l,k)}, H_{\text{ad},a(l,k)}) \\ c^{\text{vol}}_{a(l,k)}(q_{a(l,k)}, \rho^{\text{in}}_{a(l,k)}, n_{a(l,k)}) \end{pmatrix},$$

$$0 \leq c_{\mathscr{I},a(l,k)}(x_{a,l}, x_{a,l+1}, x_{a(l,k)}) = c^{\text{limit}}(p_{a,l}, p_{a,l+1}, M_{a(l,k)}),$$

and the variables

$$x_{a(l,k)} = (x^{\text{base}}_{a(l,k)}, M_{a(l,k)}, n_{a(l,k)}).$$

10.1.10.3 ▪ Compressor drives

Drives deliver the power required by compressor machines. The drive d associated with a compressor machine $a(l,k)$ is given by a mapping σ from the set of compressors to the set of drives, i.e., we have $d = \sigma(a(l,k))$ if drive d is associated to compressor $a(l,k)$. While every compressor is attached to a unique drive, some compressors may share a drive, i.e., a single drive may power several compressor machines. To simplify the exposition in this section, we describe in detail only the case where every drive powers just one compressor machine. The resulting model is easily extended to the general case.

10.1. Component models

Drives are categorized into three different types based on the energy source and the design principle: *gas turbines*, *gas driven motors*, and *electric motors*.

Electrical motors use electrical power. They form the set of *electricity consuming drives* that we call $\mathbb{D}_{el}$ in the following.

In contrast, gas turbines and gas driven motors use gas from the network as their energy source. Specifically, they take their fuel gas from the inlet of the compressor group. These drives constitute the set of *gas consuming drives*, $\mathbb{D}_{gas}$. The fuel consumption q_d^{fuel} of a drive $d \in \mathbb{D}_{gas}$ is modeled by

$$0 = c_d^{fuel}(q_d^{fuel}, b_d, m_a, H_{c,a}) = q_d^{fuel} - \frac{b_d\, m_a}{H_{u,a}};$$

see Section 2.3.5.4 for a more detailed description. Here, $H_{u,a} = c\, H_{c,a}$ denotes the lower calorific value, which differs from the (upper) calorific value $H_{c,a}$ by a constant factor c (see Cerbe (2008)). In practice, the fuel gas is taken from the network. To consider this in a model with fixed and balanced supply and demand, the required fuel gas needs to be known to adjust supply or demand accordingly. Since the required fuel gas is not known a priori, its extraction from the network is not modeled. See Chapter 5 for more details.

The maximal power $\overline{P}_d$ that a drive d can deliver is the upper bound on the power $P_{a(l,k)}$ consumed by the connected compressor machine:

$$0 \leq c_{\mathcal{G},d}(P_{a(l,k)}, \overline{P}_d) = \overline{P}_d - P_{a(l,k)}, \quad d = \sigma(a(l,k)). \tag{10.43}$$

If the drive powers several machines, $\overline{P}_d$ is the upper bound on the sum of the required compressor powers. The value $\overline{P}_d$ depends on the drive speed, which equals the speed of the connected compressor, $n_{a(l,k)}$, since compressors are directly mounted on the drive shaft. If several machines are connected, their speeds are therefore identical. In addition to (10.43), every type of drive requires for its model a specific subset of the drive variables, $x_d = (q_d^{fuel}, P_{a(l,k)}, \overline{P}_d, b_d)$, and a specific subset of the equations, see Section 2.3.5.4,

$$0 = c_d^{spec\text{-}ener}(b_d, P_{a(l,k)}) = b_d - \psi(P_{a(l,k)}; \alpha_d^{energy}),$$
$$0 = c_d^{biquad\text{-}power}(\overline{P}_d, n_{a(l,k)}) = \overline{P}_d - \chi(n_{a(l,k)}, T_{amb,a}; A_d^{max\text{-}power}),$$
$$0 = c_d^{quad\text{-}power}(\overline{P}_d, n_{a(l,k)}) = \overline{P}_d - \psi(n_{a(l,k)}; \alpha_d^{max\text{-}power}).$$

Here, $\alpha_{b_d}, \alpha_{P,d}$, and $A_{P,d}$ are the corresponding coefficient vectors or coefficient matrix of the polynomials; see Section 2.3.5.4.

Gas turbines Gas turbines are modeled by the specific energy consumption rate b_d, which depends on the power consumed by the compressor, $P_{a(l,k)}$, and a relation between the power limit $\overline{P}_d$, the compressor speed $n_{a(l,k)}$, and the constant ambient temperature $T_{amb,a}$. In terms of the above constraints the model reads

$$0 = c_{\mathcal{E},d}(x_{a(l,k)}, x_d) = \begin{pmatrix} c_d^{fuel}(q_d^{fuel}, b_d, m_a, H_{c,a}) \\ c_d^{spec\text{-}ener}(b_d, P_{a(l,k)}) \\ c_d^{biquad\text{-}power}(\overline{P}_d, n_{a(l,k)}) \end{pmatrix}. \tag{10.44}$$

Gas driven motors Gas driven motors behave like gas turbines, except that the maximal power does not depend on the ambient temperature:

$$0 = c_{\mathscr{E},d}(x_{a(l,k)}, x_d) = \begin{pmatrix} c^{\text{fuel}}(q_d^{\text{fuel}}, b_d, m_a, H_{c,a}) \\ c_d^{\text{spec-ener}}(b_d, P_{a(l,k)}) \\ c_d^{\text{quad-power}}(\overline{P}_d, n_{a(l,k)}) \end{pmatrix}.$$

Note that the difference to (10.44) is the usage of $c^{\text{quad-power}}$ instead of $c^{\text{biquad-power}}$.

Electric motors These drives consume electric power rather than fuel gas. Depending on the specific design, the ambient temperature may or may not have an influence on the power limit. An electric motor d is thus modeled by one of the following constraints:

$$0 = c_{\mathscr{E},d}(x_{a(l,k)}, x_d) = \begin{pmatrix} c_d^{\text{spec-ener}}(b_d, P_{a(l,k)}) \\ c_d^{\text{quad-power}}(\overline{P}_d, n_{a(l,k)}) \end{pmatrix}$$

or

$$0 = c_{\mathscr{E},d}(x_{a(l,k)}, x_d) = \begin{pmatrix} c_d^{\text{spec-ener}}(b_d, P_{a(l,k)}) \\ c_d^{\text{biquad-power}}(\overline{P}_d, n_{a(l,k)}) \end{pmatrix}.$$

10.1.10.4 ▪ Complete compressor group model

A generic model of an active compressor group is highly complex even though the active configuration is determined by the candidate solution of a decision approach. The model needs to incorporate several compression stages with their respective compressor machines and drives as well as the inlet and outlet resistors and the gas cooler. The constraints representing a compression stage $l = 1, \ldots, s$ can be summarized as

$$0 = c_{\mathscr{E},a,(l,l+1)}(x_{a,l}, x_{a,l+1}, x_a^{\text{base}}, x_{a,(l,l+1)})$$
$$= \begin{pmatrix} c^{\text{flow-dist}}(q_a, (q_{a(l,k)})_{k=1}^{m_l}) \\ c^{\text{mix-temp}}((q_{a(l,k)})_{k=1}^{m_l}, T_{a,l+1}, (T_{a(l,k)}^{\text{out}})_{k=1}^{m_l}) \\ c_{\mathscr{E},a(l,k)}(x_a^{\text{base}}, x_{a,l}, x_{a,l+1}, x_{a(l,k)})_{k=1}^{m_l} \\ c_{\mathscr{E},\sigma(a(l,k))}(x_{a(l,k)}, x_{\sigma(a(l,k))})_{k=1}^{m_l} \end{pmatrix},$$

$$0 \leq c_{\mathscr{I},a,(l,l+1)}(x_{a,l}, x_{a,l+1}, x_{a,(l,l+1)})$$
$$= \begin{pmatrix} c_{\mathscr{I},a(l,k)}(x_{a,l}, x_{a,l+1}, x_{a(l,k)}) \\ c_{\mathscr{I},\sigma(a(l,k))}(P_{a(l,k)}, \overline{P}_{\sigma(a(l,k))}) \end{pmatrix}_{k=1}^{m_l},$$

$$x_{a,(l,l+1)} = ((x_{a(l,k)}, x_{\sigma(a(l,k))})_{k=1}^{m_l}).$$

With these definitions, the complete model of a compressor group reads

$$0 = c_{\mathscr{E},a}(x_u, x_{a,v}, x_a) = \begin{pmatrix} c_{a,(0,1)}(x_{a,0}, x_{a,1}, x_a^{\text{base}}, x_{a,(0,1)}) \\ c_{\mathscr{E},a,(l,l+1)}(x_{a,l}, x_{a,l+1}, x_a^{\text{base}}, x_{a,(l,l+1)})_{l=1}^{s} \\ c_{a,(s+1,s+2)}(x_{a,s+1}, x_{a,s+2}, x_a^{\text{base}}, x_{a,(s+1,s+2)}) \\ c_{a,(s+2,s+3)}(p_{a,s+2}, p_v, T_{a,s+2}, T_{a:v}) \end{pmatrix}, \quad (10.45\text{a})$$

$$0 \leq c_{\mathscr{I},a}(x_a) = \left(c_{\mathscr{I},a,(l,l+1)}(x_{a,l}, x_{a,l+1}, x_{a,(l,l+1)}) \right)_{l=1}^{s}, \quad (10.45\text{b})$$

$$x_a = (x_a^{\text{base}}, (x_{a,l})_{l=1}^{s+2}, (x_{a,(l,l+1)})_{l=0}^{s+1}). \quad (10.45\text{c})$$

10.1.11 ▪ Variable bounds

Almost all variables of the NLP model have lower or upper limits (or both) resulting from physical properties, technical restrictions, and legal requirements.

More restrictive bounds are typically given by technical limitations of the network elements. For instance, the maximal pressure reduction of a control valve station depends on the technical capabilities of the control valve, and the capabilities of compressor machines usually induce lower and upper bounds on the required power, specific change in adiabatic enthalpy, and volumetric flow. Of course, the values of technical bounds are determined by the specific design of every individual network device.

Finally, security prescriptions and legal requirements may induce even tighter limits. For instance, the gas temperature is typically kept between 273.15 K and 318.15 K to prevent critical processes like hydrate formation. Pipes are usually approved for a certain maximum pressure, which induces corresponding pressure bounds at the tail and head. Moreover, high-speed gas flow causes vibrations of the pipes. To reduce the resulting noise and to prevent damage of the pipes, the gas speed is bounded by some suitable value, depending on the pipe material.

Note that even wide physical bounds must be specified explicitly in certain cases to prevent the solution algorithms from generating invalid iterates. For instance, the compressibility factor cannot become smaller than zero in reality, but the AGA formula (10.4) and Papay's formula (10.5) can yield negative values of z for physically possible pressures and temperatures outside the ranges of validity of these formulas.

10.2 ▪ Objective functions

The detailed description of individual model components is now complete. We turn to the objective function before stating a complete NLP model. Depending on the specific planning task under consideration, various objectives can be of interest; we will present a few typical ones.

For the main problem considered in this book, the validation of nominations, any feasible solution is satisfactory, and a constant objective is formally sufficient. However, since we wish to obtain additional information if the NLP solver does not succeed in finding a feasible solution, we actually use a relaxed problem formulation, where the objective consists of minimizing a suitable slack norm; see Section 10.3. Moreover, a zero objective does not lead to a well-posed formulation, so we would need some proper objective for regularization in any case. From an economic perspective, the most natural goal is to minimize the cost of network operation, which is dominated by the energy costs of gas consuming drives $\mathbb{D}_{gas}$ and electrically powered drives $\mathbb{D}_{el}$. Gas coolers and gas preheaters also consume a certain amount of fuel gas, but this can be neglected when compared to the energy consumption of compressors. The resulting objective reads

$$f^{cost}(q^{fuel}_{\mathbb{D}_{gas}}, P_{\mathbb{D}_{el}}) = \sum_{d \in \mathbb{D}_{gas}} \omega_d q^{fuel}_d + \sum_{d \in \mathbb{D}_{el}} \omega_d P_d,$$

where q^{fuel}_d and P_d denote the mass flow and power consumed by drive d, and ω_d is a cost coefficient.

If electric power is available in abundance and hence cheap, one might also be interested in minimizing only the consumption of fuel gas,

$$f^{fuel}(q^{fuel}_{\mathbb{D}_{gas}}) = \sum_{d \in \mathbb{D}_{gas}} q^{fuel}_d.$$

A more ecological goal consists of minimizing the power consumption of compressors rather than the monetary cost of this power,

$$f^{\text{power}}(P_{A_{\text{cg}}}) = \sum_{a \in A_{\text{cg}}} \sum_{l=1}^{s_a} \sum_{k=1}^{m_l} P_{a(l,k)}.$$

10.3 • Relaxations

The component models presented in the preceding sections satisfy the standard smoothness requirements stated at the beginning of this chapter. Hence standard NLP solvers can be applied, and the models are well suited to validate candidate solutions of the decision approaches. If this yields a feasible solution, the given solution candidate has been verified to be a valid approximation and has been adjusted to a more detailed model of gas physics and technical devices. Unfortunately, the opposite outcome does not provide much decisive information, since standard NLP solvers are *local* optimizers: if the problem is infeasible, they do not offer any hints to possible reasons, and they may fail to find feasible solutions even if they exist. On the other hand, the problem is much too hard to apply *global* solvers; otherwise we would not need the separate stages of decision approach and NLP validation in the first place.

To gain additional information in the case of unsuccessful NLP runs, we introduce slack variables σ^+ and σ^- and introduce relaxed versions of general NLP models like (10.1) instead:

$$\begin{aligned}
\min_{x, \sigma^+, \sigma^-} \quad & f^{\text{relax}}(x, \sigma^+, \sigma^-) \\
\text{s.t.} \quad & c_i(x) + \sigma_i^+ - \sigma_i^- = 0, \quad i \in \mathcal{E}, \\
& c_j(x) + \sigma_j^+ \geq 0, \quad j \in \mathcal{I}, \\
& \sigma_i^+, \sigma_i^-, \sigma_j^+ \geq 0, \quad i \in \mathcal{E}, j \in \mathcal{I}, \\
& x \in [\underline{x}, \overline{x}].
\end{aligned} \quad (10.46)$$

The new objective function f^{relax} will be defined as a suitable norm of the slack variables, so that we minimize some measure of infeasibility. In fact, the relaxed NLP is feasible as long as $\underline{x} \leq \overline{x}$, and the x component of any optimal solution to (10.46) is a feasible point of the original problem (10.1) *if and only if* the objective value of (10.46) is zero, i.e., if all slack variables vanish. If this is not the case, the nonzero slack components give some indication which constraints are hard to satisfy in a heuristic sense.

The standard choice of f^{relax} is the ℓ_1-norm,

$$f^{\text{relax}}(x, \sigma^+, \sigma^-) = \|(\sigma^+, \sigma^-)\|_1 = \sum_{i \in \mathcal{E} \cup \mathcal{I}} \sigma_i^+ + \sum_{i \in \mathcal{E}} \sigma_i^-.$$

If the objective value is positive, this choice of f^{relax} tends to produce only a small number of nonzero slack variables, which may indicate what components or areas of the gas network "cause" the infeasibility. This kind of information proves to be quite useful to practitioners (see Chapter 11, too).

A second natural choice of the objective function is the ℓ_∞-norm, which is obtained by minimizing an upper bound $\overline{\sigma}$ on all slack variables,

$$f^{\text{relax}}(x, \sigma^+, \sigma^-, \overline{\sigma}) = \overline{\sigma}, \quad \sigma_i^+, \sigma_i^-, \sigma_j^+ \leq \overline{\sigma}, \quad i \in \mathcal{E}, j \in \mathcal{I}.$$

This objective, however, does not offer any insight into possible reasons for infeasibility, since a nonzero objective value tends to produce a large number of nonzero slack values. For a discussion of the different norms see Section 11.6. The relaxation idea can be applied to our gas network model in several ways. While feasibility of (10.46) is only guaranteed when all constraints are relaxed, it is often sufficient to apply the relaxation to a suitable subset. For example, a restriction to special types of network elements, like compressor groups or pipes, can increase the chance to obtain useful information from the nonzero slack variables. Similarly, a relaxation of the flow balance equations (10.10) can be useful to detect obstructive, possibly implicit, flow restrictions.

10.4 ▪ A concrete validation model

In this section, we fix a single instantiation of the family of NLP models that have been presented in the preceding sections. This model is called ValNLP in the following and is used for the computational experiments in Chapter 12.

For choosing the concrete model variants, often a compromise between physical and technical accuracy on the one hand and solvability of the model on the other hand is necessary. Most of the model choices correspond to the component models of compressors and pipes. For pipes, we choose a model that is similar to the pipe models of the decision approaches, i.e., we use

▷ the quadratic approximation (10.21) of the pressure loss along a pipe,
▷ the mean pressure approximation, see (10.24),
▷ the smooth friction approximation (10.18), and
▷ the AGA formula (2.5).

For compressors and drives, we use a model incorporating

▷ the complete machine models with characteristic diagrams of turbo and piston compressors,
▷ the complete drive model, see Section 10.1.10.3, and
▷ a constant isentropic exponent, see (10.40).

We also choose to not include any mixing constraints for the gas parameters. In particular, this concrete model does not handle mixing of calorific values, but is applied to flow nominations derived from power nominations assuming an average calorific value as described in Section 5.3.3.

To be precise, ValNLP is given by

$$\min_{x,\sigma^+,\sigma^-} \quad f^{\text{relax}}(x,\sigma^+,\sigma^-) \tag{10.47}$$

$$\begin{aligned}
\text{s.t.} \quad & (10.10) && \text{for all } u \in V, \\
& (10.25), (10.26) && \text{for all } a \in A_{\text{pi}}, \\
& (10.28) && \text{for all } a \in A_{\text{lin-rs}}, \\
& (10.29) && \text{for all } a \in A_{\text{nl-rs}}, \\
& (10.30) && \text{for all } a \in A_{\text{va}}(a \text{ open}), a \in A_{\text{sc}}, a \in A_{\text{cg}} \cup A_{\text{cv}}(a \text{ in bypass}), \\
& (10.31) && \text{for all } a \in A_{\text{va}}(a \text{ closed}), a \in A_{\text{cg}} \cup A_{\text{cv}}(a \text{ closed}), \\
& (10.34) && \text{for all } a \in A_{\text{cv}}(a \text{ active}), \\
& (10.45) && \text{for all } a \in A_{\text{cg}}(a \text{ active}), \\
& x \in [\underline{x},\overline{x}].
\end{aligned}$$

The variable vector x consists of all variables required by the constraints. In addition, all nonlinear constraints of (10.47) are relaxed, leading to according slack variable vectors σ^+ and σ^-. Finally, we remark that we use the ℓ_1-norm as the measure of infeasibility in the objective function $f^{\text{relax}}(x,\sigma^+,\sigma^-)$.

Chapter 11
What does "feasible" mean?

Imke Joormann, Martin Schmidt, Marc C. Steinbach,
Bernhard M. Willert

Abstract *The main goal of our efforts described in this book consists of solving the problem of validation of nominations in gas networks, i.e., deciding whether a feasible solution exists for a given set of boundary conditions represented by a nomination. However, it turns out that the meaning of "feasible" is not self-evident. This is due to a multitude of reasons, ranging from the accuracy of problem data over subtle differences between our models to tractability of optimization problems. In the current chapter we elaborate on these points and try to clarify precisely in what sense and how well the mathematical methodology presented can distinguish feasible from infeasible solutions.*

This chapter treats a number of topics addressing feasibility issues. Following the process from modeling the gas network and its operation to a solution in our models, we highlight the different stages where feasibility might be a concern. We start with Section 11.1 by discussing important differences between reality and our mathematical optimization models, as well as the practical interpretation of the results. Section 11.2 then addresses the interrelation of modeling accuracy and requirements on the availability and accuracy of model data. We then turn to the various mathematical difficulties concerning feasibility arising in our planning problems and how they are addressed by considering a hierarchy of abstract and computational models. This is discussed in Section 11.3 with a focus on the relations of these models to each other. We then present selected results comparing our nonlinear program (NLP) validation model with the commercial simulation package SIMONE in Section 11.4. In Section 11.5, we explain how infeasible results of the ValNLP (see Chapter 10) are postprocessed in order to give the user hints for possible reasons of the infeasibility. Finally, the last section is devoted to provably infeasible problem instances: Section 11.6 presents techniques for analyzing infeasibility in relaxed models.

11.1 • Feasible network operation

"What is feasible network operation?" is one of the most fundamental questions of this book. This question needs to be answered in order to define the capacity of a gas network. It is thus behind every modeling and algorithmic decision that has been made. There is a precise (and trivial) answer/definition: a network operation is feasible if all technical, legal, and contractual conditions of the given gas network are satisfied. In this section we highlight several issues arising from this definition and give connections with the different parts of this book.

The first issue is that, naturally, the above definition implicitly includes a load flow situation and external conditions like temperatures, etc. In short-term planning, the load flow and external conditions must be known, or sufficiently accurate forecasts must be available, to generate suitable network operation plans. Since we concentrate on long-term planning in this book, the above data are not available in our context: we consider fictitious future *nominations*, where load flows and external conditions are assumed to be constant over a fixed time period. For the evaluation of capacities, a complete probability space of such nominations has to be generated using statistical data from the past (see Chapter 13) in combination with methods to include contractual data and forecasts for points for which no statistical data are available (see Chapter 14); see also Chapter 4 for a description of the approach currently used in practice.

An important second issue is that network operation is transient in practice, i.e., dynamic over time. The most detailed evaluation of capacities would thus consider transient computations. However, this would require us to replace the probability space of nominations with a probability space of dynamic load flow situations, each consisting of

▷ a dynamic profile of load flows and external conditions over the considered time period,
▷ an initial state of the system, and
▷ an optimization engine that can handle the transient computations.

It is entirely unclear whether it is possible to avoid the "curse of dimensionality" in such an attempt and whether sufficiently reliable probabilities can be generated from the data available. Moreover, the network sizes currently solvable by transient methods that handle discrete decisions are small (see, e.g., Domschke et al. (2011)); simulation and optimization without discrete decisions can currently handle larger sizes (see Ehrhardt and Steinbach (2005); Steinbach (2007)), but would need improvements as well. See also the discussion in Section 5.3.1.

As already discussed in Section 5.3.1, we therefore concentrate on stationary models and computations. While this is currently the only way to go, it has the drawback that certain temporal effects cannot fully be represented in stationary models. Examples are as follows:

▷ In contrast to a stationary model, entry and exit flows do not have to be balanced.
▷ It is sometimes possible to operate a turbo compressor with a working point left to its surgeline. This is realized by redirecting part of the compressor outflow back to its inlet, which results in a circular gas flow whose temperature is increased repeatedly; see Section 2.3.5.1. There is no adequate stationary model for this process: it requires an artificial relaxation of the stationary operating range of the turbo compressor, while the dynamic operating range is never violated in practice.
▷ Temporarily storing gas in a pipeline, the so-called linepack, cannot be represented in stationary models. It is, however, of great practical importance for gas network operation, since the pipes themselves can store a significant amount of gas and thus act as a large buffer to smoothen short-term demand peaks.

Consequently, stationary models are usually considered as being more conservative, since transient operation offers more flexibility. Moreover, a stationary network state is a special case of a transient state in which the load flows and external conditions do not change over time and the network reaches a steady state. However, it is possible that a sequence of nominations, whose corresponding stationary models are all feasible, cannot dynamically be operated. Usually, however, stationary planning provides practically reliable results. It is an important goal of future research to develop transient methods for short-term planning.

In order to focus our discussion of feasibility, we from now on consider the case of a stationary computation with a given nomination.

The third issue connected with the feasibility definition concerns the choice of the model for computations. All considered models are based on the physical laws, usually the Euler equations, which describe gas behavior and flow. However, as is obvious from the discussion in Chapter 2, several approximations and simplifications have to be made. For instance, the choice of one-dimensional (1d) Euler equations instead of a 3d fluid dynamical model is essential to be able to perform computations on networks larger than a few pipelines. However, the 1d Euler equations are usually considered as reasonably exact for practical computations on larger networks. Other model choices are based on computational reasons—this is discussed in detail in Chapters 6–10. Note that these facts imply that in practice there is nothing like a natural "master model" for the whole network from which one can then derive approximations.

As a consequence, any approach has to deal with the *model gap* between reality and the actually chosen model. This, of course, has implications on the definition of feasibility, which is with respect to the chosen model. This model can describe reality more or less accurately. In this book, we take the approach to define a base model, against which we validate all other models. Thus, we define a network operation as feasible if we can find a solution of the validation NLP (ValNLP, see Section 10.4) using computed/given discrete decisions that has a zero objective value. See Chapter 12 for more information of the performance with respect to this definition.

As a final issue with respect to feasibility, we mention the limited numerical accuracy with which all computations are performed. Although not easy in detail, the handling of this issue is standard.

11.2 • Availability and accuracy of model data

Theoretically it would be possible to formulate a microscopic network model that accounts for the full physics of all the individual chemical species contained in natural gas and for the full three-dimensional geometry of the pipeline network, including special devices such as valves, compressors, control valves, filters, measurement instruments, etc.

Clearly, such an extremely detailed model would be of little practical use. First, it is impossible to gather reasonably accurate values of all the required physical and technical data: just think of the spatially varying heat conductance of the soil around the pipelines, or of technical network parameters that change slowly over time, because of usage conditions or wearout, like the roughness of inner pipe walls or the efficiency of compressors. Second, even if accurate and complete physical and technical data could be provided for (typically large) real-life networks, a numerical simulation or optimization matching the accuracy of such a model would be far beyond the capabilities of today's software and algorithms.

Fundamental considerations like these apply to most complex technical systems: modelers always face the difficulty of providing a *reasonable* degree of detail (or model accuracy) in the sense that the value of information drawn from the model by computational simulation and optimization outweighs the cost of setting up the model (complete with required data) plus the cost of performing the computations. In the case of gas networks, it turns out that

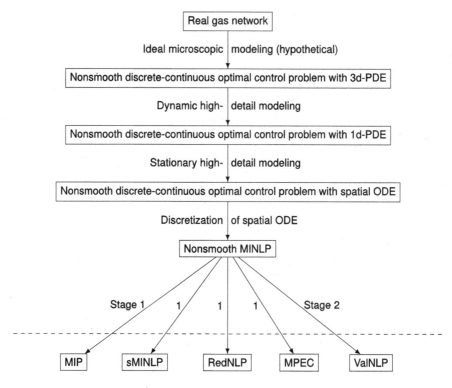

Figure 11.1. *Overview of optimization models: abstract models are above the dashed line, computational models below the dashed line.*

▷ it is rather difficult to find a properly balanced degree of modeling detail;
▷ the degree of detail depends heavily on the intended use (on-line simulation, short-term planning, mid-term planning, ...) and on the algorithms used;
▷ even models used for on-line simulation are much coarser than the "ideal" microscopic model mentioned above;
▷ providing and maintaining the data of a real gas network requires a substantial effort even for relatively coarse models;
▷ the accuracy of results is typically limited by the accuracy of the data, not the accuracy of algorithms, even for relatively coarse models.

The discussion in this chapter is primarily focused on *deterministic* network data. All mid-term and long-term planning problems involve inherently *stochastic* data such as future supplies, discharges, and temperatures, which are the subject of Chapter 13.

11.3 ▪ How "feasible" are solutions of our models?

When we speak about feasibility of solutions, we have to consider both feasibility with respect to a certain model and feasibility with respect to reality. For a proper understanding, we need to take a closer look at the interrelations of the models in our hierarchy (see Figure 11.1). Our models basically need to address the following combination of difficult mathematical aspects:

▷ nonlinearity,
▷ nonconvexity,

11.3. How "feasible" are solutions of our models?

▷ nonsmoothness,
▷ PDEs or ODEs for gas dynamics,
▷ continuous controls,
▷ discrete decisions.

All these aspects arise in the hypothetical microscopic network model mentioned before, which would have the form of a (nonlinear and nonconvex) nonsmooth discrete-continuous control problem with a system of gas dynamics PDEs in three spatial dimensions. All aspects still arise in a practical "reference" model, the next-lower level in Figure 11.1, which differs from the ideal model in that gas dynamics are only modeled in the pipes and only in one spatial dimension, and in that macroscopic technical models are used for all other network elements. As a side issue we would like to point out that even for such a transient one-dimensional model, theoretical results on existence and uniqueness of solutions for arbitrary networks are still unavailable to the best of our knowledge; for some of the most general studies see Colombo, Herty, and Sachers (2008), Colombo et al. (2009), and Gugat et al. (2012).

In the *stationary* case considered here, the 1d PDE actually reduces to a spatial ODE. Even in gas network *simulation* (see Králik et al. (1988); Záworka (1993)) one typically uses a priori discretizations of the PDEs (or ODEs), which in our context leads to a nonsmooth mixed-integer nonlinear program (MINLP) model, the lowest level in the upper part of Figure 11.1. This is still an abstract model, in the sense that we do not actually use it directly for computations. However, all our computational models are derived from it by further approximations, simplifications, and smoothings of the above mathematical difficulties; see lower part of Figure 11.1. (If applicable, further intermediate models in the derivation are described in the context of each computational model.) The first four computational models feature substantial simplifications to obtain tractable formulations of the overall discrete-continuous optimization problem; see Section 5.6 for an overview and Chapters 6–9 for details. In contrast, the validation NLP model ValNLP (Section 10.4) does not include discrete decisions but retains as much of the nonlinear physics as deemed necessary to check real-life feasibility with sufficient accuracy. For our purpose, this means that the degree of detail and the resulting solution accuracy are comparable to that of existing simulation packages (see Section 11.4 for a comparison), which have been trusted for years by practitioners to be reliable approximations of reality.

The fundamental difficulty in assessing real-life feasibility is that the different models in the hierarchy are not *quantitatively* comparable: although some of the differences are proper mathematical approximations for which error estimates exist (such as discretizations of the spatial ODE), other model differences just consist of different equations defined on different spaces (i.e., sets of variables) for which natural error measures do not even exist. The same point applies to comparisons of the four first-stage models with each other and with the ValNLP model: they all live on different spaces of variables, possibly including discrete variables, so that a natural difference measure for solutions is not clear. Thus, for comparing the results of our models with each other and for interpreting their implications in real life, we need to rely on experience with these models in practical application.

For these reasons, we consider a candidate solution from any of the four approaches as feasible with respect to our mathematical framework if fixing its discrete decisions and adding variables of the (finer) validation NLP model ultimately leads to a feasible point: a ValNLP solution with slack norm zero; see Section 10.3 and Section 5.6. Of course, there remains some unavoidable "grey area": a chance that an NLP-feasible point is slightly infeasible in real life. Such a grey area exists in simulation-based planning as well. Since the comparison with state-of-the-art simulation software is the best possible benchmark available to us, we basically try to produce decisions that are close to practically accepted simulation-based decisions.

If all four approaches fail to produce a candidate solution, due to divergence in the RedNLP and MPEC approaches or due to timeout in the MILP and sMINLP approaches, we cannot make a decision on feasibility and the runs are repeated with different parameter settings.

Note that the MILP and sMINLP approaches are actually capable of detecting *infeasibility* of their respective models, i.e., we may obtain a decisive *negative* answer to the feasibility question. This will be discussed in the last section of this chapter.

11.4 ▪ NLP validation vs. network simulation

The fundamental difficulty that our abstract and computational models are not quantitatively comparable (see Section 11.3) also applies to the relation of the ValNLP model with simulation models employed in commercial packages.

Ideally, the ValNLP model should reflect the real network behavior as closely as possible. In principle, this can be achieved by setting up an *inverse problem*, i.e., an optimization problem that estimates the model parameters from the mismatch between the given network model and measurements of the real system. However, measurements of *stationary* network states cannot be obtained during regular operation, and are therefore very expensive. On the other hand, to estimate parameters from *dynamic* measurements, we would have to develop a proper transient network model and to solve a corresponding dynamic inverse problem, both of which are way beyond the scope of the work described here. As an artificial reference substituting the real system we have hence chosen a particular simulation package, namely SIMONE version 5.73 [SIMONE], which is frequently used by Open Grid Europe GmbH (OGE) and other network operators.

In the following we shall demonstrate that, even when our NLP uses the model equations given in the SIMONE documentation, small but nonnegligible differences between the results of the NLP and SIMONE occur. These differences can be attributed to several factors:

▷ the overall numerical solution algorithms differ,
▷ differential equations are discretized using different schemes or grids,
▷ model variants like the pipe friction model may differ in minor aspects,
▷ data handling and the accuracy of data processing differ, and
▷ SIMONE possibly performs floating point computations in single precision arithmetic whereas all our computations are performed in double precision.

To test how well SIMONE and the various submodel variants of our NLP match, in particular the submodel variants used in ValNLP, we have performed a large number of comparisons on individual network elements. The tests cover all element types of Chapter 2, with multiple sets of design parameters and flow situations for each of them. An illustrative subset of these tests is discussed below.

For every comparison we fix all inflow quantities of the network element under consideration. Additionally, we fix all control quantities in case of active devices. It can be shown that all other quantities are thereby uniquely determined.

As it turns out, choosing suitable test sets of design parameters and flow situations is hard. Due to the nonlinearity of the model equations, small differences of intermediate values caused by the above-mentioned factors can get amplified during computation. Even worse, for the pipes one can create extreme cases with almost arbitrarily large (relative)

Table 11.1. *Parameters of test set for pipes.*

Quantity	Tested values	Unit
L	0.01, 0.1, 0.9, 46.0	km
D	150, 310, 405, 1185	mm
k	0.006, 0.02, 0.1, 0.5	mm
h_{out}	−500, 0, 500	m
p_{in}	3.9, 15.3, 53.8, 74.4	bar
T_{in}	288.15, 298.15, 308.15, 318.15	K
Q_0	50, 250, 500, 750	1000 Nm^3/h

differences in the final results. This happens, e.g., when considering the outflow pressure of very long pipes with small diameter.

We wish to perform a comparison that covers real-life network elements and flow situations as well as possible on the one hand and that avoids extreme cases on the other hand. In case of testing pipes we therefore proceed as follows. For the northern H-gas network of OGE (see Section 1.6), we take the distributions of pipe length, diameter, and roughness, and choose as test parameters the respective quantiles at 10%, 35%, 65%, and 90%. For the outlet elevation h_{out}, we choose the values −500 m, 0 m, and 500 m, the inlet elevation being fixed at 0 m. The resulting sets of individual parameters are given in Table 11.1. The entire test set is obtained by taking all 12 288 combinations of the individual pipe and flow parameters. Test sets of other network elements are generated in a similar manner. In addition to pipes, we will also address control valves here. More detailed results including all other element types are given in Schmidt, Steinbach, and Willert (2014).

For every type of network element, several quantities of the computed solutions are compared, such as outflow pressure and temperature. For every quantity x, we measure the absolute deviation between the NLP and SIMONE results x_{NLP} and x_{Sim}, the relative deviation, and (if applicable) the relative deviation with respect to the quantity's change from inflow to outflow, $\Delta x = x_{in} - x_{out}$:

$$d_{abs}(x) = |x_{NLP} - x_{Sim}|,$$

$$d_{rel}(x) = \frac{|x_{NLP} - x_{Sim}|}{|x_{Sim}|},$$

$$d_{rel}(\Delta x) = \frac{|\Delta x_{NLP} - \Delta x_{Sim}|}{|\Delta x_{Sim}|}.$$

If multiple model variants exist in both the NLP and SIMONE, several choices are compared. For instance, in the case of pipes we obtain four model variants by combining the AGA formula (10.4) and Papay's formula (10.5) for the compressibility factor $z(p,T)$ with two variants for the Euler equations: the approximation (10.25) and discretized ODE (10.20). Each of these four variants is used in all 12 288 test cases. Finally we exclude all test cases where one of the methods fails to produce a result and all combinations with an impossible outlet elevation, i.e., with $|h_{out}| > L$. Average deviations are then computed on the remaining test cases. In all instances, the settings (like compressibility factor or friction model) of the NLP and SIMONE are matched as closely as possible. Note that this will yield poor matchings in certain cases: while both the NLP and SIMONE offer the AGA formula and Papay's formula for the compressibility coefficient,

Table 11.2. *Average deviations of pipe outflow pressure (bar).*

NLP model choice	$z(p,T)$	Sample size	$d_{\text{abs}}(p_{\text{out}})$	$d_{\text{rel}}(p_{\text{out}})$	$d_{\text{rel}}(\Delta p)$
Approximation	AGA	2419	0.11	0.86%	5.8%
Approximation	Papay	2651	0.10	0.94%	5.9%
ODE	AGA	2658	0.046	0.31%	3.2%
ODE	Papay	2659	0.050	0.33%	3.3%

Table 11.3. *Average deviations of pipe outflow temperature (K).*

NLP model choice	$z(p,T)$	Sample size	$d_{\text{abs}}(T_{\text{out}})$	$d_{\text{rel}}(T_{\text{out}})$	$d_{\text{rel}}(\Delta T)$
Approximation	AGA	2419	1.1	0.38%	16%
Approximation	Papay	2651	1.0	0.35%	15%
ODE	AGA	2658	0.22	0.075%	4.6%
ODE	Papay	2659	0.15	0.051%	2.9%

for instance, SIMONE always uses an ODE model for the pipe flow. The approximation model of our NLP (which is actually used in ValNLP) has no counterpart and is expected to be less accurate than the ODE model of SIMONE. In cases like this, the measured average deviations provide a reasonable indication of the absolute quality of the less accurate model.

The computed average deviations are listed in Table 11.2 and Table 11.3; Figure 11.2 displays logarithmically scaled histograms of the absolute deviations of pressure and temperature for the model variant with ODE discretization and Papay's formula. The standard deviation of the distribution of absolute pressure differences is 0.24 and the leftmost bin contains 2574 of 2659 samples (97%). The standard deviation of the distribution of absolute temperature differences is 0.29 and the leftmost bin contains 2330 of 2659 samples (88%).

The deviations between the NLP and SIMONE with discretized ODEs are smaller than with the approximating model. This is to be expected, since the ODE discretization is more accurate than the approximation in both cases. Specifically, SIMONE applies an implicit integration method for the partial differential equations (10.15), based on algorithms presented in Královič et al. (1988) and Záworka (1993). Nevertheless, the absolute deviation of pressures of 0.1 bar using the approximation models is still sufficiently accurate for most of the planning tasks addressed in this book. In these cases, one can use the approximation models in order to achieve reduced computation times, as is being done in the ValNLP. If, however, this accuracy was expected to be insufficient, one could switch to the discretized ODE model.

Some pressure and temperature profiles along an exemplary pipe are plotted in Figure 11.3 and Figure 11.4. These graphs show the outflow pressure and temperature depending on the normal volumetric flow along the pipe for an inflow pressure of 74.4 bar at 318.15 K with a soil temperature of 284.15 K. The figures on the right show enlarged details of the most interesting area of the figures on the left, as indicated by the dashed box.

In Figure 11.3 we see that the ODE discretizations of the NLP and SIMONE agree quite well for the pressure profile while the less accurate approximation model of ValNLP is qualitatively similar, but with a larger absolute difference, as expected. This observation is in line with the average deviations given in Table 11.2 and Table 11.3.

11.4. NLP validation vs. network simulation

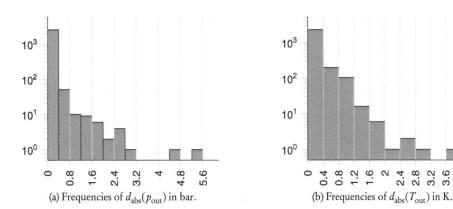

(a) Frequencies of $d_{abs}(p_{out})$ in bar.

(b) Frequencies of $d_{abs}(T_{out})$ in K.

Figure 11.2. *Logarithmic histograms for the tested pipes and the model choices "ODE discretization" and "Papay's formula."*

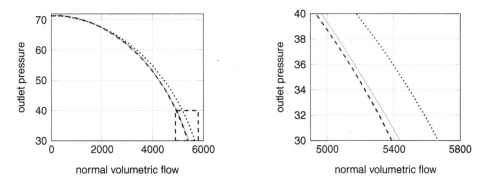

Figure 11.3. *Outlet pressure (bar) of a pipe ($L = 46$ km, $D = 1185$ mm, $k = 0.006$ mm, slope $s = 0.01$) vs. normal volumetric flow (1000 Nm³/h), computed using Papay's formula in SIMONE v5.73 (——), ODE discretization (- - -), and ODE approximation ($\cdots$). Figure on the right is a zoom of the dashed frame in the left figure.*

In Figure 11.4 we see that the ODE discretizations of the NLP and SIMONE agree almost as well for the temperature profile, except for excessive deviations at very small flow values: here SIMONE yields drastically increasing temperatures that cannot be physically correct, which indicates numerical problems or a bug in the tested version. Spot tests suggest that this has been fixed in the release 5.83 of SIMONE. The approximation model of ValNLP is again qualitatively similar. It shows larger absolute differences, but only for medium and large flow values. At small flow values up to 400 000 Nm³/h, some curvature information gets apparently lost when approximating the energy equation (10.17c), and the convex section of the temperature graph does not show up as with the ODE discretization. This could be a reason for the larger average deviations between the approximation model of ValNLP and SIMONE as reported in Table 11.3. Note, however, that even these larger average deviations of about 1 K lie within the range of data accuracy, since more accurate forecasts of environmental and soil temperatures required for planning processes are rarely available. Moreover, a small number of tests that we performed with two different versions of SIMONE indicate that the deviations between these two releases of the same software lie also in the same order of magnitude.

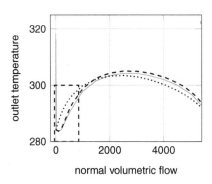 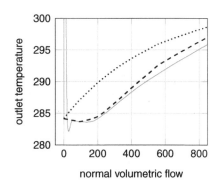

Figure 11.4. *Outlet temperature (K) of a pipe (L = 46 km, D = 1185 mm, k = 0.006 mm, slope s = 0.01) vs. normal volumetric flow (1000 Nm³/h), computed using Papay's formula in SIMONE v5.73 (——), ODE discretization (- - -), and ODE approximation (······). Figure on the right is a zoom of the dashed frame in the left figure.*

Table 11.4. *Average deviations of control valve outflow temperature (K).*

$z(p,T)$	Sample size	$d_{\text{abs}}(T_{\text{out}})$	$d_{\text{rel}}(T_{\text{out}})$	$d_{\text{rel}}(\Delta T)$
AGA	54	0.74	0.25%	34%
Papay	54	0.094	0.032%	4.1%

Let us finally consider control valves. Here we fix the inflow pressure, inflow temperature, flow value and controlled pressure reduction, and we compare outflow temperatures. Again, the compressibility factor can be modeled by the AGA formula (10.4) or Papay's formula (10.5). In both the NLP and SIMONE, the temperature loss is computed by the detailed model of Joule–Thomson; see (10.8). Note that there exists a model difference in the case of zero flow, where the "outflow" temperature is set to the soil temperature in SIMONE, whereas it is set to the inflow temperature in the NLP. Thus, we consider only test cases with nonzero flow. In total, there are 108 test cases, all of which produce results with both software packages. The resulting average deviations are listed in Table 11.4.

The differences between the AGA and Papay models are striking: despite the fact that all other quantities and model aspects are identical, the mean deviations with the AGA formula are about eight times as large as the values with Papay's formula. Moreover, the standard deviation of the absolute error for Papay's formula is less than 10^{-2} whereas the standard deviation for AGA is 0.63. This strongly indicates differences in the implementation of the AGA formula.

Although the pressure reduction is fixed, there also exists an average absolute deviation of 5×10^{-5} bar between the NLP and SIMONE. The NLP reproduces the fixed value correctly; hence, we must conclude that in this case either the internal accuracy of SIMONE is lower or the API of SIMONE rounds output values to a precision of roughly 10^{-4} bar. The second explanation seems more likely.

In summary, the validation procedures that we have carried out confirm that our models actually match the level of detail and accuracy provided by SIMONE (and presumably also by other commercial simulation packages). However, the NLP results must always be expected to differ from simulation results even if the models are based on the same equations, since different algorithms and different implementations are involved. Moreover, there is no systematic way of eliminating the remaining differences since the source codes of the simulation packages are inaccessible. Our numerical experiments have shown that the same observations also apply to the comparison of different releases of SIMONE. Results are always roughly comparable, but often not exactly; see Willert (2014) and Schmidt, Steinbach, and Willert (2014).

11.5 • The interpretation of ValNLP solutions

Our general workflow for solving the problem of validation of nominations consists of two stages (see Section 5.6). The decision approaches produce solution candidates with discrete decisions and an approximative description of the pressure-flow situation in the network. This discrete-continuous solution is used to fix the discrete decisions of all controllable network devices. In addition, the given solution is used to initialize the variables of the ValNLP model. For this, the situation can be divided into two distinct cases. First, there are variables (like pressure on vertices and flow on arcs) that appear both in the decision approaches and in the validation NLP. For these variables, the solution candidates are used directly to initialize the ValNLP variables. Second, the ValNLP model has additional variables that do not appear in the models of the decision approaches. These variables are initialized in a heuristic way depending on the given pressure and flow values of the solution candidate, the constraints of the ValNLP model, and suitable constants.

As it is described in Section 10.3, the original ValNLP is relaxed by using slack variables for a certain (sub)set of constraints, and the ℓ_1-norm of the slack variable vector is minimized. When solving the relaxed ValNLP model, three different outcomes are possible:

1. The ValNLP model is solved to (local) optimality with a slack norm value of zero. In this case, we have provably found a feasible solution to the underlying nonsmooth and nonlinear mixed-integer feasibility problem.
2. The ValNLP model is not solved to (local) optimality.
3. The ValNLP model is solved to (local) optimality with a slack norm value larger than zero.

The last two cases require further consideration. If no (locally) optimal solution could be found at all, we can only state that the solution candidate could not be validated by the ValNLP model. This can happen for various reasons, such as improper starting points, poor problem scaling, ill-conditioning, etc.

If we have found a (locally) optimal solution with a slack norm value larger than zero, there is more information available. Obviously, we have *not* validated the given solution candidate from the decision approach. But in contrast to the second case, we can interpret the infeasibility by taking a closer look at the slack variables that do not vanish in the ValNLP solution. In practice, minimizing the ℓ_1-norm of the constraint violation typically leads to a small number of nonvanishing slack variables, i.e., the set

$$\mathcal{N} := \{i \in \mathcal{E} \mid \sigma_i^+ > 0 \text{ or } \sigma_i^- > 0\} \cup \{j \in \mathcal{I} \mid \sigma_j^+ > 0\}$$

has small cardinality (see Section 11.6.1). The sets $\mathcal{E}$ and $\mathcal{I}$ denote the index sets of equality constraints and inequality constraints; see Chapter 10. With the indices in $\mathcal{N}$, we can then identify the network elements $u \in V$ or $a \in A$ at which the constraint violations appear. This gives the practitioner a first idea where reasons of infeasibility might be located. However, the slack variable *value* should be used for the infeasibility analysis, too. Unfortunately, the pure value does not give a reasonable tool for the analysis if we do not know the unit in which it should be interpreted. For instance, a slack variable value of 10 is of different importance if it is interpreted in bar or in MW. Finally, the situation becomes even more complicated by the fact that many constraints in the ValNLP model are reformulated in order to obtain better numerical behavior. For instance, the original constraint for the pressure loss at resistors (see 10.1.5),

$$p_u - p_v = \frac{8\zeta_a}{\pi^2 D_a^4} \frac{q_a |q_a|}{\rho_{a,\text{in}}},$$

might be reformulated as

$$\rho_{a,\text{in}}(p_u - p_v) = \frac{8\zeta_a}{\pi^2 D_a^4} q_a |q_a|,$$

to avoid the density variable in the denominator of the right-hand side. A relaxed version might then take the form

$$\rho_{a,\text{in}}(p_u - p_v) - \frac{8\zeta_a}{\pi^2 D_a^4} q_a |q_a| + \sigma^+ - \sigma^- = 0, \tag{11.1}$$

having the unit bar kg/m³ which cannot be interpreted by a practitioner in a direct manner.

These considerations are the reason why we reinterpret the nonvanishing slack variable values in order to allow an infeasibility analysis based upon the ValNLP solution, as follows:

1. definition of a set of physical quantities together with units in which the ValNLP solution should be interpreted, e.g., pressure in bar and flow in kg/s;
2. identification of vertices and arcs whose constraints are not satisfied exactly;
3. reinterpretation of nonvanishing slack variables and reporting the reinterpreted values to the practitioner.

An intuitive way for the reinterpretation is to compute a slack value with the unit chosen in step 1 by rearranging the unrelaxed form of the violated constraint until a term with the appropriate unit is isolated. Relaxation variables are then reapplied and, based on the solution of the ValNLP, values for the relaxation variables are computed. In constraint (11.1), for instance, the recomputed values $\bar{\sigma}^\pm$ of the slack variables satisfy the equation

$$p_u^* - p_v^* - \frac{8\zeta_a}{\pi^2 D_a^4} \frac{q_a^* |q_a^*|}{\rho_{a,\text{in}}^*} + \bar{\sigma}^+ - \bar{\sigma}^- = 0,$$

yielding $\bar{\sigma}^\pm = (\sigma^\pm)^* / \rho_{a,\text{in}}^*$ (where superindices "$*$" denote values of the ValNLP solution). These slack values have the unit bar, which can be directly interpreted by a practitioner. In the following, this approach is referred to as the *direct approach*.

The direct approach has some significant drawbacks. First, it is not always possible to apply it to an arbitrarily chosen quantity in step 1. This is the case for implicitly stated constraints. Second, consider the ValNLP model of pipes. As it is also the case for other network elements, the model involves a *vector* of constraints, $c: \mathbb{R}^n \to \mathbb{R}^m$, i.e., $c(x) = (c_i(x))_{i=1}^m$. Thus, it might be the case that there are nonvanishing slack variables σ_j^+ or σ_j^- for different constraints $c_j(x)$, $j \in \mathcal{J} \subset \{1, \ldots, m\}$, of the pipe model. If we apply the direct approach, we obtain a reinterpreted slack value for every violated constraint $c_j(x)$, $j \in \mathcal{J}$, and all reinterpreted values can have different units. For the modeler of the problem, this information might be useful since it corresponds to a level of detail that the modeler can handle. However, this is not the case for a practitioner who has no

11.5. The interpretation of ValNLP solutions

knowledge about the concrete model formulation. Thus, the practitioner is interested in a *single* interpretable value for each network element that has a constraint violation. For instance, the practitioner might ask the question: "What is the error in terms of gas flow at the arc, i.e., how much more or less gas has to flow through it, in order to achieve a feasible solution?" For taking such requirements of practitioners into account, we do not apply the direct approach. Rather, we solve small optimization problems for every network element with nonvanishing slack variables in a post-processing step. In the following, we describe these optimization problems for arcs $a = (u, v)$. The concrete formulations for vertices are completely analogous.

Let $c_{a,\mathscr{E}}(x_a, x_u, x_v)$ and $c_{a,\mathscr{I}}(x_a, x_u, x_v)$ be the component model of the network element a described in Chapter 10. That means, $c_{a,\mathscr{E}}$ and $c_{a,\mathscr{I}}$ are all constraints required to model the arc a and x_a, x_u, and x_v are the vectors of variables required by these constraints. Thus, the variable for the quantity selected in step 1 has to be contained in x_a, x_u, or x_v. Let $\hat{x} \in \mathbb{R}$ denote this single variable. With this notation, the post-processing optimization problem for arc a reads

$$\min_{x_a, x_u, x_v} \quad |\hat{x} - \hat{x}^*| \tag{11.2a}$$

$$\text{s.t.} \quad c_{a,\mathscr{E}}(x_a, x_u, x_v) = 0, \tag{11.2b}$$

$$c_{a,\mathscr{I}}(x_a, x_u, x_v) \geq 0, \tag{11.2c}$$

$$x_u - x_u^* = 0, \quad \text{except for } \hat{x}, \tag{11.2d}$$

$$x_v - x_v^* = 0, \quad \text{except for } \hat{x}, \tag{11.2e}$$

$$x_a - x_a^* = 0, \quad \text{except for } \hat{x}, \tag{11.2f}$$

$$\hat{x} \in [\underline{\hat{x}}, \overline{\hat{x}}]. \tag{11.2g}$$

Thus, all variables of the arc and its incident nodes are fixed to the values defined by the ValNLP solution, except for the quantity $\hat{x}$ whose deviation to its value in the solution of the validation NLP is minimized. The constraints (11.2d) and (11.2e) fix the tail and head variables x_u and x_v, respectively, except for the quantity represented by $\hat{x}$ if it is part of the node variable vectors. In analogy, constraint (11.2f) fixes the arc variables except for $\hat{x}$, if the latter is an arc variable. The objective consists of finding the minimum deviation of the variable $\hat{x}$ with respect to its value $\hat{x}^*$ in the ValNLP solution. For reasons of clarity, let us again take a look at problem (11.2) for the concrete case of a being a pipe and the case in which $\hat{x}$ is chosen to be the flow $Q_{0,a}$ through the pipe. The constraints (11.2b), (11.2c), and simple bounds (11.2g) are the same as for the pipe in the ValNLP model and constraints (11.2d)–(11.2f) fix all variables of the post-processing problem to the values of the ValNLP solution except for the flow variable $Q_{0,a}$ that is part of x_a. Thus, if (11.2) is feasible, the value $Q_{0,a}^{**}$ in the solution of the post-processing problem has the minimal distance to the flow value $Q_{0,a}^*$ of the ValNLP solution. The answer to the question *"How much more or less gas has to flow through the pipe in order to achieve a feasible solution?"* is thus $|Q_{0,a}^{**} - Q_{0,a}^*|$.

Unfortunately, the approach of solving problems of type (11.2) in post-processing may fail when the infeasibility cannot be expressed in the quantity $\hat{x}$. For example, this may happen at a compressor group, when the working point of a compressor machine lies above *and* to the right of the feasible operating range in the characteristic diagram. Thus, the required compression ratio as well as the flow through the compressor are too high with respect to the capability of the machine. Obviously, the infeasibility cannot be measured in only *one* of the quantities compression ratio or flow. In such cases, the infeasibility of (11.2) has to be reported to the practitioner with the consequence that the situation has to be analyzed by inspection.

Nonetheless, the described approach turned out to be very useful in practice, since in most of the cases it is possible to give the practitioner an interpretable value of the constraint violations and their location in the network.

11.6 • Analyzing infeasibility in a first stage model

In the previous section we have seen how to analyze near-feasible but not provably feasible solutions in the second stage, the ValNLP model. As described at the beginning of Section 11.5, we need to fix the discrete decisions to use this model. In the current section, we deal with the case where none of the first stage approaches can find a feasible discrete-continuous solution to start the ValNLP.

The mixed-integer linear program (MILP) (Chapter 6) and sMINLP (Chapter 7) models are solved by *global* methods: in exact arithmetic and with unlimited computing time, these methods either find a global minimum or they detect that no solution can exist. This offers the advantage of *proving* infeasibility of a nomination with respect to each of the two mixed-integer models—up to the chosen accuracy and if no timeout occurs, of course. In contrast to the RedNLP (Chapter 8) and MPEC (Chapter 9) approaches, which do not provide any information in case of divergence, we may thus obtain a decisive *negative* result.

Moreover, the MILP model is a *relaxation* of a certain intermediate MINLP model so that infeasibility can be proven for this abstract model whose accuracy is close to that of the nonsmooth MINLP (see Figure 11.1). Thus, we can actually detect infeasibility, again of course with some "gray area" that depends on the respective accuracy of the model.

The impossibility to find feasible discrete decisions can originate from different sources: First, the input data might be defective (as mentioned in Section 11.2) due to typing errors, outdated data, or misunderstandings in format. On the other side, there could be modeling and implementation issues. In the validation of a nomination, there might be too much flow to fit through the pipes (a "real," physical infeasibility). During the validation of a booking (see Section 3.2.1), we could encounter an infeasible nomination, but should the booking really be declined because of that? We might ask, "how far away" from feasibility is the nomination? Additionally, there is still the difference between the stationary and dynamic viewpoint (see Section 11.1), and we might be able to "repair" the nomination by pulling an interruptible contract (see Section 3.3.2) which was not incorporated from the beginning. In topology planning (see Section 15.4), we are explicitly asked how one can extend the network to turn a given infeasible nomination into a feasible one—a question for which an understanding of the location of the infeasibility is highly advantageous.

From a practitioner's perspective it is unsatisfactory to obtain a negative result without further explanation: it would be valuable to know why the problem is infeasible or, to pose it more positively, how it could be made feasible with—in some sense—minimal modifications.

There are different approaches how to analyze infeasibilities in models; for an overview in the LP case see Greenberg and Murphy (1991). The authors state

> An *isolation* means to find a portion of the model, preferably of minimal dimensions [sic], that contains the source of infeasibility. A *diagnosis* is an isolation that is meaningful—that is, consistent with the model's syntax, rather than an arbitrary portion that is as difficult to interpret as the original

problem. A diagnosis is *good* if it leads to the correction of the error in a reasonable amount of time.

Following this definition, our goal here is to develop a good diagnosis. Therefore, we present a "tool kit" with different problem-specific methods, which can give usable information not only on the model but the network. First, we could compute the smallest distance to feasibility (SDF) in a suitable but, in principle, arbitrary measure. These methods will include, on the one hand, changes of the network, and on the other hand, modifications of the nomination. The distance is determined by solving a relaxed version of the base model with an MILP solver in a black box fashion. The second approach uses "isolation": We compute the smallest part of the network which is still infeasible. This is the concept of irreducible infeasible subsystems (IISs). In the following, we will present the different possibilities and practical implementations of these approaches.

As implied above, we have to choose a model on which our analyses can be based, and, since the ability to detect infeasibility should be given, we can use the MILP or the sMINLP. Although it is possible for most of the above-mentioned analyses to transfer the key ideas to other formulations for validation of nominations, we will present our models based on the MILP for ease of description.

11.6.1 ▪ Smallest distance to feasibility

For this approach, we ask how we can change as little of the problem as possible and obtain a feasible solution. This is done by introducing slack variables to the chosen validation of nominations model. These slacks will relax different aspects of the formulation, represented by certain constraints, proposing either changes of the network or modifications of the nomination. The aim, and therefore the objective function, is to minimize the deviance from the original model. To demonstrate the principle of the models and to illustrate the flexibility of different validation of nominations models, we will use the following generic model to describe the validation of nominations problem:

$$
\begin{aligned}
\min \quad & f(q,p,r) \\
\text{s.t.} \quad & g(q,p,r) = a, && \text{(pipes)} \\
& h(q) = q^{\text{nom}}, && \text{(flow conservation)} \\
& k_{\text{active}}(q,p,r) \leq c_1, && \text{(active elements)} \\
& k_{\text{passive}}(q,p,r) \leq c_2, && \text{(further passive elements)} \\
& \underline{q} \leq q \leq \overline{q}, && \text{(flow)} \\
& \underline{p} \leq p \leq \overline{p}, && \text{(pressure)} \\
& \underline{r} \leq r \leq \overline{r}, && \text{(other variables)} \\
& r \in \mathbb{Z}^m \times \mathbb{R}^n, && \text{(partly binary)}
\end{aligned}
$$

where f, g, h, k_{active}, and k_{passive} are functions, and a, q^{nom}, c_1, and c_2 are vectors with the associated right-hand sides and of appropriate dimension. As usual, q and p denote the flow and pressure, respectively, while r is a wildcard for every other variable.

Throughout the next sections, we will present some exemplary computational results to give a better understanding of the strengths and weaknesses of the different models. These results are the means of 50 computations with different nominations, respectively, and were computed on an Intel i3 Dual-Core, 3.20 GHz, 8 GB with a time limit of 24 h and an ℓ_1 objective function, unless otherwise stated. As a solver for the models, we used GUROBI 4.6. The examined nominations have no input data errors, but are "physically" infeasible, due to rather large flow amounts. The network in our numerical experiments corresponds to the H-gas north network (see Figure 1.5). It is medium-scale and consists of 31 sources, 129 sinks, 432 inner nodes, 452 pipes, 6 compressor stations, 23 control valves, 34 valves, and 9 resistors. The topology of the network is the same as that of the test set HN-SN used in Chapter 12.

Selection of the objective function As noted above, we are looking for a diagnosis of the infeasibility, ideally with a physical interpretation. For the network changing models, this is given by the concept of an *extension* (see also Section 15.4) of the network. An extension is a new or remodeled network element which helps realize a previously infeasible nomination. Herewith, every nonzero slack value corresponds to a network element which must be altered.

In the different models, the objective function is presented as an arbitrary norm of the slack variable vector σ. We have implemented and tested models using the ℓ_0-"norm" (ℓ_0 is not homogeneous and therefore not a norm by definition) and the ℓ_1-norm; thus, we minimize the number of nonzero entries and the sum of the absolute values of the entries, respectively. It would be equally possible to use, for example, the ℓ_2-norm, although the resulting model would no longer be linear. Anyway, in most cases the desired goal should be to minimize the number of affected elements and hence the usage of ℓ_0. Unfortunately, the ℓ_0-case is NP-hard to solve; see problem [MP5] in Garey and Johnson (1979), even for a pure LP. Therefore, we also investigated the ℓ_1-norm, which leads to more easily solvable models and gives—in most cases—a decent alternative. Especially in the nomination modifying models, the ℓ_1 solution is often very sparse. In a certain sense, the ℓ_1-norm is also the best convex "relaxation" of the ℓ_0-"norm": The ℓ_p-norm, given by

$$\|x\|_p = \left(\sum |x_i|^p\right)^{1/p},$$

is nonconvex for $0 < p < 1$, but convex for $p \in [1, \infty]$, and it holds that $\lim_{p \searrow 0} \|x\|_p = \|x\|_0$. For a detailed discussion about the relation between ℓ_0- and ℓ_1-norms see Elad (2010). Besides, the ℓ_1-norm has the additional property of minimizing the value of the slack variables (in contrast to the ℓ_0-solution) which might be, depending on the application, a desired feature.

Note that (theoretically) the objective value is zero if (and only if) the original model has a feasible solution, since this means every entry in σ is zero. Unfortunately though, it might be possible that this is not true in reality: First, in every model we have to cope with numerical difficulties, and second, when changing the bounds for pressure or flow variables, this leads to different linearizations of the pipe modeling in the MILP, since the bounds are used as outer supporting points (see Section 6.2). This problem could also lead to some other undesired effects: Since we work with a given model accuracy, wider bounds mean probably many more supporting points, which in turn leads to much bigger models. So even feasible nominations could become unsolvable or we observe seemingly contradictory solutions. It is therefore advisable to test the feasibility of a nomination first, and not depend on the characterization of the objective function value.

11.6. Analyzing infeasibility in a first stage model

Flow bounds The first possibility for a network modification is a relaxation of the given flow bounds:

$$
\begin{aligned}
\min \quad & \|\sigma^q\| \\
\text{s.t.} \quad & g(q,p,r) = a, \\
& h(q) = q^{\text{nom}}, \\
& k_{\text{active}}(q,p,r) \leq c_1, \\
& k_{\text{passive}}(q,p,r) \leq c_2, \\
& \underline{q} - \sigma_1^q \leq q \leq \overline{q} + \sigma_2^q, \\
& \underline{p} \leq p \leq \overline{p}, \\
& \underline{r} \leq r \leq \overline{r}, \\
& r \in \mathbb{Z}^m \times \mathbb{R}^n, \\
& \sigma^q \geq 0.
\end{aligned}
$$

The objective function here is a suitable norm of the slack variable vector σ^q, where σ^q is the concatenation of σ_1^q and σ_2^q.

The physical interpretation of this model is, How much wider do the flow bounds need to be to operate the network? This model is fairly well solvable for our instances; unfortunately it is more or less useless in a real-world application. The explicit flow bounds are provided by rather wide technical capacities of pipes and simply not the limiting factor in a network, and therefore not responsible for the infeasibility, as the computational results emphasize: Out of 50 nominations, 50 models were still infeasible, with a mean solution time of 214.2 seconds. Thus, the only case in which this model might be useful is for detecting wrong input data.

Pressure bounds The next model relaxes the pressure bounds in a similar fashion:

$$
\begin{aligned}
\min \quad & \|\sigma^p\| \\
\text{s.t.} \quad & g(q,p,r) = a, \\
& h(q) = q^{\text{nom}}, \\
& k_{\text{active}}(q,p,r) \leq c_1, \\
& k_{\text{passive}}(q,p,r) \leq c_2, \\
& \underline{q} \leq q \leq \overline{q}, \\
& \underline{p} - \sigma_1^p \leq p \leq \overline{p} + \sigma_2^p, \\
& \underline{r} \leq r \leq \overline{r}, \\
& r \in \mathbb{Z}^m \times \mathbb{R}^n, \\
& \sigma^p \geq 0.
\end{aligned}
$$

Here, a "good" solution might be found, meaning that there is actually a pressure bound given for the validation of nominations that is too tight, since the pressure bounds are part of contracts. In this case, the interpretation of the solution would be asking the contracting party to adjust the requested pressure bound or, if the pressure is bounded by technical components, to extend their operating capacity.

In the computational experiments, 36 nominations were infeasible in the above model, 5 reached the time limit (24 hours), and 9 had a solution with a mean of 21.4 bar difference (in sum) to the original bounds and 2.6 affected nodes.

Pipes Another possibility is to target the pipes in the network:

$$
\begin{aligned}
\min \quad & \|\sigma^\pi\| \\
\text{s.t.} \quad & \tilde{g}(q,p,r,\sigma^\pi) = a, \\
& h(q) = q^{\text{nom}}, \\
& k_{\text{active}}(q,p,r) \leq c_1, \\
& k_{\text{active}}(q,p,r) \leq c_2, \\
& \underline{q} \leq q \leq \overline{q}, \\
& \underline{p} \leq p \leq \overline{p}, \\
& \underline{r} \leq r \leq \overline{r}, \\
& r \in \mathbb{Z}^m \times \mathbb{R}^n, \\
& \sigma^\pi \geq 0,
\end{aligned}
$$

where $\tilde{g}$ is a relaxed pipe modeling, depending on the slack variables σ^π. Here, the actual modeling details are a little more involved since we have some additional equations for the pipe description, as presented below. Preferably, our model will be easy to interpret, so the solution should be transferable to a network extension (e.g., a looped pipe or a different diameter of the pipe). For that reason, we allow the pressure drop on a pipe $a = (u,v)$ to be lower or higher than on the actually built pipe in the pressure-flow situation. Starting with Eq. (2.24),

$$ p_u^2 e^{-S} - p_v^2 = \Lambda \phi(q_a) \frac{e^S - 1}{S} e^{-S}, $$

we use

$$ p_u^2 e^{-S} - p_v^2 + \sigma_+^\pi - \sigma_-^\pi = \Lambda \phi(q_a) \frac{e^S - 1}{S} e^{-S}, $$

$$ \operatorname{sgn}\left(p_u^2 e^{-S} - p_v^2\right) = \operatorname{sgn}\left(\Lambda \phi(q_a) \frac{e^S - 1}{S} e^{-S}\right). $$

Note the additional constraint, concerned with the flow direction (disregarding possible differences in height): We prevent the flow from going from the lower to the higher pressure node, since this could not be realized just by adding more pipes.

In the computational experiments, the solution of 3 nominations were aborted due to memory issues and 47 were solved to optimality, with an average of 5.9 affected pipes and 72.5 difference in the pressure drop in sum (measured in bar^2). The same models but with an ℓ_0 objective function lead to the following comparison: only 8 nominations could be solved optimally, and the number of changed pipes is minimum 1, maximum 12, and average 4.4. The average computation time was 182.3 minutes (including nonoptimal solutions) and 128.1 minutes for the ℓ_1 objective function.

Combination It is possible for all previously described models that they are still infeasible, with the real-world interpretation that it is just not enough to build new pipes alone, or to widen certain pressure bounds. To overcome this issue, we can combine the models, i.e., introduce slack variables on all mentioned conditions at once. This yields a model which allows corresponding violations at the same time and with directly transferable interpretation as the mentioned ones. Unfortunately, the models would become much harder to solve, so there must be some sort of trade-off. Of course, the size of the network plays a major role. In the computational experiments, all 50 instances were not solvable due to memory issues, although for networks smaller as the one considered here, the combination might be possible.

If the models are combined, one must incorporate weights for the different slack variables in the objective function—on the one hand, to actually weight the preference for the different violations (a pressure bound violation might be cheaper to eliminate than to replace a pipe) and on the other hand to even out the different orders of magnitude of the slack variables: Since, for example, flow and pressure, and thus their bounds and the slack variables, are measured in different units, but are represented in the model by real numbers, there should be some sort of a conversion in the penalty (e.g., a violation of 1 m^3/s is not as bad as a violation of 1 bar).

Active elements The last model aimed at an addition of pipes, but an extension of the network could also mean new active elements. Therefore we introduce the following model; the idea here is to allow each nonpipe element (i.e., compressor groups, control valves, resistors, short cuts, and valves) in the network model to work as an ideal compressor or control valve (i.e., change the pressure by an arbitrary amount):

$$\begin{aligned}
\min \quad & \|\sigma^p\| \\
\text{s.t.} \quad & g(q,p,r) = a, \\
& h(q) = q^{\text{nom}}, \\
& k_{\text{active}}(q,p,r) \leq c_1 + \sigma^p, \\
& k_{\text{passive}}(q,p,r) \leq c_2, \\
& \underline{q} \leq q \leq \overline{q}, \\
& \underline{p} \leq p \leq \overline{p}, \\
& \underline{r} \leq r \leq \overline{r}, \\
& r \in \mathbb{Z}^m \times \mathbb{R}^n, \\
& \sigma^p \geq 0.
\end{aligned}$$

Again, this model should serve as a basic concept. The reason to choose the existing elements as a location for these ideal compressors and control valves is that existing elements are cheaper to extend than to build completely new elements. This remodeling is done by adding a slack variable to the particular delimiting pressure relation. For short cuts this is (6.5), for valves Eq. (6.11c) and Eq. (6.11d), for resistors either (6.8) or (6.10). Compressor groups and control valves are handled by relaxing the bypass valve (see Section 5.1.6 and Section 5.1.7, respectively).

For instance, the modeling of a valve $a = (u, v)$ becomes

$$(\overline{p}_v - \underline{p}_u)s_a + p_v - p_u - \sigma_c^p \leq \overline{p}_v - \underline{p}_u,$$
$$(\overline{p}_u - \underline{p}_v)s_a + p_u - p_v - \sigma_r^p \leq \overline{p}_u - \underline{p}_v,$$

with slack variables σ_c^p for the compressor and σ_r^p for the control valve, put in the respective correct position.

In the computational experiments, 25 nominations remained infeasible, 1 reached the time limit with a feasible solution, and 24 were solved optimally with an average of 3.1 ideal compressors and 1.7 ideal control valves.

Flow amount So far, we have considered models that would induce changes of the network; we are now switching to nomination modifying models. An application for these could arise after an unsuccessful validation of a booking: As mentioned earlier, we would ask "how infeasible" a nomination is. Thus, the next model tries to find the "closest smaller" feasible nomination with respect to the given one, by allowing a smaller supply and demand at the entries and exits, respectively:

$$
\begin{aligned}
\min \quad & \|q^{\text{nom}} - \sigma^q\| \\
\text{s.t.} \quad & g(q,p,r) = a, \\
& h(q) = \sigma^q, \\
& k_{\text{active}}(q,p,r) \le c_1, \\
& k_{\text{passive}}(q,p,r) \le c_2, \\
& \underline{q} \le q \le \overline{q}, \\
& \underline{p} \le p \le \overline{p}, \\
& \underline{r} \le r \le \overline{r}, \\
& r \in \mathbb{Z}^m \times \mathbb{R}^n, \\
& 0 \le \sigma_u^q \le q_u^{\text{nom}} && \text{for all } u \in V_+, \\
& q_u^{\text{nom}} \le \sigma_u^q \le 0 && \text{for all } u \in V_-, \\
& \sigma_u^q = 0 && \text{for all } u \in V_0.
\end{aligned}
$$

The objective function value—in relation to the total supplied flow amount—should give a pretty good idea of how to answer our question.

Of the 50 nominations, 48 were solved to optimality, while 2 were aborted due to memory limits. In the minimum case, 0.8% of the original flow amount fed into the network had to be reduced ("almost feasible"), in the maximum case 25.1% ("highly infeasible"), with an average of thereby affected nodes of 9.1.

Supplier The studied nominations are potential situations with limits given by contracts (see Section 3.2). These contracts might contain additional agreements which permit the network operator to manipulate the incoming flows. With the next model, we allow the flow at entries to be reallocated to other entries:

$$
\begin{aligned}
\min \quad & \|\sigma^q - q^{\text{nom}}\| \\
\text{s.t.} \quad & g(q,p,r) = a, \\
& h(q) = \sigma^q, \\
& k_{\text{active}}(q,p,r) \le c_1, \\
& k_{\text{passive}}(q,p,r) \le c_2, \\
& \underline{q} \le q \le \overline{q}, \\
& \underline{p} \le p \le \overline{p}, \\
& \underline{r} \le r \le \overline{r}, \\
& r \in \mathbb{Z}^m \times \mathbb{R}^n, \\
& \sigma_u^q \ge 0 && \text{for all } u \in V_+, \\
& \sigma_u^q = q_u^{\text{nom}} && \text{for all } u \in V_- \cup V_0.
\end{aligned}
$$

In our numerical experiments, 3 nominations were infeasible, 2 reached a limit, and 45 were solved to optimality with an average of 1139.6 relocated flow units (1000 Nm³/h) and 19.3 affected nodes.

There are still other models imaginable, built in a similar fashion, which can be used to analyze infeasibility. The presented ones should demonstrate that it is possible to design models tailored to particular needs and interests (e.g., the aforementioned infeasibility in the validation of a booking). So, when deciding which model to use, the first question should be, What will the obtained information be used for? Otherwise, it might be a good idea to compare the solutions from several models. It is possible (and likely) that different models have ambiguous solutions (i.e., indicating other parts of the network as the source of infeasibility), hence, we cannot speak of "the" bottleneck. However, if multiple models highlight the same part, an actual congestion for the nomination was presumably found (see also Section 4.2.3.5).

11.6.2 ▪ Infeasible subsystems

The next basic approach is to search for a subsystem as small as possible of the corresponding MILP which is still infeasible. This is the concept of an IIS of the given MILP; see Chinneck (2008) for more details. The computation of IISs for MILPs is done in a straightforward way by removing constraints of the MILPs while maintaining infeasibility, with certain selection rules; see Guieu and Chinneck (1999) for details. This algorithm is implemented in most of the commercial MILP solvers, e.g., GUROBI and CPLEX.

What is left open after the computation is that, in fact, one wants an infeasible subnetwork, not a subsystem of the MILP. Thus, we identify network elements by their describing constraints (where one element can have more than one constraint). In the next step, we build the desired infeasible subnetwork by including every element in the network for which at least one constraint or an associated variable (e.g., the flow variable on a pipe) is part of the IIS. One of the properties of this subnetwork is that it is connected (see Joormann (2013)), hence, this network can be processed further to obtain more detailed information. By doing so, we gain a considerable limitation of the search area for manual search (e.g., for defective data) and, more generally, this method could be combined with the experience of network planners, or the SDF models in Section 11.6.1. The problem is that if the IIS is large, there is almost no informative value.

Although it is not always possible for MILPs in the considered dimensions to compute an *irreducible* infeasible subsystem within reasonable time, one can get a sufficiently good reducible infeasible subsystem (IS) with the same advantages as an IIS. The disadvantage, however, in contrast to the SDF models, is that there is no statement implied on how to resolve the infeasibility, or what the "reason" for the infeasibility is.

For our numerical experiments, we could compute 20 IISs; 30 nominations reached the time limit. The contained constraints translate to a minimum of 7 edges, 8 nodes and a maximum of 468 edges, 446 nodes (compared to 524 edges and 592 nodes in the original network).

This approach is not easily transferred to a basic model other than the MILP: In any model with a nonlinear part, it is nontrivial to compute an IIS, e.g., today's local solvers could report a model to be infeasible, but by adding one constraint, the solver may find a feasible point and the status turns to feasible.

11.6.3 ▪ Identifying bottlenecks of the network

The identification of a transport congestion or bottleneck of the network would be highly beneficial for a number of reasons, e.g., for a readily given extension decision (see Section 15.4). Furthermore, bottlenecks are used in the generation of worst-case scenarios (see Section 4.2.3). Unfortunately, mathematically defining a bottleneck of a gas network is not straightforward, let alone computing it.

The standard network approach of computing a min-cut does not work here, as the computational experiments for the flow bound model (Section 11.6.1) emphasize. Besides, the problem of both presented approaches, i.e., SDF models and IISs, is the dependence on the nomination, while we are actually looking for a bottleneck of the *network*. For an arbitrary model, one always needs a right-hand side and that is precisely the nomination. We will now sketch different attempts to break this dependence and we will see that it is not easily overcome.

Intuitively, we could try to use only lower and upper bounds instead of a fixed nomination. The problem with this approach is that for real-world data, the network is not so strict, meaning that, from our experience, we always end up with a feasible model.

In a second attempt, we could perform the presented analyses for multiple nominations and compare the results. But, unfortunately, there are most likely no overlappings of the affected elements, especially due to ambiguous solutions.

Instead of one nomination, we could try to regard several nominations at once (e.g., representatives of temperature classes). The first shortcoming of this idea is that it is not applicable for IISs, and for SDF with nominations $q_1^{\text{nom}}, \ldots, q_N^{\text{nom}}$, there would arise a system

$$
\begin{aligned}
g(q_i, p_i, r_i) &= a && \text{for all } i = 1, \ldots, N, \\
h(q_i) &= q_i^{\text{nom}} && \text{for all } i = 1, \ldots, N, \\
k_{\text{active}}(q_i, p_i, r_i) &\leq c_1 && \text{for all } i = 1, \ldots, N, \\
k_{\text{passive}}(q_i, p_i, r_i) &\leq c_2 && \text{for all } i = 1, \ldots, N, \\
\underline{q} \leq q_i &\leq \overline{q} && \text{for all } i = 1, \ldots, N, \\
\underline{p} \leq p_i &\leq \overline{p} && \text{for all } i = 1, \ldots, N, \\
\underline{r} \leq r_i &\leq \overline{r} && \text{for all } i = 1, \ldots, N, \\
r_i &\in \mathbb{Z}^m \times \mathbb{R}^n && \text{for all } i = 1, \ldots, N,
\end{aligned}
$$

with copies of the constraints for every nomination (note that q_i, p_i, and r_i are still vectors). This is problematic because there already is a little luck involved if the elementary models can be solved, let alone a model N times the size.

Instead, it may be worth looking at the following empirical approach. We start with a given set of nominations (feasible and infeasible) and ask how their status changes if the network elements are modified successively (scaling up or down the pressure interval at a fixed node, changing a pipe diameter, etc.). Then, we count how many infeasible nominations became feasible and vice versa. After such computations we could state that if a smaller (pressure, diameter, etc.) value does not lead to many more infeasible nominations, then the examined element is *not* a bottleneck. Similarly, if a larger value does not lead to many more feasible nominations, the element is *not* a bottleneck (at least not alone). Finally, if a larger value does lead to many more feasible nominations, the corresponding element *is* a bottleneck.

For this last approach a lot of individual computations need to be performed. However, since these are only ordinary validations of a nomination, they can be carried out much faster than SDF computations (see Section 11.6.1 as an indication for SDF computational times and Section 12.2 for validation of nomination computations). The computational effort can be justified by the independence of only a single nomination, although it should be clear that some sort of a preselection of the regarded elements is necessary; for a real-world network it is not possible to observe every element individually (nor every combination of these).

Chapter 12
Computational results for validation of nominations

Benjamin Hiller, Jesco Humpola, Thomas Lehmann,
Ralf Lenz, Antonio Morsi, Marc E. Pfetsch, Lars Schewe,
Martin Schmidt, Robert Schwarz, Jonas Schweiger,
Claudia Stangl, Bernhard M. Willert

Abstract *The different approaches to solve the validation of nomination problem presented in the previous chapters are evaluated computationally in this chapter. Each approach is analyzed individually, as well as the complete solvers for these problems. We demonstrate that the presented approaches can successfully solve large-scale real-world instances.*

12.1 ▪ Introduction

The main topic of Part II of this book is the validation of nomination problem (NoVa). The chapters in this part provide contributions to the solution of this problem. Since one of the driving forces of this book is to provide practically applicable approaches, it is thus time to computationally evaluate them.

To this end, we first define our test set of networks and nominations. We then first evaluate the behavior of the four decision approaches for NoVa in Sections 12.2–12.5. In particular, the influence of different algorithmic choices on the performance is evaluated. Since each approach has a quite different flavor, we aim at giving insights into the algorithmic behavior of these solvers. In the next step, the behavior of the validation NLP (ValNLP) is analyzed in Section 12.6. Here, the goal is to analyze the effect of applying ValNLP: To what extent are the initial solutions arising from the previous approaches changed in order to yield a feasible solution, i.e., a slack-0 solution? Finally, all these approaches are combined to yield a complete solver for NoVa. Indeed, it turns out that this solver is successful in solving large-scale instances.

We mention that Pfetsch et al. (2015) contains extensive computational results similar to the ones used here. However, we present more details in this book and partly use different test sets.

Since all our approaches for NoVa aim at solving a nonconvex mixed-integer nonlinear problem, the question whether it is possible to solve large-scale instances with state-

Table 12.1. *Overview of the test sets.*

Test set	$|V|$	$|A_{pi}|$	$|A_{re}|$	$|A_{sc}|$	$|A_{va}|$	$|A_{cv}|$	$|A_{cg}|$	# nom.
HN-AB	661	498	9	116	33	26	7	43
HN-SN	592	452	9	98	35	23	6	4227
gaslib-582	582	451	8	96	26	23	5	4227
gaslib-582-95	582	451	8	96	26	23	5	4227
HNP	592/661	452/498	9	98/116	33/35	23/36	6/7	100
gaslib-582-P	582	451	8	96	26	23	5	100
gaslib-582-95-P	582	451	8	96	26	23	5	100

of-the-art black-box mixed-integer nonlinear program (MINLP) solvers seems to be natural. For such an investigation with the global MINLP solvers BARON (Tawarmalani and Sahinidis (2002), (2004); Tawarmalani and Sahinidis (2005)) and SCIP (SCIP (2013); Vigerske (2012)), as well as the local MINLP solvers BONMIN (Bonami et al. (2008)), ALPHAECP (Westerlund and Pörn (2002)), and KNITRO (Byrd, Nocedal, and Waltz (2006)) we refer the interested reader to Pfetsch et al. (2015). Since this examination demonstrated that almost no instance can be solved by these general-purpose solvers within four hours, we decided to omit any additional comparison with such black-box solvers in this book.

12.1.1 ▪ Evaluating our hierarchy: Our test set

To show the practical relevance of our approaches, they are applied to four test sets (HN-AB, HN-SN, gaslib-582-P, gaslib-582-95-P) of country-size real-world gas networks and corresponding nominations; see Table 12.1 for an overview. The networks for the first two test sets arise at Open Grid Europe GmbH (OGE) and the ones for the last two are publicly available versions of a similar sized real-world network.[16] We operate with four different base test sets that can be characterized as follows:

HN-AB The network of this test set is the northern German high calorific gas network of OGE; see the purple network in Figure 1.5. It has 661 nodes, 498 pipes, 116 short cuts, 33 valves, 26 control valves (all with remote access), and 7 compressor groups. This test set contains 43 difficult hand-made expert scenarios.

HN-SN The network of this test set is topologically very similar to the one in HN-AB and contains 592 nodes, 452 pipes, 98 short cuts, 35 valves, 23 control valves (all with remote access), and 6 compressor groups. This test set contains 4227 nominations generated from statistical data as described in Section 14.2.

gaslib-582 This test set is part of the publicly available library for gas transmission problems (see Gaslib (2013)). The data have been disturbed in such a way that the resulting network has no resemblance with the original network, but it is still realistic. The network contains 582 nodes, 451 pipes, 96 short cuts, 26 valves, 23 control valves, and 5 compressor groups. The test set contains 4227 nominations. Generally, these nominations represent demanding flow situations. Especially the compressors typically need to work at high levels.

gaslib-582-95 This test set is the same as gaslib-582, except that the nominations are scaled by a factor of 0.95. As a consequence, the nominations are less

---

[16]See http://gaslib.zib.de

12.1. Introduction

demanding and, in particular, compressors typically do not need to operate at their limits.

In order to present the evaluations of the different decision approaches, we use three smaller sized test sets that represent parts of the above test sets:

HNP This test set consists of 100 nominations: 43 are given by HN-AB and the remainder consists of 57 randomly selected instances from HN-SN.

gaslib-582-P We randomly selected 100 nominations from test set gaslib-582.

gaslib-582-95-P This test set contains the same 100 nominations as in the test set gaslib-582-P, but scaled by a factor of 0.95.

12.1.2 ▪ Computational setup

The computations were performed on a cluster running an OpenSuse 12.3 Linux with a gcc 4.7.2 compiler. Each node has two 64 bit Intel Xeon X5672 CPUs at 3.20 GHz with 12 MB cache and is equipped with 48 GB main memory. In all experiments, we ran only one job per node to avoid random noise in the measured run time that might be caused by cache-misses if multiple processes share common resources. Moreover, all computations were run single-threaded, except for the MILP approach. All timings include the time for reading data and setting up the model, which for some instances consumes a major part of the running time.

The presented methods have been implemented in C++, using the framework LAMATTO++ (see LaMaTTO++ (2014); Morsi (2013)). The software provides the basic data handling, as well as model setup and interfaces to the following external solvers that are used in the different approaches: GUROBI 5.1 (Gu, Rothberg, and Bixby (2013)) and CPLEX 12.5.0.1 (CPLEX (2011)) are used to solve LP and MILP problems. The constraint integer programming framework SCIP (Achterberg (2009); Vigerske (2012), (2013)) is used in a prerelease version of SCIP 3.0. A variety of solvers are used to solve the different types of NLP problems: IPOPT 3.10 (Wächter and Biegler (2006)) with MA27 (HSL (2003), (2013)) as linear solver, CONOPT 3.15C and CONOPT 4.00 (see Drud (1992), (1994), (1995), (1996)), in the following called CONOPT and CONOPT4, respectively, and SNOPT, version 7.2-12 (Gill, Murray, and Saunders (2002)).

GUROBI was used to solve the constructed problems by means of the MILP approach (see Chapter 6). The sMINLP approach (see Chapter 7) was implemented in SCIP. The LP and NLP subproblems therein were solved using CPLEX and IPOPT, respectively. NLP problems in the RedNLP approach (see Chapter 8) are solved by CONOPT, CONOPT4, and IPOPT. CONOPT4, IPOPT, and SNOPT solve the NLP problems in the MPEC approach (see Chapter 9). Since the ValNLP model (see Chapter 10) can be tackled by several NLP solvers, we sequentially tried the solvers IPOPT, CONOPT, and CONOPT4 in this order until one of them converged to a feasible point (or all had been tried). This last step could, of course, be parallelized. The RedNLP and MPEC approach, as well as the validation NLP, use GAMS 23.8.2 (GAMS) to interface the solvers.

We imposed a time limit of four hours for finding discrete settings, i.e., for each of the four decision approaches from Chapter 6, Chapter 7, Chapter 8, and Chapter 9. The validation NLP has a time limit of one hour per solver.

Many of the results are illustrated by performance diagrams: Each point on the x axis corresponds to some given measure, e.g., running time, slack value, etc. We then display on the y axis the percentage of the total number of instances that were solved using at most the given measure on the x axis. All times are reported in seconds.

12.2 ▪ Results for the MILP-relaxation approach

In Chapter 6 we formulated a mixed-integer linear programming relaxation of the mixed-integer nonlinear feasibility problem of validating a nomination. The main characteristic of this approach is that all nonlinear functions are approximated by piecewise linear functions. Using an a priori bound on the approximation error we derived a piecewise-polyhedral outer approximation and thus a rigorous mixed-integer linear programming relaxation, which can be solved by any general-purpose mixed-integer linear program (MILP) solver. The discrete aspects of NoVa on the other hand, are modeled very accurately within this MILP formulation. Moreover, due to the relaxation property, this approach allows us to reliably conclude infeasibility of a nomination, at least with respect to the underlying MINLP model used to set up this relaxation.

Our aim in this section is to analyze the results of the MILP approach, introduced in Chapter 6, for our test instances in detail. In particular, we are interested in the quality of the solutions to NoVa computed by these MILP relaxations. Of course the run time is important as well and especially whether it is possible to solve the instances within an acceptable amount of time. It is hardly surprising that quality and run time are correlated objectives, both mainly controlled by the a priori approximation tolerance parameter τ_{abs}^{plf}. A smaller value of τ_{abs}^{plf} leads to tighter MILP relaxations of the underlying MINLP and simultaneously requires more auxiliary binary variables to represent the piecewise polyhedral envelopes. Conversely, the number of auxiliary binary variables is reduced, implying a weaker MILP relaxation if the parameter τ_{abs}^{plf} is increased. Besides the accuracy parameter τ_{abs}^{plf}, we investigate the impact of the separation of supporting hyperplanes to the convex relaxation of the admissible operating range, i.e., the characteristic diagram, of a turbo compressor, as described in Section 6.2.3. Since NoVa is a pure feasibility problem, we analyze, as a third degree of freedom, the effectiveness of the artificial objective function in the MILP approach introduced in Section 6.1.9.

In summary, we analyze the impact of all these degrees of freedom, i.e.,

(i) a predescribed upper bound on the maximum approximation error τ_{abs}^{plf} to represent nonlinear relationships by piecewise polyhedral outer approximations,
(ii) the application of additional cutting planes for a more accurate representation of characteristic diagrams of turbo compressors, and
(iii) the choice of the objective function.

We apply the MILP approach to the test set HNP. For this test set, we investigate the impact of the different parameter choices and discuss their advantages and disadvantages. Based on the results for test set HNP, we derive a suitable choice of parameters that is used subsequently for the computations on the remaining test sets gaslib-582-P and gaslib-582-95-P.

For our initial investigations on test set HNP we vary the upper bound on the maximum approximation error $\tau_{abs}^{plf} \in \{0.25, 0.5, 1, 2, 4, 8, 16, 32, 64, 128\}$. The absolute unit for this error during all approximations is bar. Thus a value of, e.g., $\tau_{abs}^{plf} = 4$ within a pressure loss equation of a pipe implies that the pressure loss in the relaxation cannot deviate more than 4 bar from the underlying nonlinear pressure loss equation. We recall from the introduction of τ_{abs}^{plf} in Chapter 6 that this is just an upper bound on the true maximum approximation error and does not have to be tight; indeed, most often this upper bound is not attained. Moreover, since all error measures are done a priori and with respect to the maximum error, the final error at some feasible or optimal point to the relaxation frequently yields a much smaller deviation from the underlying nonlinearities. For all our test instances, the largest value of $\tau_{\text{abs}}^{\text{plf}} = 128$ bar may be seen as an infinite error tolerance and hence does not lead to the introduction of auxiliary binary variables. The restriction to powers of two for $\tau_{\text{abs}}^{\text{plf}}$ is just used for convenience.

Violated cutting planes for characteristic diagrams of turbo compressors are either applied whenever a new incumbent solution is found or their generation is completely turned off. Moreover, we investigate three conceivable objective functions. We minimize the distance of each active compressor to a precomputed interior point of the characteristic diagram either with respect to the one-norm or infinity-norm (see Section 6.1.9), or we explicitly use no objective function.

Before the MILP relaxation is constructed, some additional preprocessing steps are performed to improve unnecessarily large variable bounds. Since the sizes of the variable domains have a direct impact on the size of the MILP relaxation, this step is crucial. In particular, we apply the three preprocessing techniques: bound propagation based on interval arithmetic, flow bound strengthening using min-cost-flow computations, and strengthening of the bounds for the pressure drop along resistors, as described in Section 5.4. Throughout our investigations, the resulting MILP relaxations are solved by GUROBI 5.1 (Gu, Rothberg, and Bixby (2013)) using default parameter settings on 8 threads, except that dual reductions and precrush are disabled if cutting planes for turbo compressors are separated.

For the sake of completeness, we remark that other general purpose MILP solvers, especially CPLEX, behave similarly. Nonetheless GUROBI seems to perform slightly better, which is the reason why we restrict our computations to this solver. Moreover, preliminary experiments with different GUROBI parameter settings did not yield significantly different overall results. Even a limitation to a single-thread branch-and-bound solution process does lead to comparable performance results. Since every computation is executed exclusively on a machine with two quad-core CPUs anyway, we decided to accept GUROBI's default maximum number of 8 threads.

The HNP test set The results for the MILP approach in terms of solution quality for NoVa on the combined test set HNP consisting of 100 nominations for all our parameter sets are summarized by the histograms shown in Figure 12.1. Therein instances that were detected to be infeasible by the corresponding MILP relaxation are shown in black. If the discrete decision obtained from the solution of the MILP relaxation results in a feasible nomination, indicated by a zero slack sum in the second stage validation NLP, the corresponding instances are shown in dark gray. The number of successfully solved instances is the sum of those two. Instances for which a feasible solution to the MILP relaxation was found, but the resulting slack sum is strictly greater than zero, are depicted in light gray. For the remaining instances no feasible solution to the MILP relaxation could be computed within the time limit.

Table 12.2 gives a more detailed investigation of these results. Here, for each parameter combination the number of successfully solved instances are shown, i.e., the number of instances out of 100 that are validated as feasible by a slack sum of zero plus the number of those detected to be infeasible by the MILP relaxation.

From the results given in Figure 12.1 and Table 12.2 one can conclude that an approximation error bound of $\tau_{\text{abs}}^{\text{plf}} = 16$ or greater leads to MILP relaxations that are in general too imprecise to be successful. Most notably, the coarsest accuracy value of $\tau_{\text{abs}}^{\text{plf}} = 128$, which can be regarded as infinity, yields (almost) no solved instance. This is not surprising

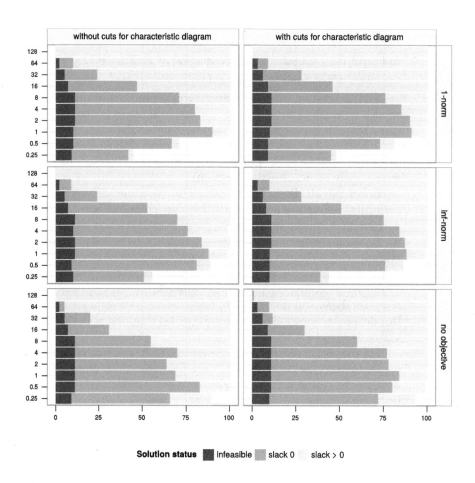

Figure 12.1. *Histogram of the solution status for the MILP approach on test set HNP.*

Table 12.2. *Successfully solved instances (slack sum zero or infeasible) for the MILP approach on test set HNP (100 instances).*

τ_{abs}^{plf}	1-norm obj.		∞-norm obj.		w/o obj.	
	w/o cuts	cuts	w/o cuts	cuts	w/o cuts	cuts
0.25	42	45	51	39	66	72
0.5	67	73	81	76	83	80
1	90	91	88	88	69	84
2	83	90	84	87	64	78
4	80	85	76	84	70	77
8	71	76	70	75	55	60
16	47	46	53	51	31	30
32	24	28	24	28	20	12
64	10	9	9	10	5	10
128	0	0	0	0	0	1

and confirms the fundamental nonlinear nature of NoVa. On the other hand, an error bound of $\tau_{abs}^{plf} = 0.5$ and smaller values lead to MILP problems that often reach the time limit, finding neither a feasible solution to the relaxation nor proving infeasibility. Counting the successfully solved instances, the error bound tolerances from $\tau_{abs}^{plf} = 1$ to $\tau_{abs}^{plf} = 4$ seem to be favorable. In fact, all of them are very reasonable.

The contribution of cutting planes for characteristic diagrams, as the second degree of freedom in our parameter analysis, is less pronounced. The separation of our cutting planes seems to be most effective in combination with a medium accuracy parameter ($\tau_{abs}^{plf} = 2$ to $\tau_{abs}^{plf} = 8$ with both norm objectives and $\tau_{abs}^{plf} = 1$ to $\tau_{abs}^{plf} = 8$ without objective) and they seem to become less important for a higher accuracy parameter ($\tau_{abs}^{plf} = 1$ with both norm objectives and $\tau_{abs}^{plf} = 0.5$ without objective, respectively).

Besides the accuracy parameter and the impact of cutting planes, the results when no objective is used in the relaxation are in general inferior to both nontrivial objective functions (in particular, if no cutting planes are used); an exception are small values of τ_{abs}^{plf}, where no objective seems to be advantageous in order to obtain some feasible solution within the time limit. Nevertheless, generally the minimization of both the one-norm and the infinity-norm distance to interior points of the characteristic diagrams is advantageous to obtain many more successfully validated instances. On the other hand, both norms seem to be quite comparable and no fundamental difference between them can be concluded.

Average sizes in terms of variables and constraints of the resulting MILP models are given in Table 12.3 for each accuracy parameter τ_{abs}^{plf}. All instances are based on the same network and thus, due to insignificant variation of the sizes for a fixed accuracy, such averaged numbers are indeed meaningful and representative. Most interesting is the number of binary variables. Starting with a model of only a few hundred initial binary decision variables ($\tau_{abs}^{plf} = 128$) up to approximately 26 000 of them are required on average for modeling the piecewise linear relationship with the most accurate relaxation parameter $\tau_{abs}^{plf} = 0.25$. While this number grows moderately with a doubling in precision for coarse relaxations (e.g., with a factor slightly above one from $\tau_{abs}^{plf} = 128$ to $\tau_{abs}^{plf} = 64$), the ratio approaches two for increasing accuracy.

At this point we again remark that a logarithmic increase in the number of binary variables would be possible if the underlying incremental model for piecewise linear functions is replaced by a logarithmic formulation (Vielma and Nemhauser (2011)). Nonetheless it has been shown computationally in the context of gas network optimization problems that this reduction in the number of binary variables is not always reflected by the run times (Geißler (2011); Morsi (2013)); in fact, the performance with a logarithmic model is in general even worse for NoVa. One explanation for this effect, given in the above references, is that due to the formation of holes in the feasible set after branching, the linear programming relaxations of the logarithmic model are in general not as strong as for the incremental formulation.

Besides the primary goal to obtain a high quality solution candidate from the MILP relaxation for the detailed second stage validation NLP, the run time required to solve the MILP problem is obviously not neglectable. Based on the quality results of the solutions, the majority of parameter choices are definitely inappropriate for our purposes, although the resulting problems might be solved within a few seconds; examples are too coarse relaxations, which do not reflect the nonlinear nature of the problem, and too accurate relaxations, which often cannot be solved within reasonable time. An overview of the

Table 12.3. *Average model sizes of the MILP relaxations on test set HNP.*

τ_{abs}^{plf}	Variables	Binaries	Constraints
0.25	56907.98	26618.55	87377.00
0.5	30488.54	13408.83	47747.84
1	17294.10	6811.61	27956.18
2	10729.20	3529.16	18108.83
4	7475.58	1902.35	13228.40
8	5890.42	1109.77	10850.66
16	5130.16	729.64	9710.27
32	4781.54	555.33	9187.34
64	4651.94	490.53	8992.94
128	4609.68	469.40	8929.55

Table 12.4. *Geometric means of run times (including NLP validation) for the MILP approach on test set HNP.*

	1-norm obj.		∞-norm obj.		w/o obj.	
τ_{abs}^{plf}	w/o cuts	cuts	w/o cuts	cuts	w/o cuts	cuts
0.25	7843.48	7867.97	7061.11	7126.53	1844.80	1826.59
0.5	4714.44	5166.57	3714.19	4240.92	467.89	535.99
1	1541.92	2044.27	937.05	1403.68	137.76	156.16
2	381.71	610.58	219.75	352.03	56.46	61.66
4	172.39	202.68	104.56	129.60	39.18	39.48
8	96.30	117.82	55.35	86.79	31.18	32.57
16	68.69	82.41	46.52	68.00	28.84	29.02
32	45.39	56.64	37.49	53.40	27.30	27.63
64	34.90	44.68	33.14	44.17	26.96	27.31
128	27.07	27.03	26.69	26.97	26.74	26.93

geometric means of the run times for each choice of parameters is presented in Table 12.4. In Figure 12.2 these run times are illustrated and compared against the success rate of the respective parameter set. Within this plot, "w. cuts" and "w/o cuts" indicate whether user-defined cutting planes are applied or not. The parameters 1-norm-obj., inf.-norm-obj., and no-obj. refer to the choice of the objective function.

As a final conclusion from the parameter investigations on this test set, we choose an upper bound on the approximation error bound of $\tau_{abs}^{plf} = 2$ in the following. This value is used because it yields a good balance between run time and success rate. Moreover, since the application of cutting planes in general does not significantly harm the run times and at least sometimes leads to an improvement of the quality, we decide to enable them from now on. Both nontrivial objective functions improve the number of successfully validated instances. Hence, we choose such a distance minimization to interior points of the characteristic diagrams. Since the results for both norms are almost indistinguishable, we arbitrarily pick the infinity-norm for our computations on the subsequent test sets.

The gaslib-582-P test set The results for test set gaslib-582-P, based on the parameter choice from the detailed analysis on test set HNP ($\tau_{abs}^{plf} = 2$, application of our

12.2. Results for the MILP-relaxation approach

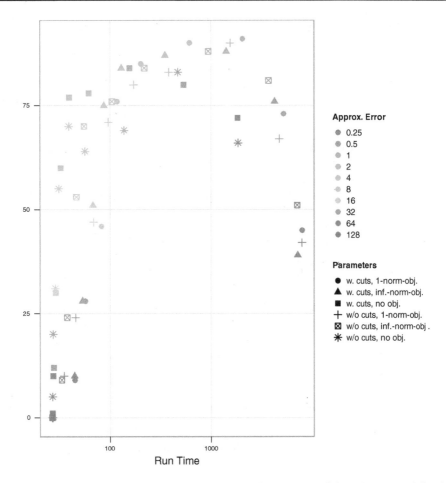

Figure 12.2. *Geometric means of run times (including NLP validation) vs. successfully solved instances for the MILP approach on test set HNP.*

cutting planes, and distance minimization based on the infinity-norm) in terms of solution quality, are as summarized follows. Altogether 68 instances out of 100 were solved successfully. Among these, 14 MILP relaxations were detected to be infeasible, and for 54 instances the respective solution candidate from the MILP relaxation leads to a slack sum of zero in the validation NLP. For the remaining 32 instances, the solution candidate merely yields a nonzero slack sum. On average a MILP relaxation has 4929.92 variables, 1617.58 binary variables, and 7870.74 constraints. The geometric mean of the run times for the MILP relaxations is 222.84 s.

The `gaslib-582-95-P` test set For this test set, again with the parameter settings resulting from the discussion of test set HNP, 85 instances out of 100 were successfully solved. Of these, 10 MILP relaxations were proved to be infeasible, and 75 solutions provide a solution candidate that is validated with slack-0 by the second stage NLP. The remaining 25 solutions to the relaxations only lead to a nonzero slack sum in the validation NLP. On average, a MILP relaxation of this test set has 4879.64 variables, 1592.44 binary variables, and 7795.42 constraints. The geometric mean of the run times for the MILP relaxations is 272.45 s.

12.3 • Results for the specialized MINLP approach

In the following, we document the results of the sMINLP approach, which is described in Chapter 7. First, we provide some additional implementation details of how and when the convex reformulations are solved. Then, we evaluate variants of the approach. To this end, we analyze the influence of the number of hyperplanes used to discretize the feasible operating ranges of active elements on the solvability and feasibility of the models. We show that the algorithm works faster and has consistent results if the elements are not discretized at all. In this case, using our approach does not yield a global solution, since we do not obey the theoretical restrictions developed in Chapter 7. As we face an existence problem, the only problematic case that may occur is the detection of infeasibility of a feasible problem. We provide computational evidence showing that this does not occur in practice.

The same line of argumentation is used for the consideration of slopes in the pressure loss equations of pipes; see Eq. (7.2). For the case that pipe slopes are approximated or ignored, our approach is in line with the theory. We show that it is practically justified to incorporate the respective coefficients in the model. Finally, we compare our problem specific approach to plain spatial branch-and-bound as described in Section 7.1.1. In all the experiments mentioned so far, we focus on our model and the different solution approaches. From these experiments, we conclude that our approach can be used in practice as an approximation using a model without discretization and including slopes. This configuration is finally evaluated with respect to the detailed ValNLP model from Chapter 10.

12.3.1 • Implementation details: Solving the convex reformulation

Recall that in the sMINLP approach, a convex reformulation (model (7.15)) of the problem is solved whenever all discrete decisions are fixed. In this section, we describe the implementation of the solution procedure for these nonlinear optimization problems in more detail.

As a first step preprocessing as provided by SCIP is applied to the reformulation. This preprocessing applies several techniques, such as bound tightening, aggregation of variables, removal of redundant constraints, etc., to make the problem more amenable for solution. Next, we solve the linear programming (LP) arising from the relaxation of the preprocessed problem. Both steps might prove infeasibility of the problem, in which case we cut off the node from the branch-and-bound tree. Otherwise, we solve the resulting NLP by IPOPT (Wächter and Biegler (2006)) as described in Section 7.2. Any resulting slack-0 solution is a valid solution candidate for NoVa. For solutions with positive slacks, we proceed as described at the end of Section 7.2 by solving (7.17). It turns out that this case does not occur in our computational experiments. Thus we did not implement the analysis of the NLP solution.

Furthermore, we solve the NLP (7.15) whenever all integer variables attain integral values in the LP solution of a node, even if they are not fixed. This sometimes yields a solution, but the node cannot be cut off, since other solutions might still exist in the subtree.

12.3.2 • The influence of discretization

In Section 7.3.3 and Section 7.3.4 we described a discretization of the operating range of active control valves and compressor groups. The discretization constraints (7.12) either yield a large number of hyperplanes or a strong restriction of the feasible solution space.

12.3. Results for the specialized MINLP approach

Therefore we analyze the impact of the discretization level in the following. We consider different numbers of hyperplanes, as well as neglecting these constraints in total. Using fewer hyperplanes yields a reduction of the model size by elimination of discrete variables, which indicate the particular hyperplanes, and thus leads to faster computation times. Moreover, we thereby circumvent the determination of an appropriate discretization level that a feasible instance requires to be solved to feasibility.

We therefore performed computational runs for different discretization levels $n \in \{9, 17, 33, 65, 129, 257, 513, 1025\}$ on the test set HNP. Here, each discretization level includes the hyperplanes of the coarser discretizations. Thus, the solution status of a particular instance can only change from infeasible to feasible along the discretization levels with increasing number of hyperplanes. Figure 12.3 shows the solution status and the run time of each instance per line.

The results reveal consistency between "no discretization" and "with discretization" in that the infeasible instances of the model without discretization are also infeasible for all discretization levels.

Concerning the solution status (left Figure 12.3), all instances have been solved to (in)feasibility without discretization, whereas for each discretization level there are instances that could not be solved within the time limit. Especially at discretization levels, where the solution status changes, e.g., $n = 129$, instances could not be solved. On coarse discretization levels, such as $n \in \{9, 17, 33\}$, infeasibility is detected very quickly, mostly in presolving.

Frequently, "no discretization" yields shorter run times than using discretization levels for instances that are not trivially infeasible; see right Figure 12.3. This might be due to the fact that additional branching decisions do not have to be taken into account without discretization. Even though, the run time of the solved instances is generally very short.

Further consistency with respect to the solution status, as described above, justifies a reduction of the model size by neglecting the discretization of the feasible operating range of the active elements.

12.3.3 • The influence of slopes

In this section we analyze the impact of the approximated slope coefficient in the pressure loss equation (7.11) on the feasibility of the problems. Recall that the basis of our solution approach is a special structure of the pressure loss equation for pipes.

We solve test sets HNP and gaslib-582-95-P with three configurations of the model and algorithm. Configuration *SB* refers to the model with exact pressure loss equation and pure spatial branching. This model yields a complete classification of the solution status of an instance. Configurations *exact* and *approximated* both apply the algorithm described in Section 7.2. The pressure loss equation used is exact and approximated, respectively. Table 12.5 gives an overview of the number of feasible and infeasible instances for different configurations.

We compare feasibility of individual instances between configurations *SB* and *exact*. All instances of HNP and gaslib-582-95-P that are feasible in *SB* are also feasible in *exact*, while there is one instance of gaslib-582-P that is feasible in *SB* but reached the time limit in *exact*. There are six instances in HNP where the time limit is reached with *SB*. Of these, only one of the instances is infeasible in *exact*. For another three instances from gaslib-582-P that hit the time limit in *SB*, two are infeasible, and one is feasible in *exact*. We conclude that the two configurations *SB* and *exact* yield similar results in all of the 200 instances.

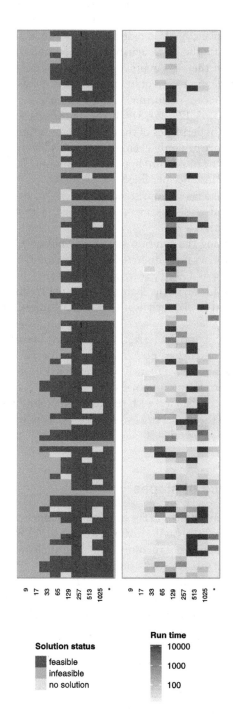

Figure 12.3. *Solution status and run time in seconds (excluding NLP validation) for different discretization levels for the sMINLP approach on test set HNP; "*" refers to the case without discretization.*

Table 12.5. *Solution status for different configurations of model and algorithm.*

Test set	Configuration	Feasible	Infeasible	Unknown status
HNP	SB	84	10	6
	exact	89	11	0
	approximated	88	12	0
gaslib-582-P	SB	41	50	9
	exact	41	52	7
	approximated	40	51	9
gaslib-582-95-P	SB	83	17	0
	exact	83	17	0
	approximated	83	17	0

We continue by comparing the configurations *exact* and *approximated*. Within test set gaslib-582-95-P, the results are always consistent, i.e., instances are feasible for *exact* if and only if they are feasible for *approximated*. There is one instance in test set HNP that is feasible for *exact* but infeasible for *approximated*. In test set gaslib-582-P, there are four instances, where either *exact* or *approximated* reached the time limit. We conclude that the configuration *approximated* yields almost equivalent results in practice compared to the configuration *exact*. Only one instance in 200 was declared infeasible although a feasible solution exists.

12.3.4 ▪ Choice of the reference inlet pressure for compressor modeling

In Section 7.3.4, we described how the feasible region of compressor machines is represented by polyhedral cones. Part of the construction is a conversion of measurements in a characteristic diagram to rays in the space of (q, p_u, p_v) variables. The conversion formula (7.22) assumes a constant compressibility factor z. In order to fix this value, we use an estimation of the expected (reference) inlet pressure p_u. Roughly speaking, the capacity of the compressor machine, in terms of mass flow and pressure ratio, increases slightly with p_u.

In this section, we investigate the impact of different choices for this value on the solution quality of the approach. We tested the range $\{1, 40, 50, 60, 70, 80\}$ bar, where we ensured that we only use values within the valid bounds w.r.t. the corresponding network data.

Setting a value that is too large results in a relaxation that is too weak and cannot detect infeasible instances. Values that are too small, on the other hand, declare infeasibility for instances where solutions actually exist. Three types of events are defined whose occurrence is to be avoided. First, we can never accept *false negative* results, i.e., our model is detected to be infeasible, while a validated (zero slack) solution is known. Secondly, we would like to minimize the number of *false positive* results, i.e., our model is feasible (but cannot be validated with zero slack), while infeasibility was proven otherwise. Lastly, there is the case where a validated solution is known, but the solution of the sMINLP approach cannot be validated. We call this situation *unhelpful*. To evaluate the above situations, we use results of the MILP approach from Section 12.2. More specifically, we use the settings $\tau_{\text{abs}}^{\text{plf}} = 2$, with cutting planes for turbo compressor enabled and the objective of minimizing the distance to the centroid of the characteristic diagram, w.r.t. to the infinity-norm.

Table 12.6. *Occurrence counts of unwanted results in the* HNP *test set.*

p_u	1	40	50	60	70	80
False negative	0	0	0	0	0	0
False positive	2	2	2	2	2	3
Unhelpful	13	16	15	16	19	23

Table 12.7. *Occurrence counts of unwanted results in the* `gaslib-582-P` *test set.*

p_u	1	40	50	60	70	80
False negative	18	1	0	0	0	0
False positive	0	0	0	0	3	3
Unhelpful	16	36	15	30	25	27

Table 12.6 counts the occurrence of these errors in the training test set HNP. Here, we never have false negative results, and the number of false positive results does not change noticeably with varying p_u. The number of unhelpful results grows from 13 to 23, as p_u increases, so the best result is achieved with a value of 1 bar (possibly adjusted to the lower bound of the individual compressor station). Therefore, we choose 1 bar as default value in all subsequent computations.

Overall, we see that the choice of p_u does not have a strong impact on the outcome within the test set HNP. Presumably, the compressors need not be operated at their limit in these instances, and solution values are not near the boundaries.

When we now evaluate the test set `gaslib-582-P`; however, we can see that this choice does not yield the expected results. As shown in Table 12.7, there are 18 occurrences of false negatives in test set `gaslib-582-P`. In fact, the choice of 1 bar is worst, while only values between 50 and 80 bar are acceptable. Of these, only 50 and 60 bar have no false positive results, while 50 bar has fewer instances with unhelpful results.

The analysis of test set `gaslib-582-95-P` is shown in Table 12.8. Here, the choice of 1 bar wins, with neither false negative nor false positive results, agreeing with the training test set. Similarly to test set HNP, the choice of p_u barely changes the outcome. The bad performance in test set `gaslib-582-P` might be due to higher throughput demands that lead to solution values near the boundary of the feasible region.

12.3.5 ▪ Comparison against spatial branch-and-bound

Consider the nonlinear optimization problem that is obtained by fixing all discrete decisions of our NoVa model. As described in Section 7.2, this problem is a passive pipe network problem that can be solved in two different ways. We either apply spatial branching (SB, see Section 7.1.1) or we proceed by solving the NLP reformulation mentioned in Theorem 7.5 (sMINLP, see Section 7.2) instead. Setting branching priorities to all our integer variables within SCIP, we ensure that spatial branching is only applied when all binary decisions are fixed. This is not necessary in general, but the discrete decisions affect the topology of the network to a great extent. A different topology usually leads to different flow values. Thus, branching on flow values is reasonable only if the topology is fixed.

Table 12.8. *Occurrence counts of unwanted results in the* `gaslib-582-95-P` *test set.*

p_u	1	40	50	60	70	80
False negative	0	0	0	0	0	0
False positive	0	1	2	2	1	3
Unhelpful	15	14	16	19	15	20

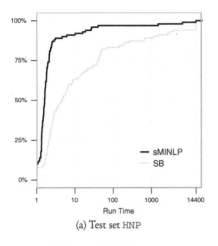
(a) Test set HNP

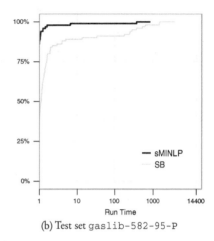
(b) Test set `gaslib-582-95-P`

Figure 12.4. *Run time comparison between spatial branching (SB) and specialized MINLP (sMINLP) (excludes NLP validation time).*

We analyzed both solution approaches concerning their solution time needed for solving the NoVa model and the number of nodes used in the branch-and-bound tree. According to Section 12.3.2, we perform this analysis without the discretization constraints (7.12) for active elements. We use the "exact" model for pipes as described in Section 12.3.3. A comparison of the run time of both approaches for our test sets HNP and `gaslib-582-95-P` is shown in Figure 12.4.

The run time could be reduced drastically following the sMINLP approach in comparison to spatial branching. The corresponding quantiles are shown in Table 12.9 and Table 12.10. This is also confirmed by the corresponding number of branch-and-bound nodes. In Figure 12.5 the differences in the solution time are shown in more detail.

Nevertheless, there is still an instance left that could not be solved by both approaches for the HNP test set within the time limit. But there are several instances running into the time limit when using spatial branching that could be solved by our sMINLP approach. For some instances the solving time does not show any differences. These cases turned out to be infeasible and every passive pipe network problem was infeasible by the LP relaxation. Thus no spatial branching and no NLP reformulation was used in the branch-and-bound process.

12.3.6 ▪ Performance on all test sets

As part of the comparison of all approaches in this chapter, we use the sMINLP algorithm as described in Chapter 7 (*not* the general algorithm based on spatial branching). The discretization of active elements is disabled, as discussed in Section 12.3.2. Following

Table 12.9. *Comparison of solver time and nodes for the* HNP *test set (excludes NLP validation time).*

	Run time		B & B nodes	
	SB	sMINLP	SB	sMINLP
Minimum	0.1	0.1	0	0
1st quartile	2.1	1.3	227	30
Median	4.7	1.6	598	47
Mean	1015.0	227.2	240 700	17 030
3rd quartile	40.3	2.1	7302	121
Maximum	14 400	14 400	6 095 000	1 465 000

Table 12.10. *Comparison of solver time and nodes for the* gaslib-582-95-P *test set (excludes NLP validation time).*

	Run time		B & B nodes	
	SB	sMINLP	SB	sMINLP
Minimum	0.0	0.0	0	0
1st quartile	0.9	0.7	98	25
Median	1.1	0.8	139	28
Mean	55.6	4.4	23 210	923
3rd quartile	1.5	0.9	287	32
Maximum	1521	363	779 200	88 930

Section 12.3.3, no approximation of the pressure loss constraint for pipes (7.11) is employed. Furthermore, we set higher branching priorities on the binary variables, compared to the continuous variables. Finally, the solver framework SCIP is configured to use the node selection strategy restartdfs with a restart frequency of 50.

HNP test set All 100 instances in the HNP test set can be solved within the time limit. They divide into 89 feasible and 11 infeasible problems for the sMINLP model; the solution is always found by solving a nonlinear subproblem with all discrete decisions fixed after branching. Among the 89 feasible instances, 69 are validated with zero slack in the ValNLP model (see Section 10.4). The infeasibility of the remaining 11 instances is shown during the preprocessing stage 8 times. For the three other infeasible instances, 120, 241, and 913 350 nodes are solved using 2.15 s, 2.77 s, and 11 909.05 s, respectively.

gaslib-582-P test set The gaslib-582-P test set consists of 100 instances, of which 93 are solved within the time limit. Among the 41 feasible instances, only 22 lead to a zero slack solution of the ValNLP. Again, the solution is discovered after solving a nonlinear subproblem in SCIP. The infeasible instances are either detected during preprocessing (13 times) or in the tree search (39 times).

gaslib-582-95-P test set Of the gaslib-582-95-P test set, all 100 instances are solved within the time limit. The 83 feasible instances lead to zero slack solutions of the ValNLP in 65 of the cases. Again, the solutions are discovered after solving a nonlinear subproblem in SCIP. The infeasible instances are either solved during preprocessing (11 times), at the root node (once), or in the tree search (5 times).

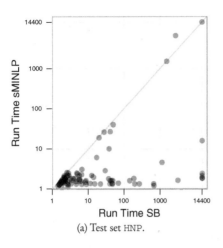 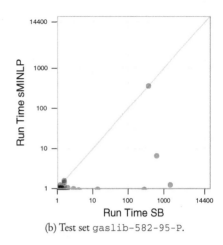

Figure 12.5. *Run time comparison between spatial branching (SB) and specialized MINLP (sMINLP). Both axes show the solver run time for the NoVa model in seconds (excludes NLP validation time).*

12.4 • Results for the reduced NLP heuristic

This section provides information about the results of the model described in detail in Section 8. Regarding the pressure loss over pipes, the model is based on the approximation given in Eqs. (2.24)–(2.29). For the ease of elimination of variables, differences in geodetic height are neglected. Compressor groups are modeled approximately as detailed in Section 8.2.1.

The RedNLP approach uses a priori fixing of binary decisions, which is done by two different methods. First the transshipment heuristic described in Section 8.5.1 is used for the current nomination. This way, promising binary decisions are fixed on the basis of the given flow situation at the exit and entry points and a set of rules. If the RedNLP problem resulting from this choice is not solved successfully, the algorithm is continued by testing a set of given binary configurations which successfully solved other nominations.

The resulting NLP could be solved by any general-purpose NLP solver. Due to the nonconvexity of the problem we often observed that the solution process can get stuck in an infeasible region or converge to an infeasible point. Thus it turned out to be favorable to interrupt the run by an iteration limit and to warm start a second, different NLP solver from the iteration point with least primal infeasibility in case no feasible point was found in the first run. Because we try to solve a nonlinear and nonconvex problem, the described technique is often necessary for tackling NoVa.

If a feasible solution to the RedNLP model is found, this solution is used as a starting point for the detailed ValNLP. The solution is called confirmed if the ValNLP then converges to a solution with zero slack.

In the process of deriving promising switching decisions using the transshipment heuristic and solving the RedNLP model we use the following adjustable parameters:

1. Transshipment problem
    (a) choice of the (non)linear programming solver,
    (b) values for the fictitious costs for valves, control valves, and compressor stations and for connecting arcs within stations.

250 Chapter 12. Computational results for validation of nominations

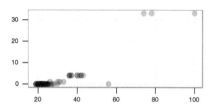

Figure 12.6. *Profile of the run time in seconds for the RedNLP approach on test set* HNP *including NLP validation; y axis: number of used configurations (total: 33). Zero refers to switching decisions given by the transshipment problem.*

2. RedNLP
   (a) choice of a nonlinear programming solver as first solver;
   (b) choice of a nonlinear programming solver as second solver;
   (c) use of an objective function;
   (d) tolerances for Eqs. (8.13)–(8.15), (8.16), (8.17), (8.20), and (8.22); these are modeled as *softened equalities*, i.e., as double-sided inequalities within a given tolerance;
   (e) choice of a factor ω in the objective function Eq. (8.21).

Extensive computational experiments showed that none of the tested settings clearly dominated the other settings; some parameter settings are better for a part of the nominations on one network, while other settings work better for other nominations or networks. All checked variants solved more than two-thirds of the test instances in comparable time.

The results shown below are achieved by the following parameter settings:

1. Transshipment problem
   (a) Solver: CONOPT4
   (b) Fictitious costs
       ▷ valves: 10^{-4};
       ▷ control valves forward: 10^{-5}, backward: 10^2;
       ▷ compressor stations forward: 10^{-11}, backward: 10^1;
       ▷ connection arc: 10^{-10}.

2. RedNLP
   (a) First solver: CONOPT.
   (b) Second solver: IPOPT.
   (c) Objective function: Minimize pressure increase over compressor stations and compressor decrease over control valves in addition to the cycle flow with $\omega = 0.001$.
   (d) Tolerances for
       ▷ bypass constraints Eq. (8.16), Eq. (8.17): 10^{-4},
       ▷ resistor constraints Eq. (8.20): 10^{-5},
       ▷ constraints for active elements Eq. (8.22): 10^{-6},
       ▷ constraints for compressor stations Eqs. (8.13)–(8.15): 10^{-3}.

If the first solver CONOPT is not able to find a feasible solution for the RedNLP model within the given iteration limit of 3500 iterations, the second solver IPOPT is used, starting at the iteration point from CONOPT with least infeasibilities.

12.4. Results for the reduced NLP heuristic

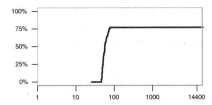

Figure 12.7. *Profile of the run time in seconds for the RedNLP approach on test set* HNP *including NLP validation; y axis: percentage of confirmed instances (total: 100).*

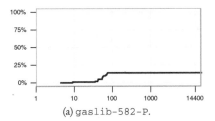 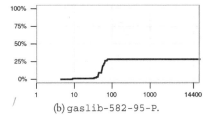

(a) gaslib-582-P.   (b) gaslib-582-95-P.

Figure 12.8. *Profile of the run time in seconds for the RedNLP approach on tests set* gaslib-582-P *and* gaslib-582-95-P *including NLP validation; y axis: percentage of confirmed instances (total: 100).*

For the first test set (HNP), the RedNLP model can solve 93 nominations, 78 of which were found using the switching decisions derived from the solution of the transshipment problem. In Figure 12.6 the connection between time spent and the number of used configurations is shown. The second solver found a solution for 34 instances. For 77 of the 93 solutions, ValNLP converged to a solution with slack 0.

The total run times for the RedNLP range from 18 to 97 s with an average of 25.58 s and a median of 23 s. Figure 12.7 shows the results for this test set. The transshipment problem for this network has 198 variables and 108 equality constraints. The size of the RedNLP depends on the chosen switching decisions. The number of variables varies between 1408 and 1441, the number of equality constraints between 1318 and 1346, and that of inequality constraints between 60 and 85.

Out of the 100 nominations in the second test set (gaslib-582-P), a solution was found for 29 nominations. The second solver found a solution for 3 instances. 14 of the 29 solved nominations were confirmed, which is shown in Figure 12.8(a). The spreading of configurations that led to a feasible solution is shown in Figure 12.9(a). The run times range from 9 to 57 s with an average of 40.41 s and a median of 39.5 s.

We did not use the transshipment heuristic for this test set, because this heuristic only works well if you have some knowledge about the network or if you spend some time becoming acquainted with the network. We instead use 21 fixed configurations. This is the reason why the solutions for the test set gaslib-582-P are worse than for the test set HNP.

The results are slightly better for the test set (gaslib-582-95-P). Here a solution was found for 33 of the 100 instances, out of which 28 were confirmed. The second solver found a solution for 3 instances. The run times range from 9 to 136 s with an average of 40.66 s and a median of 43 s.

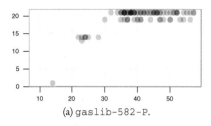
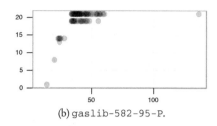

(a) gaslib-582-P.   (b) gaslib-582-95-P.

Figure 12.9. *Profile of the run time in seconds for the RedNLP approach on test sets* gaslib-582-P *and* gaslib-582-95-P *including NLP validation; y axis: number of used configurations (total: 21).*

12.5 ▪ Results for the MPEC based heuristic

In this section, we investigate the computational properties of the mathematical program with equilibrium constraints (MPEC) based heuristic.

12.5.1 ▪ Solving MPECs as (regularized) NLPs

In Chapter 9 we discussed NoVa as a *nonsmooth nonlinear mixed-integer feasibility problem* (nMINLP). Since this model is too hard to be solved directly in practice, we presented a cascade of reformulations leading to a smoothed mathematical program with equilibrium constraints (sMPEC). For this, all discrete aspects of nMINLP are reformulated by complementarity constraints whereas all nonsmooth aspects are approximated by general or problem-specific smoothings. Since our principal aim is to set up a fast primal heuristic for the underlying nMINLP, we follow an NLP driven approach, i.e., we develop a technique to solve the problem heuristically with general-purpose NLP solvers like IPOPT, CONOPT, or SNOPT. See Wächter and Biegler (2006); Drud (1992), (1994), (1995), (1996); and Gill, Murray, and Saunders (2002).

However, the reformulated and complementarity constrained model sMPEC has the significant drawback that it lacks standard constraint qualifications (CQs) like the Mangasarian–Fromowitz CQ (MFCQ) or the linear independence CQ (LICQ). The reader interested in an introduction to CQs is referred to Nocedal and Wright (2006). To be more precise, it can easily be shown that these CQs do not hold for *any* feasible point of the MPEC (see Ye and Zhu (1995)), for an early proof). As a consequence, standard NLP solvers cannot be applied to MPEC models directly without losing their theoretical convergence results. Due to this fact, an active field of research in the last decades has been to investigate techniques for solving MPECs, and mainly two branches of solution strategies evolved. On the one hand, one tries to develop MPEC tailored solution strategies that tackle the MPEC directly. On the other hand, one tries to develop reformulation techniques leading to sequences of NLPs such that every NLP of these sequences is regular (in the sense of a reasonable CQ) and such that the limit points of the sequences of optimal solutions converge to some type of MPEC stationary points (see Scheel and Scholtes (2000), for MPEC tailored stationarity concepts).

As we already mentioned, we want to exploit standard NLP solvers for tackling the MPEC problem. Thus, we have to apply regularization techniques transforming sMPEC to rsMPEC (see Chapter 9). As it is often the case in nonlinear programming, the *best* approach strongly depends on the concrete problem at hand. Thus, we tested several reformulation techniques on large test sets:

▷ The *relaxation approach* (see Scholtes (2001)): the complementarity constraints

$$\phi_i(x)\psi_i(x) = 0$$

are approximated by relaxations

$$\phi_i(x)\psi_i(x) \leq \tau, \quad \tau \geq 0. \tag{12.1}$$

Remember that $\phi_i(x), \psi_i(x) \geq 0$ holds in an MPEC standard form.

▷ The *smoothing* (or *NCP function*) *approach* (see Fischer (1992); Sun and Qi (1999)): the complementarity constraints are replaced by NCP functions like the perturbed Fischer–Burmeister function or smoothings of the min function.

▷ The *penalization approach* (see Hu and Ralph (2004)): the complementarity constraints are completely removed from the set of constraints and their violation is penalized by additional objective function terms $\pi(\phi_i(x), \psi_i(x))$ (see Section 9.2).

▷ The *direct approach*: even though one loses convergence theory, it might be possible to apply standard NLP solvers to the MPEC directly (see Fletcher and Leyffer (2004); Rodrigues and Monteiro (2006)).

In addition to the choice of the regularization approach, the type of NLP solver is crucial, too. In Fletcher and Leyffer (2004), numerical results are discussed that show that the SQP code FILTERSQP (Fletcher and Leyffer (2000)) outperforms the interior-point codes LOQO and KNITRO. However, interior-point methods that are especially tailored for MPECs came up subsequently to Fletcher and Leyffer (2004) (e.g., Benson, Shanno, and Vanderbei (2002); Leyffer, López-Calva, and Nocedal (2006); Liu and Sun (2004); Raghunathan and Biegler (03/2005)), and can today be assumed to be competitive with SQP methods or other algorithmic concepts of nonlinear optimization.

Since, in preliminary tests, the use of NCP functions did not offer good results for our model, we concentrate on the *relaxation* and *penalization approach* in the following. In addition, we compare them with the *direct approach*, in which an NLP solver is applied to the original complementarity constrained problem without any regularization.

12.5.2 ▪ Discussion of the numerical results

The MPEC is divided into two problems as described in Section 9.3: the first stage problem models the discrete states of the active elements by complementarity constraints, while the second stage focuses on the active configurations of the compressor groups.

As described in Section 9.2, we do *not* solve a series of problems with decreasing parameters τ. In case of the penalization approach, we choose $\tau = 10^{-6}$, which proved to be a good choice in preceding parameter tests. In case of the relaxation approach, a constant positive value does not require the complementarity constraints to hold, which leads to uninterpretable situations in practice. Instead, we handle the regularization parameter τ as a variable and minimize it in the objective function. The initial value of this variable is chosen to be 100.

When testing our approaches, we often recognized that additional *artificial* gas may flow in cycles of subnetworks around active elements in solutions of the rsMPEC model. This does not contradict the flow balance equation (9.2) but should be avoided, since this behavior cannot take place in reality and because it makes the second stage of the MPEC heuristic harder to solve. As a remedy, the absolute values of flows through active elements are additionally minimized. A side effect is that the first stage tends to close active elements, when possible.

The resulting optimization problems are solved with three general-purpose NLP solvers based on different optimization techniques:

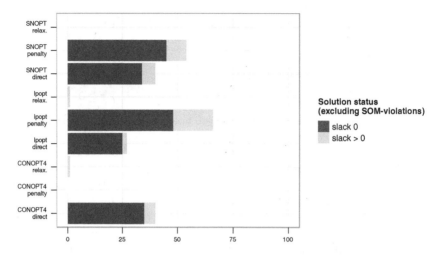

Figure 12.10. *Solution states for the tested MPEC regularizations (relaxation, penalty, and direct) and NLP solvers (SNOPT, IPOPT, and CONOPT4) on the HNP test set.*

▷ the interior-point method IPOPT (Wächter and Biegler (2006));
▷ the sequential quadratic programming (SQP) method SNOPT (Gill, Murray, and Saunders (2002));
▷ the generalized reduced-gradient (GCG) method CONOPT4 (Drud (1992), (1994), (1995), (1996)).

Since our aim here is to analyze different NLP solution strategies applied to different MPEC regularization techniques, these choices only influence the first stage of our MPEC heuristic, since the second stage does not contain complementarity constraints (see Section 9.3.2). The optimization problem of the second stage is solved by IPOPT in all cases.

The techniques described in Section 9.2 are practicable for model aspects with two (or, in some special cases, three) discrete states (see Schmidt (2013), for a more detailed discussion). Due to their general nature, subnetwork operation modes cannot be represented in the same way with a reasonable number of complementarity constraints. Thus, subnetwork operation modes are not considered by the MPEC approach and may be violated.

The HNP test set Figure 12.10 shows the solution states of all MPEC-regularization and NLP-solver combinations on the HNP test set. The computations show that the relaxation approach is drastically inferior to the penalization and direct approach. Only in the cases of IPOPT and CONOPT4 do a few instances result in feasible solutions of the validation NLP, but none of these instances are validated with vanishing slack sum value. CONOPT4 seems to have systematic problems with the penalization formulation—no solution can be found for the first MPEC stage. The best results are obtained by the penalization approach solved by IPOPT.

Since one of the approaches may find solutions of instances for which the other approaches do not find any, it is reasonable to try all regularization approaches with all solvers in parallel and finally choose the most feasible solution as a solution candidate for the NLP validation. If we do so, the ValNLP returns a feasible solution for 75 of the 100 instances, and 61 are validated with slack sum 0.

12.5. Results for the MPEC based heuristic

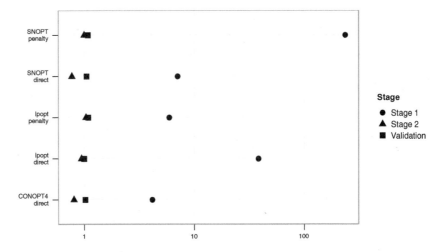

Figure 12.11. *Comparison of the average run time in seconds of successfully solved instances for Stage 1, Stage 2, and the validation NLP on the HNP test set.*

In the cases of SNOPT and IPOPT, the gain in regularity by the penalization approach is visible. Applying the penalization approximately doubles the number of validated solutions in the case of IPOPT, compared to solving the complementarity constrained model directly. In the case of SNOPT, the gain is also significant.

Figure 12.11 shows the average solver run time in seconds of all solved test instances for both MPEC stages and the ValNLP. All combinations of NLP solvers and MPEC regularization strategies for which no feasible solution can be found are missing in Figure 12.11. The run times show that solving the first MPEC stage is the hardest part of the MPEC based heuristic. The average run time of the first MPEC stage is about 380s for the combination SNOPT/penalization, about 46s in case of the direct approach solved by IPOPT, and about 6s in the three remaining cases. The second stage and the ValNLP require about a second for a successful run on average.

There are several possible reasons why the complete MPEC based primal heuristic may fail for a single instance:

▷ The chosen solver may fail to solve the first stage, i.e., the (regularized) MPEC.
▷ The solver used for the second stage may fail to find an optimal solution based on the solution of the first stage.
▷ The MPEC approach delivers a solution candidate but the solver chosen for the validation NLP fails to converge to a solution.
▷ The solver used for the validation NLP finds a solution with nonvanishing slack variables.

Our experiments show that all reasons occur in practice. Table 12.11 lists one row for every MPEC solution strategy together with the most successful solver for this strategy, i.e., the one with the most instances of vanishing slack sum value (see Figure 12.10). For this solver, we analyze the success rates of the first and second stages as well as for the validation NLP. Despite its relatively large freedom due to the regularization *variable* τ, the relaxation approach has difficulties in finding any solution when solved by IPOPT. Only one of the 12 solutions of the first stage can be processed further by the second stage—but it is not validated with a sufficiently small slack sum by the validation NLP. The penalization strategy solved by IPOPT performs better than the direct approach using

Table 12.11. *Success rates of Stage 1, Stage 2, and the validation NLP for selected combinations of MPEC regularizations and NLP solvers on the HNP test set.*

MPEC approach	Solver	Stage 1	Stage 2	ValNLP converged	ValNLP slack sum 0
Relaxation	IPOPT	12	1	1	0
Penalization	IPOPT	100	70	66	50
Direct	CONOPT4	50	41	40	33

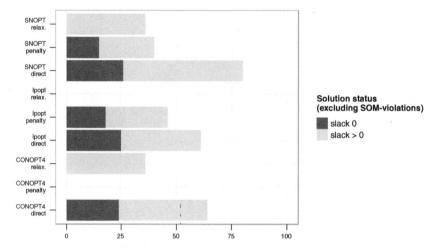

Figure 12.12. *Solution states for the tested MPEC regularizations (relax., penalty, and direct) and NLP solvers (SNOPT, IPOPT, and CONOPT4) on the* `gaslib-582-P` *test set.*

CONOPT4, since it solves 50 test cases with vanishing slack values compared to 33 in case of CONOPT4. Both settings show a very different behavior of the intermediate results. While all test cases pass the first MPEC stage in case of the penalization approach, no solution candidates could be generated for 30 instances, and another 20 are rejected by the ValNLP. Using the direct approach, the first stage appears to be much harder to solve, since only 50 instances pass this stage successful. This behavior meets our expectations, since the complementarity constraint violations are only minimized in the penalization approach and do not have to be fulfilled exactly as by the direct approach. In the second stage only 9 cases are lost in case of the direct approach, compared to 30 cases using the penalization approach. This indicates that the direct approach may be harder to solve in Stage 1, but the chance of a solution of the first stage leading to a successful validation is higher.

The `gaslib-582-P` test set The `gaslib-582-P` test set seems to differ fundamentally from the HNP test set; see Figure 12.12. It seems to be significantly harder to solve and, interestingly, the direct approach is the best choice in this case. In combination with the solver SNOPT, the direct approach finds 26 solutions with a vanishing slack sum value, while 25 cases are validated with slack sum value 0 in case of IPOPT and in case of CONOPT4 24 cases are validated. Similar to the HNP test set, the relaxation approach performs poorly for all solvers and CONOPT4 does not cope with the penalization

12.5. Results for the MPEC based heuristic

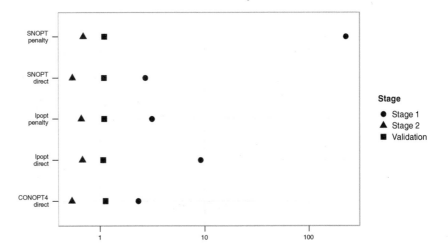

Figure 12.13. *Comparison of the average run time in seconds of successfully solved instances for Stage 1, Stage 2, and the validation NLP on the* `gaslib-582-P` *test set.*

Table 12.12. *Success rates of Stage 1, Stage 2, and the validation NLP for selected combinations of MPEC regularizations and NLP solvers on the* `gaslib-582-P` *test set.*

MPEC approach	Solver	Stage 1	Stage 2	ValNLP converged	ValNLP slack sum 0
Relaxation	CONOPT4	99	63	36	0
Penalization	IPOPT	99	92	46	18
Direct	SNOPT	98	86	80	26

approach. When combining all approaches, the ValNLP converges for all of the 100 test cases, and 48 of all cases are validated with slack sum 0.

The average run times are comparable to those of the HNP test set. The first stage results require slightly less time on average, as do the results of the second stage; see Figure 12.13. ValNLP requires about a second for every combination.

The results of the best solver for each regularization scheme are listed in Table 12.12. Nearly all problem instances pass the first stage for all listed approaches. The gap between the first and second MPEC stages indicates the quality of the generated discrete decisions. Only a small number of instances that pass the first stage fail in the second for the penalization approach and the direct approach. In contrast, about one third of the instances that pass the first stage fail in the second stage when the relaxation approach is applied. Two differences in the model exist between these stages: the discrete decisions, except for the active configurations, are fixed in the second stage and the compressor groups are modeled in a more sophisticated way. Since the more detailed compressor model is fully relaxed, it is hardly the reason for the large number of failed instances. More probably, the relaxation parameter is not minimized to zero, resulting in bad interpretations of the relaxed complementarity constraints.

A large number of instances do not result in a vanishing slack sum of the validation NLP. A possible reason is an overestimation of the capabilities of the compressor groups that results in inapplicable discrete decisions.

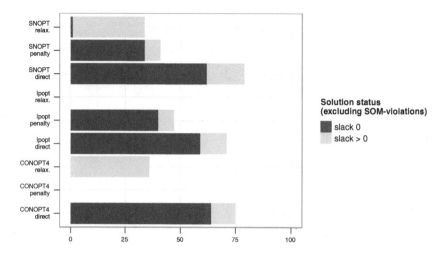

Figure 12.14. *Solution states for the tested MPEC regularizations (relax., penalty, and direct) and NLP solvers (SNOPT, IPOPT, and CONOPT4) on the test set* `gaslib-582-P` *95.*

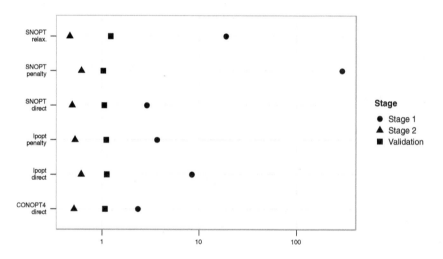

Figure 12.15. *Comparison of the average run time in seconds of successfully solved instances for Stage 1, Stage 2, and the validation NLP on the* `gaslib-582-95-P` *test set.*

The `gaslib-582-95-P` test set In this test set, the supplied and discharged flows are down-scaled to 95% of those of the `gaslib-582-P` test set. This small change has a significant impact on the results of the MPEC approach.

Figure 12.14 shows the solution states in the case of the `gaslib-582-95-P` test set. Again, the relaxation approach performs poorly and CONOPT4 has difficulties when applied on the penalization approach. As with the `gaslib-582-P` test set, the direct approach is the best choice in this case. In combination with the solver CONOPT4, the direct approach finds 64 solutions with a vanishing slack sum value, while in the case of SNOPT, 62 cases are validated with slack sum value 0. When combining all approaches, the ValNLP converges for all of the 100 test cases, and 86 of all cases are validated with slack sum 0.

12.5. Results for the MPEC based heuristic

Table 12.13. *Success rates of Stage 1, Stage 2, and the validation NLP for selected combinations of MPEC regularizations and NLP solvers on the* `gaslib-582-95-P` *test set.*

MPEC approach	Solver	Stage 1	Stage 2	ValNLP converged	slack sum 0
Relaxation	SNOPT	98	63	34	1
Penalization	IPOPT	99	92	47	40
Direct	CONOPT4	99	87	75	64

The average solver run time of all successful computations is shown in Figure 12.15. This confirms the results and their interpretation given for the HNP and the `gaslib-582-P` test sets.

Table 12.13 lists detailed success rates for three parameter choices. Again, the solver with the largest number of instances with vanishing slack values in the ValNLP run is chosen for each approach. In contrast to Table 12.11, the results for the relaxed MPECs are now discussed for SNOPT. The first stage is solved by all three combinations for at least 98 instances. The quality of these solutions can be seen in the second stage and the ValNLP run. While the penalization and the direct approaches find 92 and 87 solutions, respectively, the solutions of the relaxation approach can only be processed in the second stage for 63 of 98 instances. This indicates that the regularization parameter τ is not reduced enough in many cases and the resulting solutions cannot be interpreted as correct switching decisions. In case of the penalization approach, 47 of the remaining 92 instances converge to a solution of the ValNLP, of which 40 are validated with vanishing slack values. The results of the direct approach offer a higher quality than the results of the penalization approach, since 75 of the 87 MPEC solutions converge, of which 64 validate with vanishing slack sum value.

Compared to the unscaled `gaslib-582-P` test set, a much larger number of instances is successfully validated. Possibly, the larger flows require the compressor groups to operate near the boundary of their feasible operating ranges. Since the first stage does not incorporate the operating ranges, the first stage solution might be infeasible with respect to them, which might result in a bad starting point for the second stage.

12.5.3 ▪ Summary

There is no clear winner among the solvers and regularization schemes, since the results differ quite strongly between the test sets. Overall, a good choice is the penalization approach solved by the interior-point method IPOPT. It is the best combination in the case of the HNP test set and it solves a large number of instances in the cases of `gaslib-582-P` and `gaslib-582-95-P`.

The success rate of the regularization scheme is far behind the penalization approach or the direct approach. Compared to the original scheme (Scholtes (2001)), two differences exist. First, we do not solve a sequence of problems, but minimize the relaxation parameter in the objective. Second, we square the terms in the complementarity constraint if they can be negative, i.e., instead of (12.1) we state

$$\phi_i(x)^2 \psi_i(x)^2 \leq \tau, \quad \tau \geq 0$$

in these cases. The squared functions seem to increase the problem difficulty drastically. Alternative modifications of (12.1) are suggested in Willert (2014).

In general, the direct approach is harder to solve than the penalization approach, i.e., fewer instances pass the first MPEC stage. The quality of the first stage solutions of the direct approach is better, in the sense that more of them are validated by the ValNLP. However, in the case of the HNP test set, the increased number of results passing the first stage of the penalization approach results in a larger number of validated instances for SNOPT and IPOPT anyway.

In summary, the MPEC approach offers a fast primal heuristic for networks similar to the networks of HNP and `gaslib-582-95-P`, when no subnetwork operation modes have to be considered. The average combined solver run time of a successful run including the validation is about 10 s. In the case of the HNP test set, the interior-point method IPOPT and the SQP method SNOPT benefit from the gained regularity of the penalization approach, in contrast to the GCG algorithm CONOPT4.

12.6 • Results for the validation NLP

In this section, we present and discuss some statistics about the ValNLP model (see Section 10.4) and its solution process.

12.6.1 • Model and solution statistics

The ValNLP instances solved for the `gaslib-582-95-P` test set all contain about 10 400 variables including slack variables (see Section 10.3) and about 4710 constraints. They result in about 22 850 nonzero entries in the Jacobian of the constraints and in about 4550 nonzero entries in the Hessian of the Lagrangian. Despite the fact that all instances are computed on the same network, the exact numbers vary between the instances and solvers. The reason is that different discrete decisions determined by the decision approaches and different decisions made in preprocessing lead to slightly different instances. The largest deviation in the number of variables is 130 and the largest deviation in the number of constraints is 59.

A sequence of NLP solvers, consisting of IPOPT, CONOPT, and CONOPT4, is used to solve the ValNLP model. If a solver does not find a solution with an objective value that is below a given positive and small threshold, the computation is restarted using the next solver until all solvers have been applied. Since IPOPT is the first solver in the sequence, all other solvers are only applied if IPOPT fails. This is the reason why the number of results is only representative for IPOPT and why we only report results concerning IPOPT in Table 12.14. This table displays the run times and the number of iterations of IPOPT in dependence of the chosen decision approach. The average number of iterations is about 55, and the average solution time is about 1 s.

Summarizing, one can say that the NLP validation stage is very fast in comparison to the decision approaches. The main reasons are the relatively small model size and the good quality of the starting points which are generated based on the solution candidates coming from the solution of the decision approaches. We furthermore observe that no decision approach is preferred by the validation NLP, since the run times and number of iterations are approximately the same.

12.6.2 • Comparison of decision approach and NLP validation solutions

As it is described in Chapter 5, the solution candidates coming from the decision approaches (see Chapter 6 to Chapter 9) are used to fix the discrete decisions in the network and to initialize the variables of the ValNLP. In this sense, the solution candidates form the

12.6. Results for the validation NLP

Table 12.14. IPOPT *run times and required iterations for solving the ValNLP model on the* `gaslib-582-P` *test set.*

Decision approach	Run time in s			# iterations		
	min	max	avg.	min	max	avg.
sMINLP	0.74	1.51	1.12	35	71	56
MPEC	0.75	1.43	1.05	37	79	55
RedNLP	0.70	1.74	1.01	30	81	49
MILP	0.71	1.84	1.16	39	105	60

basis of the subsequent validation by the ValNLP model. Since all decision approaches as well as the validation NLP model the same physical and technical phenomena in different ways, the question arises whether these model differences can be expressed quantitatively and/or qualitatively. It turns out that this question is hard to answer, as we will see in the following.

As a means to approach this question, we compare the solution candidate given by some decision approach and the solution of the subsequent ValNLP model that is set up based on the solution candidate, for each instance of the test set. For doing so, we are mainly interested in the differences in pressure at nodes and in flow at arcs, since these are the most important common variables of the decision approaches and the ValNLP model. To compare these quantities, we define the absolute and relative maximum deviations in volumetric flow under normal conditions and in pressure:

$$\delta^{\mathrm{abs}}(Q_0^1, Q_0^2) := \max\left\{ \left| Q_{0,a}^1 - Q_{0,a}^2 \right| \,\Big|\, a \in A \right\}, \tag{12.2a}$$

$$\delta^{\mathrm{abs}}(p^1, p^2) := \max\left\{ \left| p_u^1 - p_u^2 \right| \,\Big|\, u \in V \right\}, \tag{12.2b}$$

$$\delta^{\mathrm{rel}}(Q_0^1, Q_0^2) := \max\left\{ \frac{|Q_{0,a}^1 - Q_{0,a}^2|}{\max\{|Q_{0,a}^1|, |Q_{0,a}^2|\}} \,\bigg|\, a \in A \right\}, \tag{12.2c}$$

$$\delta^{\mathrm{rel}}(p^1, p^2) := \max\left\{ \frac{|p_u^1 - p_u^2|}{\max\{|p_u^1|, |p_u^2|\}} \,\bigg|\, u \in V \right\}. \tag{12.2d}$$

Here, the superindex 1 represents the corresponding quantity in the solution of the decision approach, whereas superindex 2 refers to the ValNLP.

It is important to remark that no techniques are applied in the ValNLP that aim for keeping these deviations small. One way would be to add objective function terms in order to minimize the distance between the pressure-flow situation of the solution candidate and the solution of the validation NLP. Since these additional terms may conflict with the principal aim of the ValNLP (i.e., trying the find a feasible solution for the given discrete decisions), no such terms are incorporated in the ValNLP model.

Despite this fact, we try to analyze the differences defined in (12.2). Unfortunately, it turned out that it is very hard to come to interpretable results with this analysis. The reasons can be split up into reasons for flow deviations and reasons for gas pressure deviations.

Flow differences All models discussed in this book incorporate Kirchhoff-type flow balance equations at the nodes of the network. These equations mainly ensure the basic feasibility of gas flow, namely that all flow entering a node also has to leave it. However,

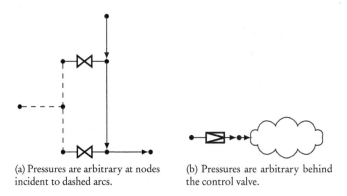

(a) Pressures are arbitrary at nodes incident to dashed arcs.

(b) Pressures are arbitrary behind the control valve.

Figure 12.16. *Solutions with arbitrary pressure values.*

none of the presented models incorporate constraints that exclude artificial gas flow in cycles, since *no* artificial flow is possible in cycles consisting of arcs affecting the pressure of gas (it would lead to arbitrary pressure increases or losses). However, artificial cycle flows can be present in a solution to our models for cycles that consist of arcs which have no influence on the pressure (e.g., open valves, short cuts, or compressor groups and control valve stations in bypass mode). These cycle flows appear both in solutions of the decision approaches and in solutions of the ValNLP model. Since these cycle flows disturb the flow measures defined in (12.2), we exclude them leading to the modified definitions

$$\delta^{abs}_{no\text{-}cycle}(Q_0^1, Q_0^2) := \max\left\{ \left|Q_{0,a}^1 - Q_{0,a}^2\right| \,\middle|\, a \in A \text{ without cycle flow} \right\}, \tag{12.3a}$$

$$\delta^{rel}_{no\text{-}cycle}(Q_0^1, Q_0^2) := \max\left\{ \frac{|Q_{0,a}^1 - Q_{0,a}^2|}{\max\{|Q_{0,a}^1|, |Q_{0,a}^2|\}} \,\middle|\, a \in A \text{ without cycle flow} \right\}. \tag{12.3b}$$

Pressure differences As it is the case for artificial cycle flows, there are also situations in which gas pressure is arbitrary (within bounds) leading to arbitrary differences between the pressures in solution candidates and in ValNLP solutions.

One example for such a situation is illustrated in Figure 12.16(a). The closed valves lead to an area in which no gas flows (everything *left* of the valves). All edges in this area are short cuts, and all nodes in this area are inner nodes or exits that do not demand any flow. If $[\underline{p}_u, \overline{p}_u]$ denotes the nonempty intersection of the pressure bound intervals of all nodes in the area under consideration, the pressures are completely arbitrary within these bounds. Thus, the used solvers have no reason to identify equal pressures.

A second situation, in which arbitrary pressure values (within bounds) occur, is illustrated in Figure 12.16(b). Here, an area, in which only exits and inner nodes appear, is supplied with flow by a single active control valve. In contrast to the first situation, the exits demand flow in this case, leading to pressure losses in the area *behind* the control valve. However, since in most of these situations the flow is very small, the pressure losses are almost negligible. Thus, the intersection of pressure bound intervals $[\underline{p}_u, \overline{p}_u]$ mainly leads to a range of pressures that are possible outlet pressures of the control valve.

The main problem with these pressure situations is that they cannot be handled a priori, since they depend on the concrete pressure-flow situation in the solution and on the given discrete decisions. The only aspect that we excluded a posteriori from our analysis is the pressure at nodes without any flow in both compared solutions. To be more precise,

12.6. Results for the validation NLP

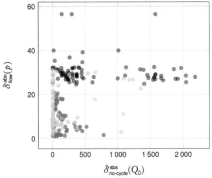
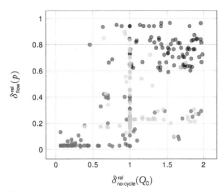

(a) Absolute maximum deviations between pressure (in bar) on nodes and normal volumetric flow (in 10^3 m^3/h) on arcs for validated instances of the `gaslib-582-95-P` test set.

(b) Relative maximum deviations between pressure (in bar) on nodes and normal volumetric flow (in 10^3 m^3/h) on arcs for validated instances of the `gaslib-582-95-P` test set.

Figure 12.17. *Maximum deviations between pressure and flow for MILP (black), RedNLP (red), MPEC (blue), sMINLP (green), and the ValNLP.*

let
$$Q_{0,u} := \sum_{a \in \delta(u)} |Q_{0,a}|$$
be the sum of all flows at node u. With this we modify the definitions of (12.2) leading to

$$\delta_{\text{flow}}^{\text{abs}}(p^1, p^2) := \max\left\{ |p_u^1 - p_u^2| \;\middle|\; u \in V \text{ with } Q_{0,u}^1 > 0 \text{ or } Q_{0,u}^2 > 0 \right\}, \qquad (12.4a)$$

$$\delta_{\text{flow}}^{\text{rel}}(p^1, p^2) := \max\left\{ \frac{|p_u^1 - p_u^2|}{\max\{|p_u^1|, |p_u^2|\}} \;\middle|\; u \in V \text{ with } Q_{0,u} > 0 \right\}. \qquad (12.4b)$$

This modified measure will address *not* the second situation, but the first one. Figure 12.17(a) and Figure 12.17(b) show scatter plots of the described measures for all validated (i.e., slack-0) instances of the `gaslib-582-95-P` test set. Thus, every dot represents a pair of solution candidate and corresponding ValNLP solution with vanishing slack variable norm.

As one can see in Figure 12.17(a) and Figure 12.17(b), there are some considerable outliers of differences between the solution candidates and the ValNLP solutions. For instance, the MILP leads to maximum deviations of up to 60 bar and up to 2.2×10^6 Nm^3/h. Altogether, the MILP approach shows the largest deviations with respect to the results of the ValNLP. This is interesting, because the MILP approach turns out to be a very successful approach for producing solution candidates (see Section 12.7). The RedNLP approach is *best* with respect to deviations in flow, followed by the sMINLP approach which leads, in contrast to the RedNLP approach, to more outliers. When looking at both pressure and flow deviations, the MPEC approach produces the smallest deviations with respect to the ValNLP solutions.

We remark that the measures used in Figure 12.17(a) and Figure 12.17(b) are based on maximum differences; see (12.3) and (12.4). Obviously, these measures are highly sensitive for outliers. However, the comparability of solution candidates and ValNLP solutions is much better as one might guess when looking at Figure 12.17(a) and Figure 12.17(b). If we replace the maximum-based measures by arithmetically averaged differences, we

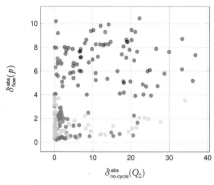 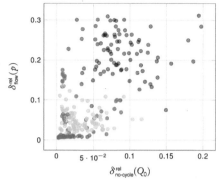

(a) Absolute average deviations between pressure (in bar) on nodes and normal volumetric flow (in $10^3 \, m^3/h$) on arcs for validated instances of the gaslib-582-95-P test set.

(b) Relative average deviations between pressure (in bar) on nodes and normal volumetric flow (in $10^3 \, m^3/h$) on arcs for validated instances of the gaslib-582-95-P test set.

Figure 12.18. *Average deviations between pressure and flow for MILP (black), RedNLP (red), MPEC (blue), sMINLP (green), and the ValNLP.*

obtain Figure 12.18(a) and Figure 12.18(b). The qualitative behavior of the approaches stays the same: Although the MILP later leads to a comparably high rate of NLP validated instances, it leads to the largest deviations both in pressure and flow. Again, the RedNLP yields the smallest differences in flow, followed by the sMINLP approach. Finally, the MPEC produces the solution candidates that are the closest to the ValNLP solutions.

We think that these results can be explained both by the model formulations as well as by the used solvers. Both the MPEC and the RedNLP are nonlinear optimization models as the ValNLP model is. Thus, they are solved using the same solver technology as the ValNLP. In addition, the model formulation of MPEC is closest to the ValNLP model; see Chapter 9. Since the sMINLP approach also uses NLP models and solvers, it is reasonable that it leads to smaller differences than the MILP approach, which is based on solving a linearized model formulation with MILP techniques.

To sum up, one can see that the comparison of solutions is very complicated and has its own intrinsic challenges. We thus have not presented a conclusive analysis of these aspects, but have to leave this as an important topic of future research. For instance, the analysis should be more sophisticated in order to exclude all appearances of situations such as in Figure 12.16 (and related types). Such an analysis should lead to an understanding of the model differences and of the differences in their results. This should help to further couple the different models and improve their validation results.

12.7 ▪ Comparison of the decision approaches and combined solver

As shown in the previous sections, each decision approach has different strengths and weaknesses. It thus makes sense to combine these four decision approaches with the validation NLP in order to obtain a combined solver for NoVa. In this section, we compare the performance of these approaches as components of this combined solver.

As described previously, we are validating the solutions of the different solvers using ValNLP. Before passing the solution to the NLP, we check whether all discrete decisions are taken in accordance with the technical restrictions described by the subnetwork operation modes. The latter check is often the reason that a solution of the MPEC approach

12.7. Comparison of the decision approaches and combined solver

Table 12.15. *Results of the solvers on test set* HN-SN *(4227 instances).*

	Slack 0	Infeasible	Slack > 0	No solution	Contradiction
MILP	3280	444	495	7	0
sMINLP	3150	541	501	21	14
RedNLP	3070	0	685	472	0
MPEC	0	0	1	4226	0

is rejected. (For the other approaches this test is satisfied by design—and only included to catch implementation errors.) For this reason, the MPEC heuristic often looks much worse than the other approaches in the following. If these constraints are ignored, the MPEC heuristic performs much better; we refer to Schmidt, Steinbach, and Willert (2013) for more details.

In this section, we call an instance for which an approach produced a solution that yields an NLP slack below the precision of the validation NLP (which allows constraint violations of at most 10^{-5}) *confirmed*. We repeat that this might exclude valid instances, since the validation NLP might only find a local optimum. Moreover, we call instances that are confirmed or the corresponding approach proved infeasibility as *solved*.

When comparing results of the four decision approaches it might happen that they give contradictory results. This happens if one approach claims that the corresponding model is infeasible, but a solution of ValNLP with 0 slack is found using some other approach. We then call the results for the particular instance a *contradiction*. This case can arise, since none of the underlying models for the decision approaches is a complete relaxation of the model used for ValNLP. In fact, the sMINLP and MILP approaches are the only ones that can claim infeasibility (the RedNLP and MPEC approaches are primal heuristics). The case of the sMINLP approach is discussed in Section 12.3.4. The MILP approach never produced a contradiction in our computations, but in principle this cannot always be guaranteed.

12.7.1 ▪ Results for test set HN-SN

The results of the decisions approaches for test set HN-SN are shown in Figure 12.19(a). Recall that each solver can report at most one solution to be validated by ValNLP and that the time limit is four hours. The ValNLP has an additional time limit of one hour per solver. Moreover, in this section the time for NLP validation is included in the statistics, since we are now interested in the combined solver.

We observe the following facts: All instances that are solved by the sMINLP approach within 20 s are infeasible and therefore no validation with the NLP model is carried out. This explains the notable "bump" in the graph of the sMINLP approach in Figure 12.19(a). The remaining instances are almost all solved within 100 s. Similarly, the RedNLP approach solves all its instances within 100 s. The MILP continuously solves instances over the whole time period. Moreover, when the MILP cannot prove optimality within the time limit, but has found a feasible solution, the best solution found is validated at the very end of the computation. This explains the "jump" of the MILP success towards the end.

Table 12.15 shows the corresponding numbers. The columns give the name of the decisions approach, the number of instances that are confirmed with a slack 0 solution, the number of instances that have been proven to be infeasible by the corresponding model, the number of instances for which a solution has been found that leads to a ValNLP

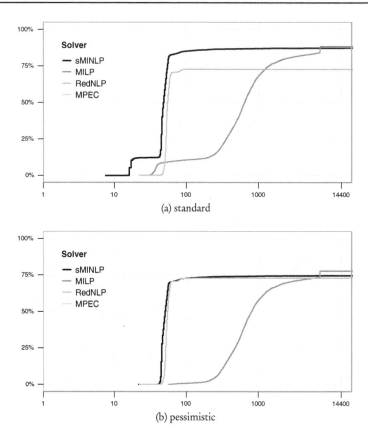

Figure 12.19. *Profile of the run times in seconds for test set HN-SN including NLP validation; y axis: percentage of confirmed or infeasible instances (total: 4227).*

solution with positive slack, the number of instances for which no solution could be found, and the number of contradictions, i.e., the number of instances for which infeasibility was detected with respect to the corresponding model, but a slack-0 solution could be found by another approach.

Since the RedNLP and MPEC approaches cannot prove infeasibility, the number of "infeasible" instances is 0. Moreover, the RedNLP approach finds fewer slack-0 solutions than the sMINLP and MILP approaches. The MPEC approach is unsuccessful, because of the reasons explained above.

We see that the sMINLP and MILP approach both solve about 88% of the instances, i.e., they find slack-0 solutions or state infeasibility, but sMINLP is usually faster. The RedNLP approach is less successful (about 73%), but also quite fast.

The sMINLP approach produces 14 contradictions (in the sense explained above). Note that we cannot detect contradictions for instances for which no slack-0 solution was found. Figure 12.19(b) therefore shows the *pessimistic* view on the same results: we only count those instances as solved for which a slack-0 solution has been found. Here, the sMINLP, RedNLP, and MILP approaches are quite close together. The MILP is slower, but also solves slightly more instances.

After this comparison of the individual solvers, the question is how they behave when combined. The results are shown in Figure 12.20(a). As can be seen from the figure, the combination of the different approaches increases the success rate to almost 96%.

12.7. Comparison of the decision approaches and combined solver

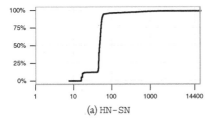
(a) HN-SN

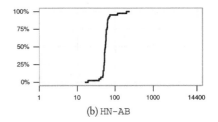
(b) HN-AB

Figure 12.20. *Profile of the run times in seconds of the combined solver for test sets* HN-SN *and* HN-AB *including NLP validation; y axis: percentage of confirmed or infeasible instances (total: 4227).*

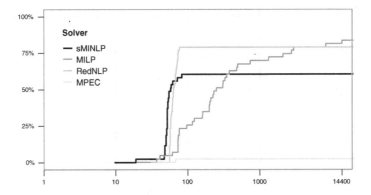

Figure 12.21. *Profile of the run times in seconds for test set* HN-AB *including NLP validation; y axis: percentage of confirmed or infeasible instances (total: 43).*

In total, 996 instances were classified as infeasible and there are 150 instances for which only a positive slack solution could be found. There are nine contradictions, all from the sMINLP approach. Taking the pessimistic viewpoint, the success rate reduces to about 72%. The combined approach is quite fast: most instances are solved within 100 s.

12.7.2 ▪ Results for test set HN-AB

The results for test set HN-AB are shown in Figure 12.21 and Table 12.16. In principle, the results are similar as for test set HN-SN, but here the RedNLP approach performs very well ($\approx$ 79% solved instances) and is fast. The sMINLP is similarly fast, but has a worse confirmation rate ($\approx$ 60%); for this test set no contradictions arise. The MILP approach is slower than the other approaches, but yields the highest success rate ($\approx$ 84%). The MPEC finds a single slack-0 solution satisfying the SOM constraints.

Figure 12.20(b) again shows the result of the combined solver. It has a success rate of 100%. All instances are solved within 200 s. The test set contains 40 confirmed instances and three infeasible ones.

12.7.3 ▪ Results for test sets `gaslib-582` and `gaslib-582-95`

The results for test set `gaslib-582` are shown in Figure 12.22(a) and Table 12.17. Here, again the MILP approach is slower than the other approaches, but solves the most number instances ($\approx$ 70% solved). The RedNLP-approach is again fast, but only solves about 14% of the instances. The MPEC is again not performing well, but finds 45 solutions with

Table 12.16. *Results of the solvers on test set* HN-AB *(43 instances).*

	Slack 0	Infeasible	Slack > 0	No solution	Contradiction
MILP	33	3	7	0	0
sMINLP	25	1	17	0	0
RedNLP	34	0	6	3	0
MPEC	1	0	0	42	0

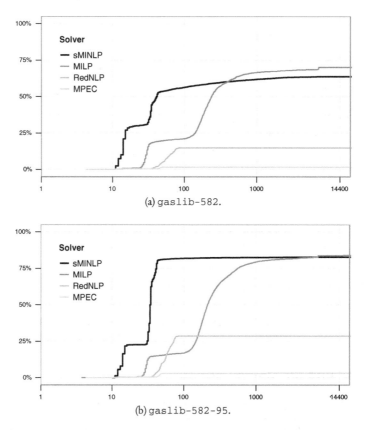

(a) gaslib-582.

(b) gaslib-582-95.

Figure 12.22. *Standard evaluation: Profile of the run times in seconds for test set* gaslib-582 *and* gaslib-582-95 *including NLP validation; y axis: percentage of confirmed or infeasible instances (total: 4227).*

Table 12.17. *Results of the solvers on test set* gaslib-582 *(4227 instances).*

	Slack 0	Infeasible	Slack > 0	No solution	Contradiction
MILP	2054	909	1230	34	0
sMINLP	971	1707	830	257	443
RedNLP	602	25	585	3015	0
MPEC	45	20	87	4075	0

12.7. Comparison of the decision approaches and combined solver

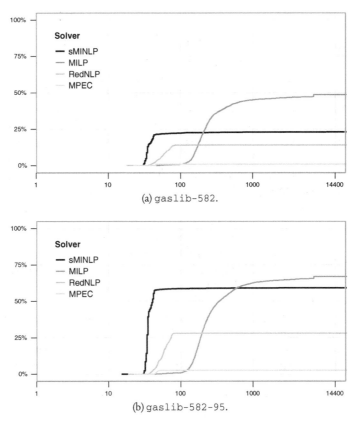

Figure 12.23. *Pessimistic evaluation: Profile of the run times in seconds for test set* `gaslib-582` *and* `gaslib-582-95` *including NLP validation; y axis: percentage of confirmed instances only (total: 4227).*

Table 12.18. *Results of the solvers on test set* `gaslib-582-95` *(4227 instances).*

	Slack 0	Infeasible	Slack > 0	No solution	Contradiction
MILP	2831	716	661	19	0
sMINLP	2504	996	706	12	9
RedNLP	1188	20	170	2849	0
MPEC	108	20	42	4057	0

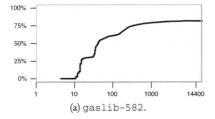
(a) `gaslib-582`.

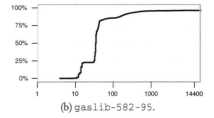
(b) `gaslib-582-95`.

Figure 12.24. *Profile of the run times in seconds of the combined solver including NLP validation; y axis: percentage of confirmed or infeasible instances (total: 4227).*

slack-0. The sMINLP approach is quite fast, but has the main drawback that it produces 443 contradictions (in the sense explained above). In order to get a safe lower bound on the number of solved instances, Figure 12.23(a) shows the "pessimistic" view on the results. Here, the MILP approach is clearly the best approach in terms of slack-0 solutions.

Note that the RedNLP and the MPEC approach classify a different number of instances as infeasible. Since both approaches are not able to detect infeasibility by themselves, all detected infeasibilities result from preprocessing (see Section 5.4), where the approaches use different parts. To be precise, the MPEC approach does not use the flow bound strengthening that is based on min-cost-flow computations since our numerical experience shows that dropping this part of the preprocessing leads to better results for the MPEC approach over all. Thus, the difference of the five instances for which the RedNLP approach detects infeasibility and the MPEC approach does not is the outcome of the min-cost-flow based preprocessing technique.

In comparison to the other test sets, `gaslib-582` seems to be hard to solve. We therefore consider the scaled version `gaslib-582-95`. The results are given in Figure 12.22(b), 12.23(b), and Table 12.18. The resulting test set is indeed easier to solve. The MILP and sMINLP approaches both solve about 84% of the instances. The sMINLP now only produces nine instances on which it contradicts the other solvers. The RedNLP and MPEC approaches also perform better than on `gaslib-582`.

The results of the combined solver are shown in Figure 12.24 for both test sets. These figures again show that the `gaslib-582` instances are very hard to solve, but the scaled versions are easier. The success rate for the first is about 82%, while it is 96% for the second. The scaled instances `gaslib-582-95` have similar characteristics to HN-SN, but 996 instances are claimed to be infeasible compared to 542 for HN-SN. The running times are about a factor of 10 larger for `gaslib-582` than for `gaslib-582-95`. For the latter, most instances are solved within 100 s, but there are some instances which take longer.

Part III
Verification of booked capacities

Chapter 13
Empirical observations and statistical analysis of gas demand data

Holger Heitsch, René Henrion, Hernan Leövey,
Radoslava Mirkov, Andris Möller, Werner Römisch,
Isabel Wegner-Specht

Abstract *In this chapter we describe an approach for the statistical analysis of gas demand data. The objective is to model temperature-dependent univariate and multivariate distributions allowing for later evaluation of network constellations with respect to the probability of demand satisfaction. In the first part, methodologies of descriptive data analysis (statistical tests, visual tools) are presented and dominating distribution types identified. Then, an automated procedure for assigning a particular distribution to the measurement data of some exit point is proposed. The univariate analysis subsequently serves as the basis for establishing an approximate multivariate model characterizing the statistics of the network as a whole. Special attention is paid to the statistical model in the low temperature range.*

The goal of our data analysis consists of evaluating historical data on gas demand at exits of some gas transportation network. The results will be used to extract statistical information, which may be exploited later for modeling the gas flow in the network under similar temperature conditions. More precisely, the aim is to generate a number of scenarios of possible exit loads, which will be complemented in several subsequent steps to complete a nomination (see Chapter 14). Such scenarios are needed for validating the gas network and for calculating and maximizing its technical capacities.

The analysis will be based on historical measurement data for gas consumption, which is typically available during some time period, and on daily mean temperature data provided by a local weather service. Due to a high temperature-dependent proportion of heating gas, the gas demand is subject to seasonal fluctuations. During the warmer season the gas consumption decreases: hot water supply for households and process gas consumption are the only basic constituents.

The method for analyzing the data should be applicable to all exits, no matter what their distribution characteristics are, and should allow for multivariate modeling to take into account statistical dependencies of different exits of the network. Therefore, the use of local temperatures as in day-ahead prediction of gas demands is less appropriate. Rather, we introduce a reference temperature which is given as a weighted sum of several local temperatures and in this way is more representative for the entire network. Due to the stationarity of the gas flow model considered throughout this book, we consider daily averages of the gas demands at all exits, based on measurement data which is mostly given for smaller (e.g., hourly) time intervals.

The statistical data analysis will consist of several steps, namely,

  (i) visualization and categorization of the available data,
 (ii) analysis of basic structural properties,
(iii) statistical tests,
 (iv) definition of temperature classes,
  (v) fitting univariate probability distributions for each exit and temperature class,
 (vi) fitting multivariate distributions, and
(vii) forecasting gas demands for low temperatures.

Issues (i)–(iii) are discussed in Section 13.1, (iv) in Section 13.2, (v) in Section 13.3, (vi) in Section 13.4, and (vii) in Section 13.5. All tables and figures illustrating the different steps are based on historical data provided by the company Open Grid Europe GmbH (OGE) for its H-gas and L-gas transportation network. Both networks contain almost 800 exit points.

13.1 • Descriptive data analysis and hypothesis testing

In this section, we discuss several methods to analyze gas demand data. These techniques form the basis for the succeeding steps to construct distributions.

13.1.1 • Visualization and categorization

The first step of data analysis typically consists of the visualization and categorization of the available data for all exit points of a given network. Then basic structural properties (trends, clusters, temperature dependence, etc.) should be analyzed before in a third step different types of univariate (exit-based) statistical distributions (normal, uniform, etc.) can be classified.

For the L-gas network and the given data set, Figure 13.1 illustrates the distribution of daily mean gas flows as a function of temperature, for selected exit points. Different years in the data set are visualized by different colors. As mentioned above, gas demands are mainly temperature dependent. The type of dependence varies between (piecewise) linear, sigmoidal and—in rare cases—other functional relations. Sometimes the exit flow data contain zero flows for certain time periods without clear temporal or temperature dependence. Furthermore, one may be faced with band- or cluster-like structures.

Table 13.1 provides an overview of a general categorization in the L-gas network. Figure 13.2 reveals that one may have to cope also with the presence of outliers and of nonstationary characteristics, such as trends and breaks, which necessitate an appropriate data preparation prior to the actual model calibration (Brockwell and Davis (2002)).

For gas demand forecast one typically uses nonlinear regression models in which the functional relationship between the daily average gas flow and the daily mean temperature or a weighted mean temperature is described using sigmoid functions; see Cerbe (2008) and Leövey et al. (2011). For an interpretation of the model parameters in the context of gas flow modeling we refer, e.g., to Geiger and Hellwig (2002). Based on the visual analysis of the data originating from the given networks, we suspect that classic assumptions such as homogeneity of variance over the entire temperature range or approximate normality of residuals (Seber and Wild (2003)) are frequently not met.

13.1. Descriptive data analysis and hypothesis testing

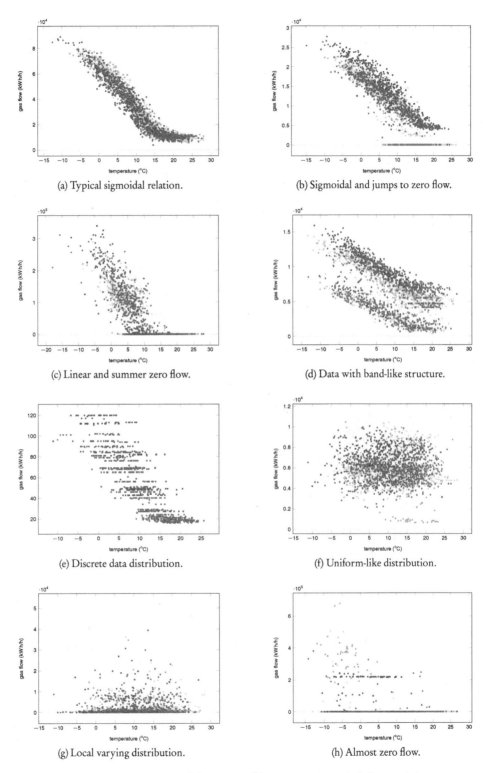

(a) Typical sigmoidal relation.

(b) Sigmoidal and jumps to zero flow.

(c) Linear and summer zero flow.

(d) Data with band-like structure.

(e) Discrete data distribution.

(f) Uniform-like distribution.

(g) Local varying distribution.

(h) Almost zero flow.

Figure 13.1. *Overview of the variety of data containing typical characteristics.*

Table 13.1. *Categorization of the data characteristics obtained for the data of the L-gas network. The values are given in percentage w.r.t. relative frequency of occurrence. Stationarity indicates that data do not contain significant trends over time.*

Temperature dependency	Description	Stationarity yes	Stationarity no
Dependent	sigmoidal	41.21	0.70
	sigmoidal and jumps to zero flow	12.52	0.28
	piecewise linear (summer zero flow)	17.30	0.42
	band- and cluster-like structures	6.05	0.56
	nonregular cluster-like structures	1.55	0.56
	discrete data distribution	3.52	0.00
	other noncategorized relations	4.50	2.81
Independent	completely positive	4.50	0.70
	jumps between zero and positive flow	0.14	0.00
	band-like structures and jumps	0.98	0.84
	uniform-like distribution	0.84	0.98
	local varying distribution	0.00	0.42
	discrete data distribution	0.70	0.14
	other noncategorized relations	2.53	0.56
	almost zero flow	3.66	0.00

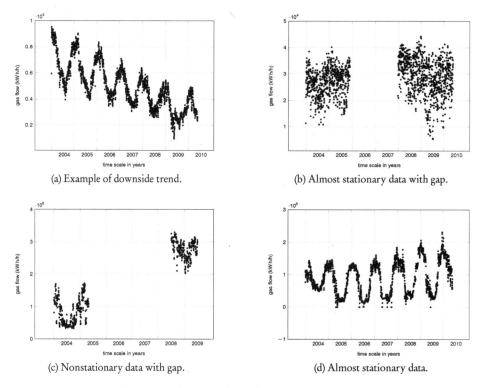

(a) Example of downside trend. (b) Almost stationary data with gap.

(c) Nonstationary data with gap. (d) Almost stationary data.

Figure 13.2. *Illustration of varying data behavior w.r.t. gaps and trends over time.*

13.1.2 • Statistical tests

In order to corroborate the latter observation one may apply suitable statistical tests. As far as homogeneity of variance for residuals of adjacent temperature intervals is concerned, the methodology of analysis of variance (ANOVA) can be employed. Here, a variety of statistical tests is at our disposal, e.g., Bartlett's test for normally distributed data or robust (distribution-free) alternatives such as the Siegel-Tukey test. For further details we refer to the comprehensive description in Hartung (2005) and Fahrmeir et al. (2007). The application of ANOVA to the available data showed that the assumption of homogeneous variances over the entire temperature interval is violated for a majority of exit points in the H-gas and L-gas network.

As far as the normality assumption is concerned, there exist numerous tests which can differ in their design and properties; compare D'Agostino and Stephens (1986), Hartung (2005), and Thode (2002). Probably the most prominent among these is the classical Kolmogorov–Smirnov goodness-of-fit test. It checks whether a random variable obeys some fixed distribution.

To give a short description of the *Kolmogorov–Smirnov test*, let F_ξ denote the distribution function of a real random variable ξ on some probability space $(\Omega, \mathcal{F}, \mathbb{P})$, i.e.,

$$F_\xi(t) := \mathbb{P}(\xi \leq t) \quad \text{for all } t \in \mathbb{R}.$$

The *Kolmogorov distance* between the distributions $\mathbb{P} \circ \xi^{-1}$ and $\mathbb{P} \circ \eta^{-1}$ of two real random variables ξ and η is given as the uniform distance of the corresponding distribution functions, i.e.,

$$\mathbb{D}_K(\mathbb{P} \circ \xi^{-1}, \mathbb{P} \circ \eta^{-1}) := \sup_{t \in \mathbb{R}} \left| F_\xi(t) - F_\eta(t) \right|. \tag{13.1}$$

The observed data $\{x_1, \ldots, x_N\}$ induce a new random variable ξ^N, whose distribution (called *empirical distribution*) is defined by the empirical distribution function

$$F_N(t) := \frac{1}{N} \#\{x_i \mid x_i \leq t\}. \tag{13.2}$$

Assuming that the measurement data are ordered ($x_1 \leq \cdots \leq x_N$) the Kolmogorov–Smirnov test statistic is given by the Kolmogorov distance between the distributions of ξ and ξ^N, i.e.,

$$D_N = \sup_{t \in \mathbb{R}} \left| F_\xi(t) - F_N(t) \right| = \max_{1 \leq i \leq N} \left\{ F_\xi(x_i) - \tfrac{i-1}{N}, \tfrac{i}{N} - F_\xi(x_i) \right\},$$

under the null hypothesis that the measurement data $\{x_1, \ldots, x_N\}$, are drawn from the distribution of ξ. Significant advantages of the Kolmogorov–Smirnov test statistic are that it can be exactly determined and that it does not depend on the underlying distribution if the distribution function is continuous.

Moreover, the sequence $\sqrt{N} D_N$ converges in distribution to the random variable $\sup_{t \in \mathbb{R}} |B(t)|$, where B is the so-called Brownian bridge. In particular, it holds

$$\mathbb{P}(\sqrt{N} D_N) > x \leq 2 \exp(-2x^2) \quad (N \in \mathbb{N})$$

and

$$\lim_{N \to \infty} \mathbb{P}(\sqrt{N} D_N \leq x) = H(x) = 1 - 2 \sum_{k=1}^{\infty} (-1)^{k-1} \exp(-2k^2 x^2) \quad (x \in \mathbb{R}),$$

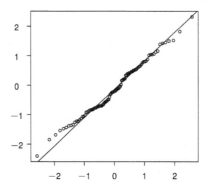

Figure 13.3. *Statistical visualization of the data set: Q-Q plot for normal distribution.*

where the estimate is the so-called Dvoretsky–Kiefer–Wolfowitz inequality with the specific leading constant 2 due to Massart (1990) and the limit is due to Kolmogorov.

The goodness-of-fit test is based on the latter argument, namely, the null hypothesis is rejected at level α if

$$\sqrt{N} D_N > H^{-1}(1-\alpha).$$

The test is more sensitive in the middle of the distribution rather than at its tails.

Alternatively, the Shapiro–Wilk test has the highest power among the normal distribution tests. The test statistic used here is based on two different estimates for the variance: an estimate from the quantile-quantile plot (see below) and of the sample variance. Both estimates should be nearly identical, if the data are drawn from a normal distribution.

In the literature one may find a lot of other commonly used tests; see Hartung (2005) and Thode (2002). Applying these to the data set of the given networks, we observed that only in rare cases can a unique result be derived from the whole variety of the available tests.

Many common tests output the so-called p value of the data $\{x_1, \ldots, x_N\}$. The p value $\mathrm{pval}(x_1, \ldots, x_N)$ is defined as the greatest lower bound on the set of all α such that the level α test based on $\{x_1, \ldots, x_N\}$ rejects the null hypothesis (see, e.g., Section 3.3 in Lehmann and Romano (2005)). One tends to reject the null hypothesis if the p value turns out to be less than a certain significance level, often 0.05. The smaller the p value, the more strongly the test rejects the null hypothesis.

The automation described in Section 13.3 is exclusively based on the Kolmogorov–Smirnov test.

13.1.3 ▪ Visual checks

In addition to statistical tests, there are many visual checks to verify the distribution assumptions of individual data series. The most commonly used ones are histograms, quantile-quantile plots, and box plots; see Hartung (2005) and Fahrmeir et al. (2007). In *quantile-quantile plots* (Q-Q plots) as shown in Figure 13.3, the quantiles (see Eq. (13.3) below) of the data series are plotted against the quantiles of the expected underlying distribution. In case of assuming the correct distribution, the points lie approximately on the bisector. With clear deviations from this reference line, the distribution has been specified wrongly. Strong deviations of the last points only indicate outliers. In box plots, location and dispersion differences between two data series as well as outliers are made

13.1.4 ▪ Outlier detection

Another component of data analysis is the detection of outliers and specifications on how to deal with them. Outliers are individual values strongly deviating from the other data. If there are no systematic errors in the data collection, one can check by suitable tests whether a suspected outlier should be removed from or retained in the sample. A simple criterion for the identification of outliers, in particular for symmetric single-peaked distributions, is provided by the so-called sigma range. In case of a normal distribution data beyond the 2.5σ range (covering realization of the random variable with a probability of 99%) are declared to be outliers.

For arbitrary distributions, Chebyshev's inequality yields conclusions about the portion of the data outside of the $\ell\sigma$ range. More exactly, it holds for arbitrary random variables with mean $\mathbb{E}[\xi] = \mu$ and finite variance $\text{Var}(\xi) = \sigma^2$ that

$$\mathbb{P}(|\xi - \mu| < \ell\sigma) \geq 1 - \frac{1}{\ell^2} \quad \text{for all } \ell > 0.$$

Robust limits for the detection of outliers for many distribution types, especially for skewed distributions, can be derived on the basis of the quartiles and the interquartile range. The latter is defined for a random variable ξ with distribution function $F_\xi(t)$ as $\text{IQR} := x_{0.75} - x_{0.25}$. Here x_p denotes the p quantile,

$$x_p := F_\xi^{-1}(p) = \inf\{t \in \mathbb{R} \mid F_\xi(t) \geq p\}. \tag{13.3}$$

There exists a variety of outlier tests, e.g., the David–Hartley–Pearson test, the Grubbs' test for normally distributed samples and average sample sizes, and the Dean–Dixon test for very small sample sizes. For further details, we refer to the relevant literature; see Hartung (2005). Removal of outliers has to be carried out very cautiously because, especially for the multivariate analysis, the sample size should be kept as large as possible. The automation in Section 13.3 and Section 13.4 refrains from outlier deletion.

13.2 ▪ Reference temperature and temperature intervals

As mentioned earlier, daily mean gas flows at some exit point depend on the local temperature. However, a numerically tractable model for this functional dependence is not available in general. An alternative approach consists of removing this dependence by introducing (i) a reference temperature for the H-gas and L-gas network, and (ii) a subdivision of the temperature range into sufficiently small and properly sized intervals in order to arrive at statistically relevant univariate distributions at each exit point. More precisely, we propose to proceed as follows:

1. Start with historical data for gas demands at the exits of the gas network on the one hand and for temperatures at certain preselected measuring stations on the other hand. Calculate for each day d of the historical time period a *reference temperature* $T^{\text{ref}}(d)$ by using a weighted average and the daily mean demand $D(n,d)$ for each exit point n.
2. The relevant temperature range (for the H-gas and L-gas network from $-15\,°\text{C}$ to $30\,°\text{C}$) is subdivided into intervals $(T^i, T^{i+1}]$, $i = 1, \ldots, I$, which are

▷ small enough to neglect the temperature dependence of demands within the interval and
▷ large enough to contain a sufficient amount of data required for statistical modeling.

(For the H-gas and L-gas network, intervals of two degrees Celsius were selected for the interior of the temperature range and $(-15\,°C, -2\,°C]$, $(20\,°C, 30\,°C]$ for its boundaries to have a reasonable amount of data available.)

3. For each temperature interval and each exit point of the network the data of gas demands for days with reference temperature belonging to the given interval are filtered:

$$S(i,n) := \{D(n,d) \mid T^{\text{ref}}(d) \in (T^i, T^{i+1}]\} \quad \forall i \, \forall n.$$

In this way the temperature dependence of gas demand is modeled for the whole gas network rather than for local parts of it. Instead of modeling this temperature dependence by an explicit formula, a subdivision into small intervals is used which allows us to establish univariate statistical models with homogeneous variance within each interval at each exit. Moreover, whenever possible, correlations in gas demand between different exit points can be taken into account for the multivariate statistical model of each temperature interval (see Section 13.4). In order to avoid confusion with respect to reference or local temperature, these intervals will be called *temperature classes* from now on. In addition to these temperature classes, the filtering procedure described above is also carried out with respect to *day classes* in order to model significantly different behavior of gas demand for specific days, namely: *working day*, *weekend*, and *holiday*.

13.3 ▪ Univariate distribution fitting

The next step consists of finding a univariate statistical model for the distribution of gas demands in each class $S(i,n)$. This is a two-step procedure where in the first step an appropriate class of distributions (e.g., normal distribution) has to be found and in the second step the associated parameters (e.g., mean value, standard deviation) for this class have to be determined. For the H-gas and L-gas network, the following selection of distributions turned out to be relevant as a result of our prior descriptive data analysis (statistical tests, visual inspection): *normal distribution, lognormal distribution, uniform distribution, Dirac distribution*.

The need to incorporate the Dirac distribution arises mainly from frequent observations of zero demands as mentioned above, but occasionally also of constant positive demands. Visualization of the data suggested the presence of mixtures of these distribution classes, e.g., Dirac and normal distribution. We then speak of *shifted* normal, uniform, etc. distributions. Therefore we extend our statistical model to positive combinations of the distribution classes above, such that the determination of weights in this combination is also part of the modeling process.

To provide an example, consider the case of the shifted normal distribution: Assume that the relative frequency of zero loads in the data sets equals $w \in [0,1]$, whereas the remaining (positive) elements of the data set follow a normal distribution with mean μ and standard deviation σ. Then, the overall distribution function F for the data set would be modeled as

$$F = wF_\delta + (1-w)F_{\mathcal{N}(\mu,\sigma)}, \tag{13.4}$$

where

$$F_\delta(t) = \begin{cases} 0 & \text{if } t < 0, \\ 1 & \text{it } t \geq 0, \end{cases}$$

13.3. Univariate distribution fitting

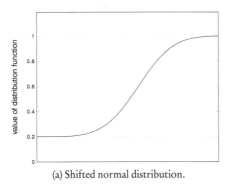
(a) Shifted normal distribution.

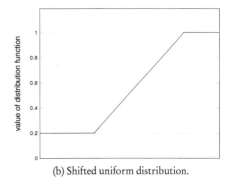
(b) Shifted uniform distribution.

Figure 13.4. *Illustration of shifted distribution functions (see Eq. (13.4) in the case of the normal distribution), where in this example the shift is given by $w = 0.2$.*

denotes the distribution function of the Dirac distribution at zero and

$$F_{\mathcal{N}(\mu,\sigma)}(t) = \frac{1}{\sigma\sqrt{2\pi}} \int_{-\infty}^{t} \exp\left(-\frac{1}{2}\left(\frac{x-\mu}{\sigma}\right)^2\right) dx$$

denotes the distribution function of the normal distribution $\mathcal{N}(\mu,\sigma)$. Figure 13.4(a) illustrates the distribution function of such shifted normal distribution, where the shift with respect to a pure normal distribution function is reflected by the intercept on the y axis. Similar shifted forms are considered for the other distributions, e.g., shifted uniform distribution on some interval; see Figure 13.4(b).

We aim at an automated procedure for statistical modeling. This means that both the assignment of a specific distribution class (e.g., shifted normal) to a data set $S(i,n)$ and the parameter estimation for this class (e.g., w, μ, σ) have to be carried out by a numerical procedure. As an assignment criterion one may use the Kolmogorov distance Eq. (13.1) which is justified, for instance, by the Kolmogorov–Smirnov test discussed above, but also by stability results in stochastic optimization (see Römisch (2003)).

More precisely, we assign to a given data set $S(i,n)$ that distribution from our portfolio which realizes the smallest Kolmogorov distance to the empirical distribution function in our data set. This idea is illustrated in Figure 13.5: The points $\{x_1,\ldots,x_N\}$ distributed along the x axis represent the measurement data $S(i,n)$ for some exit point n and temperature class i. These points induce an empirical distribution function as defined in Eq. (13.2); see the staircase function in Figure 13.5. Assume that we have two candidate distributions which we want to assign to the given data set, e.g., normal distribution or uniform distribution. Then our choice will be made according to the candidate realizing a smaller Kolmogorov distance to the empirical distribution function (see vertical bars in Figure 13.5, indicating the positions at which the largest deviation of the corresponding distribution functions occur).

The numerical computation of the Kolmogorov distance between a distribution with distribution function F and the empirical distribution function S_N associated with the measurement data $\{x_1,\ldots,x_N\}$ can be carried out according to the simple expression

$$\max_{i=1,\ldots,N} \max\left\{\left|F(x_i) - \frac{1}{N}\#\{x_j \mid x_j \leq x_i\}\right|, \left|F(x_i) - \frac{1}{N}\#\{x_j \mid x_j < x_i\}\right|\right\}. \quad (13.5)$$

For the computation, it has to be taken into account that the candidate distributions themselves depend on parameters (e.g., mean and standard deviation for the normal

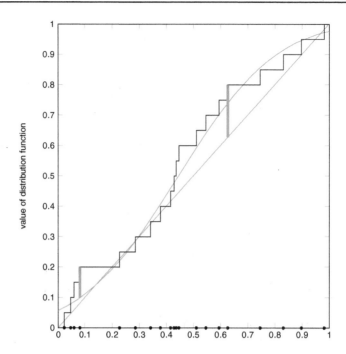

Figure 13.5. *Kolmogorov distance between the empirical distribution and the best fitting normal and uniform distribution, respectively. The empirical distribution function (staircase function) is induced by 20 sample points. Due to smaller Kolmogorov distance, in this example preference would be given to the normal distribution.*

distribution or interval limits for uniform distribution). One possibility to identify these parameters would rely on estimating them from the given measurement data (e.g., arithmetic mean and empirical standard deviation for the normal distribution, and minimum and maximum measurements for the interval limits of the uniform distribution). It appears, however, more natural to subordinate the parameter identification to the same criterion of minimum Kolmogorov distance as discussed above. More precisely, we are led to solve the following optimization problem:

$$\min_{p} \max_{i=1,\ldots,N} \max \left\{ \left| F(p, x_i) - \frac{1}{N} \#\{x_j \mid x_j \leq x_i\} \right|, \right. \quad (13.6)$$
$$\left. \left| F(p, x_i) - \frac{1}{N} \#\{x_j \mid x_j < x_i\} \right| \right\}.$$

Here, $F(p, \cdot)$ refers to the distribution function of the specific class in the portfolio depending on the parameter vector p (e.g., $p = (\mu, \sigma)$ for the normal distribution). According to Eq. (13.5), the minimization problem above identifies the parameter vector p in a way that the best fit to the empirical distribution function is realized within the given class. Similar optimization problems are solved for other distribution classes (uniform, log-normal, shifted versions, etc.). The choice of the final statistical model is then made according to the class realizing the smallest of these optimal values along with the associated parameter vector p. Evidently, Eq. (13.6) is equivalent to the following nonlinear

13.4. Multivariate distribution fitting

Table 13.2. *Percentage of univariate distributions for the H-gas network for all temperature classes: normal (ND), shifted normal (shND), lognormal (logND), shifted lognormal (shLogND), Dirac distributed (Dirac), uniform (UD), and shifted uniform (shUD).*

Temp. Class	ND	shND	logND	shLogND	Dirac	UD	shUD
(−15 °C, −2 °C]	40.40	8.75	34.01	5.39	8.08	1.68	1.35
(−2 °C, 0 °C]	40.40	8.08	31.31	7.07	8.75	3.03	1.01
(0 °C, 2 °C]	41.41	7.74	30.30	9.76	7.41	0.67	2.69
(2 °C, 4 °C]	42.42	13.13	22.56	11.78	6.73	0.67	2.02
(4 °C, 6 °C]	41.75	12.12	21.89	14.14	7.07	0.34	2.69
(6 °C, 8 °C]	41.08	15.49	18.86	12.46	7.07	2.02	3.03
(8 °C, 10 °C]	38.72	23.57	15.15	10.10	7.07	1.35	3.70
(10 °C, 12 °C]	32.32	20.20	20.88	17.17	6.73	0.34	1.68
(12 °C, 14 °C]	26.60	17.51	24.58	19.53	7.41	0.00	4.04
(14 °C, 16 °C]	21.55	21.55	27.61	15.49	10.10	0.67	3.03
(16 °C, 18 °C]	23.23	20.20	26.26	17.85	9.09	0.00	3.37
(18 °C, 20 °C]	24.58	17.17	23.91	21.55	10.44	0.00	1.68
(20 °C, 30 °C]	24.92	17.17	25.25	15.82	14.48	1.01	1.01

optimization problem:

$$\min_{t,p} \; t$$

$$F(p, x_i) - \frac{1}{N} \#\{x_j \mid x_j \leq x_i\} \leq t \quad \text{for all } i = 1, \ldots, N,$$

$$\frac{1}{N} \#\{x_j \mid x_j \leq x_i\} - F(p, x_i) \leq t \quad \text{for all } i = 1, \ldots, N,$$

$$F(p, x_i) - \frac{1}{N} \#\{x_j \mid x_j < x_i\} \leq t \quad \text{for all } i = 1, \ldots, N,$$

$$\frac{1}{N} \#\{x_j \mid x_j < x_i\} - F(p, x_i) \leq t \quad \text{for all } i = 1, \ldots, N.$$

As starting values for the unknown parameter vector p in this problem one may use the classic empirical estimates based on the measurement data. The solution of these optimization problems for all exits of the H-gas and the L-gas network leads to the results compiled in Table 13.2 and Table 13.3. In case the true distribution type is approximately normal, the classical estimator (based on estimating mean and variance) and the optimized estimators (based on solving Eq. (13.6)) are almost identical.

13.4 ▪ Multivariate distribution fitting

So far, our statistical models describe the statistical behavior of individual exit points of the network. To capture the correlations and other relationships among the exits, the underlying multivariate distributions need to be estimated.

Estimating a multivariate distribution is closely related to an analysis of correlations between the single univariate distributions. In the context of gas networks, correlations reflect tendencies of similar or opposite behavior in gas consumption between certain groups of exit points. To give a simplified idea, for instance, households could exhibit a common behavior in gas demands while this may be uncorrelated with the behavior of industrial clients. Figure 13.6 plots the pairwise correlations between gas demands of

Table 13.3. *Percentage of univariate distributions for the L-gas network for all temperature classes: normal (ND), shifted normal (shND), lognormal (logND), shifted lognormal (shLogND), Dirac distributed (Dirac), uniform (UD), and shifted uniform (shUD).*

Temp. Class	ND	shND	logND	shLogND	Dirac	UD	shUD
(−15 °C, −2 °C]	58.01	2.55	35.32	3.97	2.70	3.26	0.43
(−2 °C, 0 °C]	51.49	3.97	40.85	3.12	2.84	3.12	0.85
(0 °C, 2 °C]	48.09	4.96	40.85	6.24	2.98	2.41	0.71
(2 °C, 4 °C]	60.00	9.36	24.82	7.38	2.41	0.85	1.28
(4 °C, 6 °C]	57.16	11.49	25.39	7.80	2.55	1.42	0.43
(6 °C, 8 °C]	55.04	18.72	20.43	8.37	2.27	0.71	0.71
(8 °C, 10 °C]	49.79	18.72	20.71	10.21	2.84	2.27	1.56
(10 °C, 12 °C]	41.70	21.84	23.69	12.77	2.98	0.99	2.27
(12 °C, 14 °C]	27.66	20.57	32.77	19.86	2.55	0.28	2.55
(14 °C, 16 °C]	27.38	18.01	32.06	22.70	3.12	0.28	2.70
(16 °C, 18 °C]	29.65	17.73	29.50	20.43	5.25	0.57	2.70
(18 °C, 20 °C]	33.48	15.74	25.67	22.27	6.10	0.57	2.13
(20 °C, 30 °C]	32.48	14.61	26.52	17.87	10.78	1.42	2.41

exits from a certain area of the L-gas network. In the respective plots, one recognizes certain substructures, namely groups of exits being relatively strongly correlated within the group but only weakly correlated with exits from different groups. It is also revealed that the correlation pattern is a function of temperature. In the L-gas network, negatively correlated exits occur more frequently than in the H-gas network (not plotted here). Presumably, this is due to functional or contractual relationships reflected in the data.

As in the univariate case, a multivariate distribution is uniquely characterized by its density (if it exists) or, more generally, by its distribution function. Determining such joint (network-related) distribution on the basis of the individual (exit-wise) distributions may be very hard or even impossible in a situation where the latter are of very different character: recall from the distribution types discussed in the previous section that we are faced not only with continuous but also with discrete or shifted distributions for which it is not evident how they would fit to a joint multivariate distribution. On the other hand, for exits obeying a univariate normal distribution, it appears natural to group them in order to establish a joint multivariate distribution. The latter is characterized by the density

$$f(x) = (2\pi)^{-k/2} (\det \Sigma)^{-1/2} \exp\left[-\frac{1}{2}(x-\mu)^T \Sigma^{-1}(x-\mu)\right], \qquad (13.7)$$

where k refers to the dimension (number of exit points), μ is the mean vector, and Σ the covariance matrix. The parameters μ and Σ can be estimated from the exit-wise mean values and variances as well as from the correlations between exits. The same procedure can be applied to exits with a lognormal distribution by simple transformations.

Summarizing, regarding gas demand data one is faced with the situation that a subset of exit points is appropriate for establishing a partial multivariate model, whereas other exits (exhibiting Dirac, uniform, or shifted distributions) are difficult to join with the previous ones to set up a total multivariate model. Therefore, a simplified approach providing such total multivariate model would consist of assuming these remaining exit points to have independent distributions with all other exits. This allows us to calculate the joint distribution of the total model as the product of the multivariate normal distribution

13.4. Multivariate distribution fitting

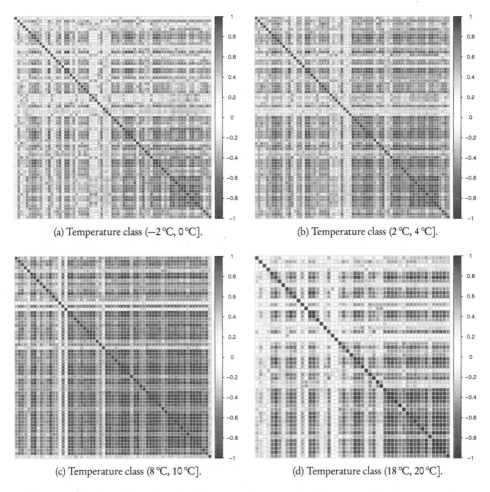

Figure 13.6. *Correlation plots for different temperature classes in an area of the L-gas network.*

with density Eq. (13.7) obtained from exit points with univariate normal or lognormal distributions on the one hand and all the univariate distributions from the remaining exits on the other hand.

A reliable estimation of the multivariate normal distribution with density Eq. (13.7) requires a sufficient amount of data. Given k exit points and a sample size N, it is known that $N \geq k$ should hold, since in case of $N < k$ the sample covariance matrix may be singular and a regularized estimator is needed. It is also known that $N = O(k)$ samples suffice (see Vershynin (2012)). As a rule of thumb, the number of samples in each exit point should exceed $3k$ to provide a stable estimate of the covariance matrix Σ. This requirement turned out not to be satisfied for the subgroup of normal or lognormal exits in the network due to the limited amount of data. As a remedy, the correlation structure as it became visible in Figure 13.6 can be exploited in order to find subgroups of these exits such that the data requirement is fulfilled for these smaller subgroups and such that the distribution between different subgroups is almost independent (indicated by small correlation between exit points of different subgroups). In this way, similar to the total model discussed before, the partial multivariate normal model is obtained as a product of smaller multivariate normal distributions each of which is estimated in a sufficiently stable manner. The analysis of block structures in the correlation matrix can be automated by means of p-median greedy heuristics based on the distance $1 - \rho_{i,j}^2$ between exit points i and j, where $\rho_{i,j}$ refers to the correlation coefficient between these exits. The greedy heuristics are also used for scenario reduction in Section 14.2.2. Evidently, the larger the correlation (positive or negative), the smaller the distance between these exits, making it more likely that they are gathered in a common group.

13.5 • Forecasting gas flow demand for low temperatures

In the following we study historical data of exit loads of the considered gas networks, in order to make a reliable and realistic prediction of the future exit loads for low temperatures. In this particular case, the prediction of gas consumption at the exit points is extremely important, since it is usually very high in cold seasons. We utilize parametric as well as semiparametric nonlinear logistic regression models to estimate the gas flow in dependence of the temperature. The relationship between load flow and temperature is closely related to empirical models for growth data, which are frequently employed in natural and environmental sciences, and sometimes in social sciences and economics. Some examples of such models and their applications can be found in Jones, Leung, and Robertson (2009), Vitezica et al. (2010), and Jarrow, Ruppert, and Yu (2004).

The thesis by Hellwig (2003), and the reports by Geiger and Hellwig (2002), and Wagner and Geiger (2005) suggest the use of sigmoidal growth models for the description of typical gas load profiles in the energy sector. An overview of methods useful for understanding the complexity of gas transportation relying on the mentioned parametric models can be found in Cerbe (2008).

Theoretically, an empirical growth curve is a scatter plot of some measure of the size of an object against a time variable x. The general assumption is, apart from the underlying random fluctuation, that the underlying growth follows a smooth curve. This theoretical growth curve is usually assumed to belong to a known parametric family of curves $f(x|\theta)$ and the aim is to estimate the parameters θ using the data at hand. The same type of models occurs when the explanatory variable x is not the time, but the increasing intensity of some other factor. We observe a change (in general a reduction) of gas consumption with increased temperatures, and seek a model with a physical basis and physically interpretable and meaningful parameters. A detailed description of growth models can be found in Seber and Wild (2003). As a more flexible alternative, semiparametric models can be utilized to tackle the problem. We choose penalized splines (P-splines), which combine two ideas from curve fitting: a regression based on a basis of B-splines and a penalty on the regression coefficients (see Wegman and Wright (1983); Eilers and Marx (2010), (1996)). This approach emphasizes a modeling based on smooth regression, where the penalty controls the amount of smoothing. We follow in particular a similar approach as proposed in Bollaerts, Eilers, and van Mechen (2006), where the authors incorporate shape constraints into the P-splines model.

Typical exit points in the considered gas networks are public utilities, industrial consumers, and storages, as well as exit points on national borders and market crossings. An important aspect to consider in the forecast is the so-called *design temperature*. The design temperature is defined as the lowest temperature at which the network operator is still obliged to supply gas without failure, and differs within Germany depending on the climate conditions in different regions. It usually lies between $-12°$ and $-16°$ Celsius. Such low mean daily temperatures are very uncommon in Germany, and we rarely encounter load flow data available at the design temperature. For this reason, prediction is usually the only way to estimate gas loads at the design temperature of the network, and we investigate several possible models suitable for the forecast.

For the aim of forecasting, we search for a fitting curve that can be interpreted as a curve of expected values in dependence of temperature. The expected values estimated in this way can be used in Chapter 14 as input parameters necessary for sampling gas load under particular distribution for low temperature intervals where nonsufficient data are available. The type of distribution assigned to these low temperature intervals containing almost no data is usually taken to coincide with the one of the neighboring temperature intervals. The neighboring interval is chosen in such a way that its estimated distribution is sufficiently reliable and stable. Thus, these temperature intervals corresponding to the exceptionally low temperatures share the same distributional properties, i.e., deviation and multivariate information as variance and correlation w.r.t. the mentioned nearest warmer temperature interval, but differ in their expected value resulting from the fitted curve. We refer to Chapter 14 for further discussion on this issue.

The data sets considered for fitting a curve of expected values in gas networks are usually large, including in some cases many periods (years) of data. Therefore for the aim of fitting a curve it is sometimes recommendable to cluster the data in convenient subintervals of temperatures and replace the data within a subinterval through its empirical average. This is done in order to reduce the size of the data set and therefore the dimensionality of the underlying optimization problems, in order to avoid very ill conditioned numerical problems. This strategy also allows the elimination of undesired outliers, by considering only subintervals where the sample size is big enough to be *significant*. To this end, we follow a criteria based on the Central Limit Theorem and define a sample size N on a subinterval to be significant if $N \geq 16\hat{\sigma}^2/(U\hat{\mu})^2$, $N \geq 2$, where $\hat{\mu}$ and $\hat{\sigma}$ are the estimated mean and standard deviation from the independent and identically distributed (i.d.d.) sample, and U is the number of deviation units measured in terms of $|\hat{\mu}|$. The latter condition means that we ask for the sample size N to be big enough such that the width of a 95% confidence interval for the standard error of $\hat{\mu}$ can be covered by fixed U units of $|\hat{\mu}|$.

13.5.1 ▪ Parametric models

In the case of parametric models, the basic assumption is that the growth curve belongs to a well known parametric family of curves. The physical interpretability of parameters in the model usually motivates the choice of the growth curve. Based on agreements between the network operators (see [KoV]), we take the following sigmoidal growth model to describe the dependence of gas consumption on temperature:

$$y_i = S(t_i|\theta) + \varepsilon_i.$$

Here y_i denotes the standardized daily mean gas flow, and the corresponding expected value (or mean) curve parameterized in $\theta = (\theta_1, \theta_2, \theta_3, \theta_4)$ is given by

$$S(t_i|\theta) = \theta_4 + \frac{\theta_1 - \theta_4}{1 + \left(\dfrac{\theta_2}{t_i - 40}\right)^{\theta_3}}. \tag{13.8}$$

The curve depends on the predictor t_i, which stands for the weighted four-day mean temperature with weights given in the following form:

$$t_i = \sum_{j=0}^{3} w_j t_{ij}, \quad w_0 = \frac{8}{15}, \; w_1 = \frac{4}{15}, \; w_2 = \frac{2}{15}, \; w_3 = \frac{1}{15}, \tag{13.9}$$

where t_{i0}, t_{i1}, t_{i2}, and t_{i3} are the temperatures corresponding to the days $i, i-1, i-2, i-3$. Finally, ε_i is an error term reflecting zero mean and constant variance.

The articles by Geiger and Hellwig (2002) and Cerbe (2008) introduce this kind of model for description of typical gas loads in dependence of temperature. According to the description of the log-logistic model provided by Ritz and Streibig (2008), the parameters θ_1 and θ_4 in Eq. (13.8) represent upper and lower horizontal asymptotes on the curve, and the other two parameters describe the shape of the decrease of the (logistic like) curve. From the point of view of the energy industry, Geiger and Hellwig (2002) discuss the meaning of parameters in the following way: θ_4 describes the constant share of energy for warm water supply and process energy, while the difference $\theta_1 - \theta_4$ explains extreme daily gas consumption on cold days. Parameter θ_2 indicates the beginning of the heating period, i.e., the change point from the constant gas loads in summer to the increasing consumption in the heating period, and θ_3 measures flexibly the dependence in the heating period.

The authors Geiger and Hellwig (2002) note that apart from the choice of the appropriate mean function $S(t_i|\theta)$, the adequate aggregation of mean daily temperatures to be included in the explanatory variable t_i is essential. Physical properties of buildings play an important role here. The four-day mean temperature is motivated by the fact that typical buildings in Germany accumulate the heat up to 85 hours, and the use of weights as in Eq. (13.9) is suggested. The weights given by Eq. (13.9) are obtained from the standardized geometric series with basis 2 applied to the temperature of the last four days, i.e.,

$$t_i = \frac{t_{i0} + \frac{t_{i1}}{2} + \frac{t_{i2}}{4} + \frac{t_{i3}}{8}}{1 + \frac{1}{2} + \frac{1}{4} + \frac{1}{8}} = \sum_{j=0}^{3} w_j t_{ij}.$$

Based on these facts, German gas transportation companies agreed to use the *sigmoidal function* $S(t_i|\theta)$ defined in Eq. (13.8) with the explanatory variable t_i given by Eq. (13.9) to describe the dependence of gas loads on temperature and to forecast the gas consumption at the design temperature.

Several generalizations of the basic sigmoid model have been considered in order to improve the forecasting of gas loads (see Friedl, Mirkov, and Steinkamp (2012); Leövey et al. (2011)). The sigmoid model and its generalizations so far represent a rough characterization of the mean gas load. The resulting forecast usually underestimates the mean responses for low temperatures. An alternative approach based on semiparametric models has been also considered in order to achieve a more accurate forecast for low temperatures, and will be described in the following section.

13.5.2 • Semiparametric models

The nuances missed by the sigmoidal models, as well as the numerical difficulties that arise in the resulting nonconvex nonlinear optimization problems, motivates the search for alternative and, in some sense, simpler models that overcome these difficulties. One possibility is to use semiparametric models, such as locally weighted regression (Cleveland (1979)) or spline models. Unfortunately, locally weighted regression models are not suitable for prediction. Many authors propose some variant of spline regression for this kind of problems, see, e.g., Jones, Leung, and Robertson (2009), Vitezica et al. (2010), Jarrow, Ruppert, and Yu (2004), Mackenzie, Donovan, and McArdle (2005), Cadorso-Suárez et al. (2010), and Riedel and Imre (1993).

In our case, we choose the penalized splines (P-splines) approach, based on Wegman and Wright (1983), Eilers and Marx (1996), Eilers and Marx (2010), and Bollaerts, Eilers, and van Mechen (2006). This choice is motivated by its simplicity and flexibility. The optimization problems resulting from the fitting problem are convex and solutions can be obtained through solving a system of linear equations. The advantage of P-splines over B-splines is the easy control of smoothness as well as the simple way to handle spline knots, i.e., their number and their positions. As emphasized by Jarrow, Ruppert, and Yu (2004), another advantage of the P-splines method is that knots can be chosen automatically. The number of knots should be sufficiently large to accommodate the nonlinearity of the underlying data. Additional shape constraints lend even more flexibility to the model, since the shape of the curve on boundaries can be adjusted if necessary.

We use the following model to describe the dependence of the gas loads y_i on temperature t_i:

$$y_i = S_\Delta(t_i) + \varepsilon_i,$$

where y_i is usually taken to be a daily mean gas flow at a particular exit point of the network, and t_i stands for the weighted four-day mean temperature, with the weights w described in Eq. (13.9). The function $S_\Delta(t)$ is given by the linear combination of basis spline functions $B_j, j = 1, \ldots, m$, on the mesh Δ, given by

$$S_\Delta(t_i) = \sum_{j=1}^{m} a_j B_j(t_i),$$

and $\varepsilon_i, i = 1, \ldots, n$, are random noise terms reflecting zero mean and constant variance. The functions B_j are basis functions of the B-spline of degree q. The mesh Δ results by taking first an equidistant grid with $m - q$ segments extended over the set of data that includes only subintervals with significant sample size, and then we modify the grid by replacing the least and maximal knots with knots at the design temperature $t_{\min}$ and at the maximal temperature $t_{\max}$ correspondingly. The resulting mesh Δ contains $m - q - 1$ inner knots.

If we introduce a smoothing penalty parameter λ, then instead of minimizing a least squares criterion like

$$\sum_{i=1}^{n} (y_i - S_\Delta(t_i))^2,$$

the objective function to be minimized is the penalized residual sum of squares

$$\sum_{i=1}^{n} (y_i - S_\Delta(t_i))^2 + \lambda \sum_{j=k+1}^{m} (\delta^k(a_j))^2,$$

where $\delta^k(a_j)$ denotes the kth order finite differences of the coefficients a_j of the corresponding B-splines. For the work described here, we use in particular the second-order finite difference $\delta^2(a_j) := a_j - 2a_{j-1} + a_{j-2}, j = 3, \ldots, m$.

The first-order shape constraints can be added to the P-splines approach, following Bollaerts, Eilers, and van Mechen (2006) or by imposing linear constraints on the spline coefficients a_j and solving the resulting constrained least-squares optimization problem. Several available commercial optimization solvers can be used for this purpose. In our approach, we include first-order boundary derivative constraints

$$\frac{\partial S_\Delta}{\partial t}(t_{\min}) = \frac{\partial S_\Delta}{\partial t}(t_{\max}) = 0,$$

aiming to simulate a constant consumption beyond the design temperature $t_{\min}$ and the considered maximal temperature $t_{\max}$. Positivity constraints where also added to ensure

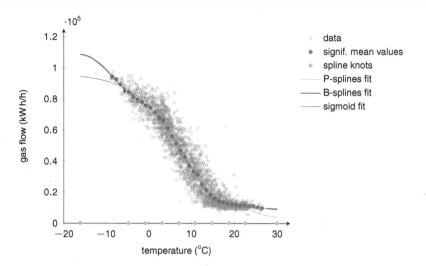

Figure 13.7. *Shown are the P-splines, B-splines ($\lambda = 0$) and sigmoid fitting mean curves with 9 cubic splines based on significant mean values on 1 °C width subintervals, with $U = 0.4$, $t_{min} = -16$, $t_{max} = 30$.*

that the fitted mean curve remains positive. Different criteria for a robust choice of the smoothing parameter λ can be founded in Lee and Cox (2010). We adopted the strategy called absolute cross validation (ACV) by Lee and Cox (2010), Section 3.2, which yields very good results for the forecasting of the mean curve in the considered gas networks.

Based on data from a typical statistical exit point of the gas network, Figure 13.7 compares P-splines, B-splines ($\lambda = 0$), and sigmoid fitting mean curves.

Chapter 14
Methods for verifying booked capacities

Benjamin Hiller, Christine Hayn, Holger Heitsch,
René Henrion, Hernan Leövey, Andris Möller,
Werner Römisch

Abstract *We formalize the problem to verify the legal requirement that transport situations arising from booked capacity rights shall be technically feasible. In particular, we propose a stochastic version of the problem of verifying booked capacities together with two heuristic solution methods. These methods have been designed as decision-support tools for real-world usage by transmission system operators (TSOs). Our approach is based on combining a stochastic model with an adversarial model to an overall model for the transport situations requested by the transport customers. The first method is based on sampling to capture the stochastic information, whereas the second method uses multivariate quantiles for that purpose. Both methods generate a set of nominations that are checked for technical feasibility to arrive at an overall conclusion.*

As described in Chapter 3, gas transmission system operators (TSOs) sell capacity rights to transport customers. Booking, i.e., buying, capacity rights entitles a transport customer to inject gas at entry points and/or withdraw gas at exit points of the gas network. In particular, TSOs are obliged to offer as much capacity as possible as freely allocable capacity (FAC), which enables transport customers to use entry and exit capacities independently (see Section 3.2.2 for details). However, a TSO may only sell capacity rights for which it can guarantee that each "likely and realistic" [GasNZV 2010] load flow complying with the capacity rights booked by all transport customers can technically be realized. Thus a TSO needs a way to check this requirement.

This chapter presents methods that (heuristically) reduce this problem to checking a (potentially large) set of possible load flows for technical feasibility. In particular, we discuss how a suitable set of load flows may be obtained and how conclusions can be drawn from the corresponding feasibility tests. Methods for performing these feasibility tests are addressed in great detail in Chapters 6–10. The methods presented are, due to the high complexity of the real-world problem, heuristical in several aspects which are discussed in more detail in the final section of this chapter.

14.1 ▪ Motivation and outline of the approach

In order to formalize the problem of verifying that the load flows corresponding to a set of booked capacity contracts are technically feasible, we introduce some terminology (for types of capacity contracts, see Sections 3.2 and 3.3.2). Recall from Section 3.2.1 that using a gas transmission network is a two-step process: First, one has to book, i.e., buy, capacity contracts from the TSO. The day before the actual transmission is going to take place, one has to nominate the amount of gas that will be injected or withdrawn, according to the limitations of the capacity contract.

We define a *booking* B to be the set of all capacity contracts booked with the network operator at a certain point in time. We will see in Section 14.4 how such capacity contracts can be modeled in detail; for now it is sufficient that they prescribe certain conditions on load flows in the gas network. Let V_+ and V_- denote the set of entry and exit points of the gas network, respectively, and denote by $V_\pm$ the union of these two sets. We represent load flows in the gas network by load flow vectors. A *load flow vector* is a vector $P = (P_u)_{u \in V_\pm}$ that specifies, for each entry and exit, a load flow. Throughout this chapter, we will assume that load flows are specified in terms of thermal power. As explained in Section 5.3.3, the load flows specified in power may be converted to ones specified as mass flows for checking their technical feasibility. We call a load flow vector *booking compliant* if it satisfies all conditions that are related to the capacity contracts in a booking B.

A *nomination* is a load flow vector $P = (P_u)_{u \in V_\pm}$ that is balanced, i.e., satisfies the condition

$$\sum_{u \in V_+} P_u = \sum_{u \in V_-} P_u.$$

Finally, we call a nomination *technically feasible* if the gas network can be controlled such that the gas flow specified by the nomination is realized.

It is important to observe that this concept of nomination is an idealization of the process of nominating in two ways (see Section 3.2.1): First, load flows in a real gas network do not have to be balanced at any point in time, but only for longer balancing periods, for instance 24 hours. Second, strictly speaking load flows are only nominated at a subset of the points, e.g., at storages or the virtual trading point (VTP).

We introduce the following sets of load flow vectors:

$$\mathcal{B}_= = \text{set of booking-compliant nominations,}$$
$$\mathcal{T} = \text{set of technically feasible nominations.}$$

Then, the task of the gas network operator can be formalized by asking whether the inclusion

$$\mathcal{B}_= \subseteq \mathcal{T} \tag{14.1}$$

holds, i.e., (14.1) means that any booking-compliant nomination should be technically feasible. This is the *deterministic* version of the *verification of booked capacities*.

Checking this inclusion in a mathematically exact sense seems to be hopeless in practice, since even checking a single booking-compliant nomination for technical feasibility requires solving a nonconvex mixed-integer nonlinear program (MINLP), whose combinatorial part is already NP-hard (Szabó (2012)). Under some additional assumptions, it would be sufficient to check the relation $P \in \mathcal{T}$ for a finite number of nominations P. For instance, given a polyhedral structure of the set $\mathcal{B}_=$ and assuming convexity of the set $\mathcal{T}$ (which does not hold true in general), the verification of (14.1) can be done by checking $P \in \mathcal{T}$ for the finitely many vertices of the polytope $\mathcal{B}_=$: If $P_1 \in \mathcal{T}$ and $P_2 \in \mathcal{T}$, so is the line $[P_1, P_2]$ by convexity of $\mathscr{T}$. Thus the fact that all vertices of $\mathscr{B}_=$ are in $\mathscr{T}$ implies that $\mathscr{B}_= \subseteq \mathscr{T}$. However, even for a moderate number of entry and exit points in the network, the number of vertices becomes astronomical. For instance, the polytope arising by intersecting the d-dimensional unit hypercube with one hyperplane given by the equation $\mathbb{1}^T x = k$, $k \in \mathbb{Z}$, has at least $\binom{d}{k}$ vertices if $2 \leq k \leq d-1$. In the simplest case, $\mathscr{B}_=$ might be given by a box $[\ell, u] \subseteq \mathbb{R}^d$ for $\ell, u \in \mathbb{R}^d$ and a balancing equation $\mathbb{1}^T P = c$ with $\|\ell\|_1 < c < \|u\|_1$, which is affinely equivalent to the polytope just mentioned. Thus one cannot benefit from finiteness in a computationally relevant sense.

Given the impossibility of checking (14.1) exactly, the question arises of how to find a finite subset of testing nominations $\{P^1, \ldots, P^N\} \subset \mathscr{B}_=$ such that the relations $P^i \in \mathscr{T}$, for $i = 1, \ldots, N$, provide a reliable substitute for the verification of (14.1). Of course, sampling of $\mathscr{B}_=$ should take into account historical information about nomination behavior. Moreover, as explained in Section 3.3.1, this is also required by current regulation rules. Depending on the considered gas network, the load flow at a subset of the points may be modeled stochastically. For instance, this is the case for exits belonging to public utilities where gas is usually used for heating and the load flow thus depends on the ambient temperature, which may be modeled stochastically as explained in Chapter 13. At these points, it is therefore justified to use probability distributions for the load flow estimated on the basis of historical data to predict future load flow patterns. We collect the points for which this is reasonable in the set V_{stat} and call them *statistical points*. We assume in the following that a stochastic model for the load flows of the statistical points is available. More precisely, we assume there is a random vector $\xi : \Omega \to \mathbb{R}^{V_{\text{stat}}}$ defined on a suitable probability space $(\Omega, \mathscr{A}, \mathbb{P})$, e.g., derived by the data analysis from Chapter 13. We will call the random vector ξ of the loads at the statistical points the *random load vector*. An element of $\mathbb{R}^{V_{\text{stat}}}$, and thus in particular a realization of ξ, is called a *statistical load scenario*.

However, there are also points for which a stochastic model is not appropriate. For instance, the behavior of entries and storages is mainly market driven, hence difficult to model in a stochastic way. This lopsided constellation suggests to consider a substitute for inclusion (14.1) which takes into account that a stochastic model for the statistical points is used. To be more precise, let π denote the projection of a load flow vector $P \in \mathbb{R}^{V_\pm}$ onto the statistical points, i.e., $\pi(P) = (P_u)_{u \in V_{\text{stat}}}$. Moreover, let $P_{\text{stat}} \in \mathbb{R}^{V_{\text{stat}}}$ be any load flow vector at the statistical points. Then, (14.1) is equivalent to the partitioned inclusion

$$\pi^{-1}(P_{\text{stat}}) \cap \mathscr{B}_= \subseteq \mathscr{T} \quad \text{for all } P_{\text{stat}} \in \mathbb{R}^{V_{\text{stat}}}, \tag{14.2}$$

stating that all booking-compliant nominations extending any P_{stat} are technically feasible. This requirement, however, is much too restrictive and unrealistic, because the given inclusion always bears the risk of being violated by some extreme but very unlikely load flow vectors P. At this place, the exploitation of stochastic information makes sense. Taking into account the stochastic character of the random load vector ξ, we relax the strict "for all" relation (14.2) by a probabilistic condition of the form

$$\beta := \mathbb{P}(\pi^{-1}(\xi) \cap \mathscr{B}_= \subseteq \mathscr{T}) \geq \alpha, \tag{14.3}$$

requiring that (14.2) holds with a specified probability $\alpha \in (0, 1)$ only. This is the *stochastic version of verification of booked capacities* on which we will focus for the remainder of this chapter. We call the probability β the *validity probability* of the booking B. With the terminology introduced above, inequality (14.3) expresses the condition that with at least probability α, every booking-compliant nomination extending the random load vector ξ will be technically feasible. Of course, the choice of which points to treat as statistical points and the particular choice of a concrete probability level $\alpha \in (0,1)$ (typically close to one) has to be agreed upon, based on experience, governmental rules, or common sense.

In order to explain the proposed methods to numerically check condition (14.3), we introduce two functions $\Gamma \colon \mathbb{R}^{V_{\text{stat}}} \to \mathscr{P}(\mathbb{R}^{V_\pm})$ and $\nu_{\mathscr{T}} \colon \mathscr{P}(\mathbb{R}^{V_\pm}) \to \{0,1\}$, where $\mathscr{P}(\mathbb{R}^{V_\pm})$ denotes the power set of $\mathbb{R}^{V_\pm}$. The function Γ maps a statistical load scenario s to its set of booking-compliant extending nominations, i.e., we have $s \mapsto \pi^{-1}(s) \cap \mathscr{B}_=$, whereas $\nu_{\mathscr{T}}$ indicates whether *all* nominations of a set of nominations are technically feasible. Moreover, we define $\Upsilon_B := \nu_{\mathscr{T}} \circ \Gamma$. With this notation, we can rewrite condition (14.3) as

$$\beta = \mathbb{P}(\Upsilon_B(\xi) = 1) \geq \alpha. \tag{14.4}$$

Assume for the moment that we could evaluate $\Upsilon_B(s)$ for a given statistical load scenario s. One natural way to check condition (14.4) is to compute the probability on the left hand side. We can rewrite this probability as

$$\mathbb{P}(\Upsilon_B(\xi) = 1) = 1 \cdot \mathbb{P}(\Upsilon_B(\xi) = 1) + 0 \cdot \mathbb{P}(\Upsilon_B(\xi) = 0) = \mathbb{E}_\mathbb{P}[\Upsilon_B] = \int_\Omega \Upsilon_B \, d\mathbb{P}.$$

Thus the computation of the probability can be seen as the numerical estimation of high-dimensional integrals. In Section 14.2 we use techniques like quasi–Monte Carlo methods and scenario reduction to construct a discrete set of statistical load scenarios $\{s^1, \ldots, s^N\}$ with associated probabilities $p_1, \ldots, p_N$. This set is the basis of an estimator for the above integral which evaluates Υ_B once for each s^i, $1 \leq i \leq N$.

A conceptually different approach described in Section 14.3 relies on so-called multivariate quantiles to check condition (14.4) directly. The basic idea is to construct a special statistical load scenario $\hat{s}$ that satisfies

$$\mathbb{P}(\xi \leq \hat{s}) \geq \alpha,$$

i.e., that dominates ξ with high probability. Assuming a certain monotonicity condition, $\Upsilon_B(\hat{s}) = 1$ then implies $\mathbb{P}(\Upsilon_B(\xi) = 1) \geq \alpha$.

It remains to discuss how to evaluate $\Upsilon_B(s)$ for a given statistical load scenario s. Observe that this task can in principle be achieved by a procedure that either generates a booking-compliant nomination extending s that is *not* technically feasible (certifying $\Upsilon_B(s) = 0$) or establishes that no such nomination exists (certifying $\Upsilon_B(s) = 1$). Given the complexity of checking technical feasibility, it is unlikely to develop a practically efficient method for this purpose. Therefore, we propose an adversarial heuristic that tries to construct a small set of extending nominations that are challenging for the gas network, i.e., which are likely to be technically infeasible if the validity probability β of the booking is too low. This adversarial heuristic is based on a model for the set $\mathscr{B}_=$ of booking-compliant nominations, presented in Section 14.4. The heuristic itself is based on sampling again and takes into account expert knowledge from network planners. It is explained in Section 14.5.

Section 14.6 puts all the pieces together and describes the two resulting overall methods, one based on sampling, the other using multivariate quantiles to capture the stochasticity of the load flows at the statistical points. From a high-level viewpoint, both methods work as follows:

1. Generate a set of statistical load scenarios that carry the stochastic information and provide load flows at the statistical points.
2. Use the adversarial heuristic to compute a set of booking-compliant nominations extending each statistical load scenario.

3. Check these nominations for technical feasibility and draw an overall conclusion from these outcomes.

In Section 14.7 we then show computational results on real-life test networks and discuss some possible improvements.

Apart from the adversarial heuristic, there are more heuristic aspects involved in these methods. These are discussed in Section 14.8, which also concludes the chapter.

14.2 ▪ Sampling statistical load scenarios for verifying booked capacities

As described in the previous section, the load flow of a substantial part of the points of the considered gas network can be modeled stochastically. For these statistical points, historical data is available and a carefully selected probability distribution model $\mathbb{P}$ for load flow can be calibrated. We will refer to the dimension d as the number of considered statistical points. As mentioned before, the considered set of statistical points is large, with values of dimension d that can range into several hundreds.

For a fixed booking B, one would like to know the validity probability β under the load flow distribution $\mathbb{P}$. The completion and validation (in the sense of checking the technical feasibility) of a statistical load scenario s^i is a process that can be interpreted as a (measurable) validation function $\Upsilon_B \colon \mathbb{R}^{V_{\text{stat}}} \to \mathbb{R}$. In the simplest case, Υ_B assigns to every generated statistical load scenario s^i, $1 \le i \le N$, the value of "1" in case of feasibility of all extending booking-compliant nominations, or "0" otherwise. In a more general case, we can consider a validity function $\Upsilon_B^* \colon \mathbb{R}^{V_{\text{stat}}} \to [0,1]$, where now the validation process can return values between 0 and 1. In this generalized setting, the task is defined to be the estimation of $\mathbb{E}_{\mathbb{P}}[\Upsilon_B^*]$ in order to take a decision of accepting or rejecting a booking, which is a high-dimensional integration problem. In our case, and as it is usual in many practical high-dimensional problems in simulation, the statistical points have been modeled with distributions over $\Omega = \mathbb{R}^d$ that allow us to consider a bijective transformation $\Phi \colon \mathbb{R}^d \to (0,1)^d$ changing the original problem into an integration problem over the unit cube. The integration problem takes now the form

$$\mathbb{E}_{\mathbb{P}}[\Upsilon_B^*] = \int_{\mathbb{R}^d} \Upsilon_B^* \, d\mathbb{P} = \int_{[0,1]^d} \Upsilon_B^*(\Phi^{-1}(x)) \, dx. \tag{14.5}$$

The latter equality is valid since an arbitrary extension of Φ^{-1} to the boundary of $[0,1]^d$ can be carried out, because the zero-measure boundary set does not influence the value of the resulting integral.

In the following we describe high-dimensional integration methods for approximating the desired expectation in (14.5) starting with Monte Carlo methods, moving to quasi–Monte Carlo and finally to a hybrid method, namely, randomized quasi–Monte Carlo. The choice of the transformation Φ is of essential importance for the problem of sampling in high dimensions. It usually has a strong influence in what is called the *effective dimension* of the problem. We will discuss this issue in Section 14.2.1, where we also argue why the class of randomized quasi–Monte Carlo methods is preferable for our application.

In the classical Monte Carlo (MC) approach (see Niederreiter (1992)) one tries to estimate (14.5) by generating statistical load scenarios pseudorandomly. Starting with a finite sequence of independent identically distributed (i.i.d.) samples $S_N = \{s^1, s^2, \ldots, s^N\}$, where the points s^i, $1 \le i \le N$, are uniformly distributed in $[0,1]^d$, the average of a given target function f

$$Q_N(f) := \frac{1}{N}\sum_{i=1}^{N} f(s^i)$$

is taken as an approximation of a desired integral $I_d(f) = \int_{[0,1]^d} f(x)\,dx$. The resulting estimator $Q_N(f)$ is unbiased, and the error can be approximated via the central limit theorem, assuming that f is square-integrable. The variance of the estimator $Q_N(f)$ is given by

$$\mathrm{Var}[Q_N(f)] = \frac{\sigma^2(f)}{N} = \frac{1}{N}\left(\int_{[0,1]^d} f^2(x)\,dx - \left(\int_{[0,1]^d} f(x)\,dx\right)^2\right).$$

The resulting integration error associated to the MC approach is then of order $O(N^{-\frac{1}{2}})$. The quality of the MC samples relies on the selected pseudorandom number generators of uniform samples in $[0,1]^d$. Good accessible generators to this end are for example the *Mersenne twister* from Matsumoto and Nishimura (01/1998), and *MRG32k3a* from L'Ecuyer (01/1999). MC is in general a very reliable tool in high-dimensional integration, but the order of convergence is in fact poor. Since validation of nominations in real-world gas network can be a very time consuming procedure, the search for good tools improving the accuracy of estimation, or reducing the amount of samples needed to reach a desired accuracy, is essential.

In contrast to MC methods, *quasi–Monte Carlo (QMC) methods* are deterministic methods based on sequences of points that are more regularly distributed than the pseudorandom points from MC (see L'Ecuyer and Lemieux; Novak and Woźniakowski; Dick and Pillichshammer; Kuo, Schwab, and Sloan (2005; 2010; 2010; 2011)). Using QMC, one can expect in many practical situations with high-dimensional integrands an error convergence of order $O(N^{-1})$, if the integrands are sufficiently smooth. Typical examples of QMC are modern shifted lattice rules and low-discrepancy sequences. To define what we mean by "regularly distributed", we now introduce the classical notion of discrepancy (see Niederreiter (1992)) of a finite sequence of points S_N in $[0,1)^d$.

Definition 14.1. *Let $S = \{s^1,\ldots,s^N\}$ be an arbitrary set of points in $[0,1)^d$. The discrepancy of S w.r.t. to an interval $[0,a) \subseteq [0,1)^d$ is measured by the function*

$$\mathrm{disc}(S,a) = \sum_{j=1}^{N} \chi_{[0,a)}(s^j) - \prod_{i=1}^{d} a_i,$$

where $\chi_{[0,a)}(\cdot)$ is the characteristic function of $[0,a)$. Let $D = \{1,\ldots,d\}$ and define $\hat{x}_I$ for any $x \in [0,1]^d$ and $I \subseteq D$ by

$$\hat{x}_I = \begin{cases} x_i & i \in I, \\ 1 & \text{otherwise.} \end{cases}$$

Then

$$\mathbb{D}_r(S) = \left(\sum_{\emptyset \neq I \subseteq D} \int_{[0,1]^{|I|}} |\mathrm{disc}(S,\hat{x}_I)|^r\,dx_I\right)^{\frac{1}{r}}$$

is called L_r-discrepancy of the point set S, $r \in [1,\infty]$, with the obvious modification for $r = \infty$. The L_∞-discrepancy is also called star discrepancy and denoted by $\mathbb{D}^*(S)$.

For a given finite point the star discrepancy gives a measure of the worst difference set $S = \{s^1,\ldots,s^N\}$, between the uniform distribution and the sampled distribution in $[0,1)^d$

14.2. Sampling statistical load scenarios for verifying booked capacities

attributed to the set S. In the context of QMC, a sequence of points in $[0,1)^d$ is called a low-discrepancy sequence if $\mathbb{D}^*(S) = O(N^{-1}(\log(N))^d)$ for all truncations of the sequence to its first N terms.

The usual way to analyze QMC as a deterministic method is to choose a linear normed space F of functions on $[0,1)^d$ with norm $\|\cdot\|$ and an associated discrepancy $\mathbb{D}(S_N)$ for the point sequence S_N. Then, the deterministic integration error can be estimated by

$$|Q_N(f) - I_d(f)| \leq \mathbb{D}(S_N)\|f\|$$

for all functions $f \in F$. Such estimates are called Koksma–Hlawka type inequalities due to the classical Koksma-Hlawka inequality (see Niederreiter (1992)), where $\mathbb{D}(S_N)$ is taken to be the star discrepancy of the point sequence S_N and $\|f\|$ is the variation in the sense of Hardy and Krause of f.

In modern QMC error analysis, one often considers weighted reproducing kernel Hilbert spaces (RKHS) as function spaces (see Kuo, Schwab, and Sloan (2011)). In this context one obtains an error bound in the above form, where $\mathbb{D}(S_N)$ represents a weighted L_2-discrepancy. If the considered weights satisfy some particular decay conditions, describing a decay of importance of the variables or group of variables, then the discrepancy $\mathbb{D}(S_N)$ can be reduced at a rate $O(N^{-1+\delta})$, $\delta \in (0, \frac{1}{2}]$, with a constant δ independent of the dimension d, in a tractable way with specially constructed shifted lattice rules and low-discrepancy sequences (see Kuo, Schwab, and Sloan (2011)).

In practice, randomly shifted lattice rules and low-discrepancy sequences are both competitive techniques of QMC. Our choice for generation of statistical load scenarios using *digital sequences*, namely, Sobol' sequences, is a special case of low-discrepancy sequences that are included in the category of (t, m, d)-nets and (t, d)-sequences (Dick and Pillichshammer (2010)). In some sense, shifted lattice rules are adaptive in the way that they allow using information of the target integrand to fix the generating vector of the lattice, if the given integrand is smooth enough (Griewank et al. (2013)). On the other hand, digital sequences focus more on other features as the discrepancy, thus they are constructed independently of the integrand at hand. In our case for the validation of statistical load scenarios, we do not know explicitly how the integrand looks like. We do know that in practice the validity function can be taken of the form $\Upsilon_B: \mathbb{R}^{V_\pm} \to \{0,1\}$, presenting discontinuity jumps.

There are some practical advantages in retaining the probabilistic scheme of the sampling, while using these nice deterministic constructions called digital sequences. Therefore we have focused on hybrid methods permitting us to combine the best features of MC and QMC together. Randomization is an important tool in high-dimensional integration if we want to estimate the error of our approximation $Q_N(f)$ to the desired integral. One goal is to randomize the deterministic point set S_N generated by QMC in a way that the randomized points in the set $\tilde{S}_N$ have the uniform distribution over $[0,1)^d$. Thus the resulting estimator $Q_N(f)$ preserves unbiasedness. The second goal is to preserve the better equidistribution properties of the deterministic construction. The simplest form of randomization applied to digital sequences seems to be the technique called *digital b-ary shifting*, see (L'Ecuyer and Lemieux (2005), Section 5.2]) and the references therein.

We choose $b = 2$, i.e., we use random digital binary shifting to obtain our randomized QMC method that works as follows. To obtain a final point $\tilde{u} \in [0,1)^d$, we generate a point $u \in [0,1)^d$ from the underlying Sobol' sequence and a pseudorandom vector Δ uniformly distributed in $[0,1)^d$. We then consider the binary expansions of each component

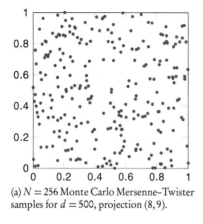 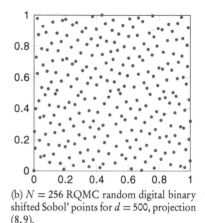

(a) $N = 256$ Monte Carlo Mersenne–Twister samples for $d = 500$, projection $(8,9)$.

(b) $N = 256$ RQMC random digital binary shifted Sobol' points for $d = 500$, projection $(8,9)$.

Figure 14.1. *Comparison of Monte Carlo Mersenne–Twister and RQMC samples.*

of both u and Δ that are given by

$$\Delta_j = \sum_{l=1}^{\infty} \delta_{jl} 2^{-l} \quad \text{and} \quad u_j = \sum_{l=1}^{\infty} u_{jl} 2^{-l} \quad \text{for all } j = 1,\ldots,d.$$

The random digital binary shifted Sobol point $\tilde{u}$ is then computed by

$$\tilde{u}_j = \sum_{l=1}^{\infty} (u_{jl} + \delta_{jl}) 2^{-l} \quad \text{for all } j = 1,\ldots,d,$$

where the addition is modulo 2. A comparison of MC samples generated by the Mersenne–Twister pseudorandom generator and random digital binary shifted Sobol' points is shown in Figure 14.1.

14.2.1 ▪ Justification of using randomized QMC methods

A partial explanation to the success of QMC against MC can be given by considering the ANOVA *(analysis of variance) decomposition* of the functions at hand. If the integrands (maybe after a proper transformation) have the property that few ANOVA terms corresponding to the interaction of few variables accumulate most of the variance of the integrand, and if these important ANOVA terms exhibit enough smoothness, then we can expect that QMC will perform better than MC for integration.

Using ANOVA we decompose a function into a sum of simpler functions (see Sobol' (2001)). Let $D = \{1,\ldots,d\}$. For any subset $I \subseteq D$, let $|I|$ denote its cardinality and $D \setminus I$ be its complementary set in D. Let $x_I = (x_j)_{j \in I}$ be the $|I|$-dimensional vector containing the coordinates of x with indices in I. Now assume that f is a square-integrable function. Then we can write f as the sum of 2^d ANOVA terms:

$$f(x) = \sum_{I \subseteq D} f^I(x),$$

where the ANOVA terms $f^I(x)$ are defined recursively by

$$f^I(x) = \int_{[0,1]^{d-|I|}} f(x_I, x_{D \setminus I}) \, dx_{D \setminus I} - \sum_{I' \subsetneq I} f^{I'}(x),$$

and $f^\emptyset = I_d(f)$. The sum of the right-hand side is over strict subsets $I' \neq I$, and we use the convention $\int_{[0,1]^0} f(x) dx_\emptyset = f(x)$. Note that the ANOVA decomposition is L_2-orthogonal.

In many practical applications, one encounters functions for which the total variance is concentrated in a small portion of its ANOVA terms. The notion of effective dimension of a function was first introduced by Caflisch, Morokoff, and Owen (1997) to describe the contribution of a group of variables to the total variance.

Definition 14.2. *A function f is said to have* effective dimension d_t *in the truncation sense with proportion p, for $0 < p < 1$, if d_t is the smallest integer that satisfies*

$$\sum_{I \subseteq \{1,\ldots,d_t\}} \sigma_I^2(f) \geq p\sigma^2(f),$$

where $\sigma_I^2(f)$ denotes the variance of the ANOVA term $f^I(x)$.

It is known that the lower-order ANOVA terms of an integrand can exhibit substantially more smoothness than the integrand itself, even if the integrand presents discontinuity jumps (Griebel, Kuo, and Sloan (2010), (2013); Heitsch, Leövey, and Römisch (2012)). Therefore effective dimension reduction techniques based on suitable transformations of the integrand are essential. In Gaussian integration, the particular choice of matrix factorization usually has a strong influence in the effective dimension of the problem and in the performance of QMC. Principal components analysis (PCA) decomposition (Wang and Fang (2003)) is usually recommended to be applied if feasible, and this is the method of choice for generating multivariate Gaussian samples for statistical load scenarios. It is well known that PCA can reduce the effective dimension and improve the performance of QMC for many integrands considered in mathematical finance (Glasserman (2004); Wang and Sloan (2005)), and the same has been shown recently in two-stage linear stochastic optimization problems (Heitsch, Leövey, and Römisch (2012)).

In most examples encountered in practical applications requiring moderate or small sample sizes N, one expects that randomized QMC will work at least as well as MC. Thus, there is usually no loss in replacing MC by particular good randomized versions of QMC. We can expect in many cases even a benefit using QMC if the given integrands can be well approximated by low-dimensional smooth functions (Kuo, Schwab, and Sloan (2011)), exhibiting in many cases order of convergence close to $O(N^{-1})$.

14.2.2 ▪ Scenario reduction

Scenario reduction may be desirable in many situations when the underlying scenario models already happen to be large scale and the incorporation of a large number of scenarios leads to high computation times. The basic idea of scenario reduction consists of determining a (nearly) best approximation in terms of a suitable probability metric of the underlying discrete probability distribution by a probability measure with smaller support. The metric should be associated to the mathematical model in a canonical way such that the model behaves stable with respect to changes of the probability distribution. Several canonical metrics are discussed by Römisch (2003).

Since the relevant optimization and feasibility problems of this book are mixed-integer nonlinear with stochastic inputs, the so-called discrepancies (see, for example, Römisch (2003)) and, in particular, the Kolmogorov distance appear to be suitable probability metrics. We refer to the monograph by Rachev (1991) for a survey of probability metrics and their properties.

Let $\mathbb{P}$ denote a discrete probability distribution on $\mathbb{R}^d$ represented by scenarios s^i (in our context: statistical load scenarios) and probabilities p_i, $i = 1,\ldots,N$, and $\mathbb{P}_J$ a discrete probability distribution with scenarios s^j and probabilities p'_j, $j \notin J$. Hence, the support of $\mathbb{P}_J$ is a subset of the support of $\mathbb{P}$, and J denotes the index set of deleted scenarios from those of $\mathbb{P}$. The *Kolmogorov distance* between two probability distributions is defined as the uniform distance of their (cumulative) distribution functions. In our special case, we have

$$\mathbb{D}_K(\mathbb{P},\mathbb{P}_J) = \sup_{x \in \mathbb{R}^d} \left| \frac{1}{N} \sum_{i=1, s^i \leq x}^{N} p_i - \sum_{j \notin J, s^j \leq x} p'_j \right|.$$

However, the numerical results by Henrion, Küchler, and Römisch (2009) show that the problem of optimal scenario reduction with respect to the Kolmogorov distance can presently not be solved in reasonable time for higher dimensions d.

Therefore, we employ alternatively the L_1-Wasserstein or *Kantorovich distance* of $\mathbb{P}$ and $\mathbb{P}_J$ given by

$$W_1(\mathbb{P},\mathbb{P}_J) = \min\left\{ \sum_{i=1}^{N} \sum_{j \notin J} \eta_{ij} \left\| s^i - s^j \right\|_1 \,\bigg|\, \eta_{ij} \geq 0, \sum_{i=1}^{N} \eta_{ij} = p'_j, \sum_{j \notin J} \eta_{ij} = p_i \right\}.$$

Clearly, computing $W_1(\mathbb{P},\mathbb{P}_J)$ means solving a linear program of dimension nN. The problem of optimal scenario reduction then consists in determining the best approximation of $\mathbb{P}$ by a distribution $\mathbb{P}_J$ with $J \subset \{1,\ldots,N\}$ and $|J| = N - n$ for given $n < N$, i.e., it may be written as

$$\min\left\{ W_1(\mathbb{P},\mathbb{P}_J) \,\bigg|\, J \subset \{1,\ldots,N\}, |J| = N - n, p'_j \geq 0 \text{ for } j \notin J, \sum_{j \notin J} p'_j = 1 \right\}$$

or

$$\min\{D_J \mid J \subset \{1,\ldots,N\}, |J| = N - n\}, \tag{14.6}$$

where D_J denotes the minimum of $W_1(\mathbb{P},\mathbb{P}_J)$ with respect to $p'_j \geq 0$, $j \notin J$, and $\sum_{j \notin J} p'_j = 1$. The minimum may be computed as (see Henrion, Küchler, and Römisch (2009)):

$$D_J = \sum_{j \in J} p_j \min_{i \notin J} \left\| s^i - s^j \right\|,$$

and the corresponding optimal weights p'_j, $j \notin J$, are given by the (optimal) redistribution rule

$$p'_j = p_j + \sum_{i \in I(j)} p_i \quad \text{for all } j \notin J,$$

where

$$I(j) = \arg\min_{i \notin J} \left\| s^i - s^j \right\|.$$

The combinatorial optimization problem (14.6) is called the n-median problem in the literature and is known to be NP-hard (Kariv and Hakimi (1979)).

There are several approaches for the computational solution of n-median problems. First, we mention the reformulation of (14.6) as mixed-integer linear program and the possibility of applying standard software (e.g., CPLEX). For example, if $y_i \in \{0,1\}$, $i = 1,\ldots,N$, denotes the decision variable whether s^i is deleted ($y_i = 0$) or not ($y_i = 1$), problem (14.6) allows the reformulation

$$\min \sum_{i,j=1}^{N} p_j x_{ij} \left\| s^i - s^j \right\|_1$$

$$\text{s.t.} \quad \sum_{i=1}^{N} y_i = n,$$

$$\sum_{1 \leq j \leq N, j \neq i} x_{ij} + y_i = 1 \quad \text{for all } 1 \leq i \leq N,$$

$$x_{ij} \leq y_i \quad \text{for all } 1 \leq i, j \leq N,$$

$$x_{ij} \in [0,1] \quad \text{for all } 1 \leq i, j \leq N,$$

$$y_i \in \{0,1\} \quad \text{for all } 1 \leq i \leq N,$$

as a mixed-integer linear program with N^2 continuous and N binary variables. Indeed, since $y_i = 1$ for $i \notin J$, and, hence, $x_{ij} = 0$ for all $i \in J$ and $j \notin J$, one obtains

$$\frac{1}{n} \sum_{i,j=1}^{N} p_j x_{ij} \left\| s^i - s^j \right\|_1 = \frac{1}{n} \sum_{i \notin J} \sum_{j \in J} p_j \left\| s^i - s^j \right\|_1 \geq \sum_{j \in J} p_j \min_{i \notin J} \left\| s^i - s^j \right\|_1 = D_J,$$

and the lower bound is attained for

$$x_{ij} = \frac{\min_{i \notin J} \left\| s^i - s^j \right\|_1}{n \left\| s^i - s^j \right\|_1} \quad \text{for all } i \notin J, j \in J.$$

Alternatively, we mention the method based on column generation and on branch-cut-and-price algorithm in (Avella, Sassano, and Vasil'ev (2007)), which is suitable for large-scale models, approximation methods based on semidefinite programming (Peng and Wei (2007)), and a hybrid heuristic (Resende and Werneck (2004)) including randomized greedy heuristics.

A forward greedy heuristic was first studied by Cornuéjols, Fisher, and Nemhauser (1977) among other exact and approximate methods. The computational experience in Heitsch and Römisch (2003), (2007) suggests that simple forward and backward greedy heuristics with final optimal redistribution lead to good results in many situations. We thus employ such heuristics to solve the n-median problem. Figure 14.2 provides an illustrative example of the scenario reduction approach applied to a temperature depending gas flow at some typical exit point.

Finally, we mention that the scenario reduction approach is extended by Heitsch and Römisch (2007) to Kantorovich–Rubinstein type metrics (e.g., Fortet–Mourier metrics) and by Henrion, Küchler, and Römisch (2008) to rectangular and polyhedral discrepancies.

14.3 • Generating quantiles for verifying booked capacities

In the previous section, scenario-based approximations of a given probability measure have been discussed. We now want to complement this approach by another possibility of characterizing multivariate distributions, namely the generation of quantiles. To this aim, let ξ be a d-dimensional random vector. To know the distribution of ξ means to know

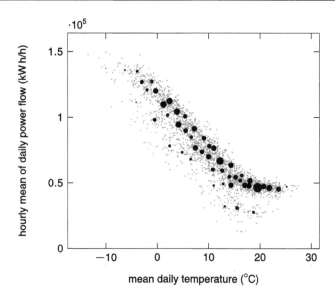

Figure 14.2. *Illustration of optimal scenario reduction from $N = 2340$ temperature depending gas load scenarios with identical probability $\frac{1}{N}$ to $n = 50$. The new probabilities after redistribution are proportional to the diameters of the points representing the remaining scenarios.*

the probabilities $\mathbb{P}(\xi \in A)$ for all Borel-measurable subsets $A \subseteq \mathbb{R}^d$. Fortunately, all these probabilities of possibly complicated sets can be recovered from probabilities $\mathbb{P}(\xi \in z + \mathbb{R}^d_{\leq 0}) = \mathbb{P}(\xi \leq z)$ of relatively simple sets, so-called cells, which are the negative orthants attached to arbitrary points $z \in \mathbb{R}^d$. The cumulative distribution function associated with ξ is defined as

$$F_\xi(z) := \mathbb{P}(\xi \leq z).$$

Therefore, F_ξ carries the whole information about the distribution of ξ. In the one-dimensional case, $d = 1$, one defines a *(univariate) p-quantile*, $p \in [0, 1]$, of the random variable ξ as the quantity

$$q_p(\xi) := \inf\{t \mid F_\xi(t) \geq p\},$$

which can be understood as the inverse of the distribution function. The benefit of univariate quantiles lies in the equivalence

$$F_\xi(z) \geq p \iff z \geq q_p(\xi), \tag{14.7}$$

which allows one to transform a probabilistic inequality into an explicit inequality. Note that univariate quantiles are easily calculated for all prominent distributions by standard software.

A multidimensional analogue of this concept is the so-called *multivariate p-quantile* (Prékopa (2012)). It is defined for a d-dimensional random vector ξ as the set

$$Q_p(\xi) := \{z \in \mathbb{R}^d \mid F_\xi(z) \geq p \text{ and } F_\xi(y) \geq p, y \leq z \text{ imply } y = z\}, \tag{14.8}$$

which is easily seen to reduce in the one-dimensional case, $d = 1$, to the classical quantile $q_p(\xi)$. If ξ has a density which is positive everywhere (as the Gaussian one), then the p-quantile is just the p-level set of the distribution function F_ξ, i.e.,

$$Q_p(\xi) = \{z \mid F_\xi(z) = p\}.$$

In contrast to the univariate case, multivariate quantiles are no longer singletons but sets (typically curved hypersurfaces in $\mathbb{R}^d$). The generalization of (14.7) now reads as follows:

$$F_\xi(z) \geq p \iff \exists q \in Q_p(\xi) : z \geq q. \tag{14.9}$$

Let us illustrate the use of multivariate quantiles in the context of the probabilistic inequality (14.3) which is central in the present chapter. We make the following *monotonicity assumption* for the feasibility of statistical load scenarios $P_{\text{stat}}, P'_{\text{stat}} \in \mathbb{R}^{V_{\text{stat}}}$ (for notation see Section 14.1):

$$P'_{\text{stat}} \leq P_{\text{stat}} \text{ and } \Upsilon_B(P_{\text{stat}}) = 1 \implies \Upsilon_B(P'_{\text{stat}}) = 1, \tag{14.10}$$

i.e., if a statistical load scenario is technically feasible in the sense of (14.2), then the same should hold true for all statistical load scenarios which are at most as large in each component. We recall that technical feasibility of P_{stat} means that all booking-compliant nominations P extending P_{stat} are technically feasible. Under the simplifying assumption (14.10), one immediately infers that the probabilistic inequality (14.3) is satisfied whenever we find a quantile $q \in Q_p(\xi)$ with $\Upsilon_B(q) = 1$. Indeed, this last relation along with (14.9) and (14.10) implies that

$$\mathbb{P}(\Upsilon_B(\xi) = 1) \geq \mathbb{P}(\Upsilon_B(\xi) = 1, \xi \leq q) = \mathbb{P}(\xi \leq q) = F_\xi(q) \geq p.$$

The generation of such a quantile can be realized by employing codes for evaluating multivariate distribution functions, see, e.g., Genz and Bretz (2009) for the examples of Gaussian and t-distributions.

Since there exists a continuous set of multivariate quantiles, the technical infeasibility of one of them does not exclude the technical feasibility of a different one. Hence, we can generate another quantile if necessary. With respect to the probabilistic inequality (14.3), this quantile-based approach has the advantage of possibly requiring only a few statistical load scenarios to be validated in contrast to a typically large number of statistical load scenarios for the sampling method of the preceding section. On the other hand, it relies on the monotonicity assumption (14.10) which is not strictly satisfied in reality and can distort the true value of the probability to be determined in (14.3).

14.4 ▪ Modeling capacity contracts

As discussed in Chapter 3, there is a sophisticated regulatory framework governing the use of gas transmission networks. Consequently, there are many different kinds of capacity contracts (see Table 3.1) and related conditions that are relevant for capacity planning in gas transmission networks. In this section we describe the typical data constituting a capacity contract and introduce a mathematical model for them. We note that apart from the capacity contracts described here, there are further types of contracts relevant for planning and operating a gas network. In particular, there are contractual limits for the pressures at entries and exits, which we will briefly discuss at the end of Section 14.5.

The model presented below not only covers capacity products that are sold to transport customers, but allows incorporating agreements between TSOs regarding the interconnection capacities between different networks or market areas. In principle, these agreements correspond to transmission capacities but they feature more complex conditions like alternative capacities that may not be used at the same time. To capture these conditions, we introduce binary variables and thus arrive at a mixed-integer linear program (MILP) instead of a linear programming (LP) model, as suggested in the introduction. The implications of this choice for the overall method are discussed in Section 14.8.

We will briefly elaborate on the modeling power of our capacity contract model at the end of this section.

A *capacity contract* defines the capacity rights of a gas transport customer, i.e., the minimum and maximum amount of gas to supply or withdraw, including additional terms and conditions. Recall that we call the set of all capacity contracts booked at a certain point in time a booking, denoted B, and that we call a nomination booking compliant if it satisfies all conditions that are related to the capacity contracts. In our model, a capacity contract c consists of one or more *capacity positions*, which is either a *freely allocable capacity (FAC) position* or a *restrictively allocable capacity (RAC) position*. The difference between FAC positions and RAC positions is in the balancing requirement: The total nomination of all entry FAC positions together has to match the total nomination of all exit FAC positions (see the discussion of market areas in Section 3.2.2). In contrast, the nominations on entry and exit RAC positions have to match *within the same contract*, which usually limits nominations to a few entry and exit points. We denote by $\mathfrak{C}^{\text{FAC}}$ and $\mathfrak{C}^{\text{RAC}}$ the sets of contracts containing FAC positions and RAC positions, respectively. Moreover, let C^{FAC} be the set of all FAC positions and C^{RAC} the set of all RAC positions.

Each capacity position c defines the capacity rights at a set of points $V(c)$ in the gas network. We require that all points in $V(c)$ are of the same type, i.e., either all entries or all exits. In most cases, $V(c)$ is just a single point, i.e., a single entry or exit. Putting more points in $V(c)$ allows one to realize zoning (see Section 3.2.2), i.e., defining a common capacity for a group of points that have to be treated as a single (virtual) point w.r.t. these capacity rights. A capacity position is usually valid for a certain time interval and possibly a restricted temperature range, since often the amount of capacity required depends on the temperature as defined in the contract. We implicitly assume that all capacity positions in B are valid for a common temperature T and date and thus define a booking situation that may occur at a single day.

The capacity of a capacity position c is given by the parameters $\underline{x}^c$ and $\overline{x}^c$, $0 \leq \underline{x}^c \leq \overline{x}^c$, which give the minimum and maximum power, respectively, that may be nominated on this capacity position. To model capacity contracts that specify alternative capacities, there is, for each capacity position c, a set $C_{\neg}^{\text{cap}}(c)$ of capacity positions that are "incompatible" with c, i.e., it is not allowed to nominate on c and another capacity position from $C_{\neg}^{\text{cap}}(c)$ at the same time. $C_{\neg}^{\text{cap}}(c)$ may contain capacity positions from any contract, not just the one to which c belongs.

We now describe a mixed-integer linear model for the set of booking-compliant nominations for a given booking B. Of course, the first condition is that the load flow vector $(P_u)_{u \in V_{\pm}}$ has to be a nomination, i.e., balanced:

$$\sum_{u \in V_+} P_u = \sum_{u \in V_-} P_u. \tag{14.11}$$

For each capacity position c and each point $u \in V(c)$, we introduce a variable $P_u^c \geq 0$ for the power nominated on c. Denoting by $C^{\text{cap}}(u)$ the set of capacity positions c at point u, the total power nominated at u is then given by

$$P_u = \sum_{c \in C^{\text{cap}}(u)} P_u^c \quad \text{for all } u \in V. \tag{14.12}$$

We set $P_u = 0$ if there is no capacity position for point u. Moreover, we have a binary variable $x^c \in \{0, 1\}$ for each capacity position c indicating whether this capacity position is used ($x^c = 1$) or not ($x^c = 0$).

For each capacity position $c \in C^{\text{FAC}} \cup C^{\text{RAC}}$, the model for booking-compliant nominations has to ensure that the load flows nominated at the points $V(c)$ are within the booked capacity limits $\underline{x}^c$ and $\overline{x}^c$, and that there is no load flow if the capacity position is not used. This is achieved by the constraints

$$\underline{x}^c x^c \leq \sum_{u \in V(c)} P_u^c \leq \overline{x}^c x^c \quad \text{for all } c \in C^{\text{FAC}} \cup C^{\text{RAC}}. \tag{14.13}$$

In the common case that $V(c) = \{u\}$, we have that P_u^c is in the interval $[\underline{x}^c, \overline{x}^c]$ if the capacity position c is used. In the case of zoning, i.e., if $|V(c)| > 1$, power may be nominated at any point in $V(c)$ as long as the sum is within the given capacity limits. Moreover, we have to ensure that capacity position $c \in C^{\text{FAC}} \cup C^{\text{RAC}}$ and each incompatible capacity position $\bar{c} \in C_\neg^{\text{cap}}(c)$ are not used simultaneously:

$$x^c + x^{\bar{c}} \leq 1 \quad \text{for all } c \in C^{\text{FAC}} \cup C^{\text{RAC}}, \bar{c} \in C_\neg^{\text{cap}}(c). \tag{14.14}$$

In contrast to FAC positions, RAC positions within contract $\mathfrak{c}$ are not independent of each other, but need to be balanced. Let $\mathfrak{C}^{\text{RAC}} \subseteq B$ be the subset of contracts with RAC positions and consider a contract $\mathfrak{c} \in \mathfrak{C}^{\text{RAC}}$, denoting by $C^{\text{RAC}}(\mathfrak{c})$ the set of all RAC positions in contract $\mathfrak{c}$. Balancing is then ensured by the condition

$$\sum_{c \in C^{\text{RAC}}(\mathfrak{c})} \sum_{u \in V(c) \cap V_+} P_u^c = \sum_{c \in C^{\text{RAC}}(\mathfrak{c})} \sum_{u \in V(c) \cap V_-} P_u^c \quad \text{for all } \mathfrak{c} \in \mathfrak{C}^{\text{RAC}}. \tag{14.15}$$

This model is already quite powerful. It allows us to model standard FAC and restrictively allocable capacity (RAC) products (see Table 3.1), which may be either defined for single points or entire entry and exit zones. Moreover, conditional versions of these products are also covered, as long as the conditions refer to temperature only, which is the usual case. This property is due to the fact that we verify booked capacities assuming a small temperature range (see Section 14.6 for details). We can thus evaluate temperature-dependent conditions and use the variant of the capacity contract that applies. The same reasoning applies to capacity contracts that are valid for a limited period only. It is also possible to model more complex conditions arising from interconnection agreements between different TSOs. These are typically based on the condition that certain hybrid points are used as an entry or as an exit point. Using the exclusion mechanism, one can ensure that either the capacities applying for entry usage or the ones for exit usage are used for nomination. So far, the model assumes that all capacities are firm capacities. It is possible to include interruptible capacities as well. However, deciding whether or not nominations should be interrupted has to be done during the nomination validation step since this is an operational measure. As the nomination validation methods presented in this book do not yet support this, we focus on firm capacities only.

14.5 ▪ An adversarial heuristic for generating booking-compliant nominations from statistical load scenarios

Assume that we obtained a statistical load scenario $s = (s_u)_{u \in V_{\text{stat}}}$ and we now want to check whether all extending booking-compliant nominations of s are technically feasible (i.e., evaluate $\Upsilon_B(s)$ in the notation of Section 14.1). We propose the following adversarial heuristic to construct a small set of extending nominations that are challenging for the gas network, i.e., which are likely to be a subset of the technically infeasible extending nominations of s. This approach is similar in spirit to the one explained in Section 4.2.

First we need a means to specify which parts of a nomination are derived from the statistical load scenario s and which are provided by the adversarial model. To this end, we distinguish *substitutable capacity positions* and *nonsubstitutable capacity positions*. The idea is that load flows for the substitutable capacity positions are "substituted" from s, whereas load flows for the nonsubstitutable capacity positions are determined adversarially. Thus capacities that are assumed to be used in the future as they were used in the past may be modeled using substitutable capacity positions and the remaining capacities as nonsubstitutable capacity positions.

It is usually not possible to generate the statistical load scenario s such that it necessarily complies with the booked capacity contracts. For instance, the support of the multivariate normal distribution is unbounded. Hence, it may happen that samples exceeding the available capacities or even negative samples are generated. To deal with this issue, we adjust s in a first step such that booking-compliant extending nominations exist, obtaining the *adjusted statistical load scenario* s'. Note that this adjustment in general affects the stochastic properties of s', i.e., s' may no longer be a multivariate quantile with the same p value or its probability is different from that of s. We therefore choose s' as close as possible to s. Finally, we select some of the nominations extending s'.

We refine the model for booking-compliant nominations from Section 14.4 as follows. For every point $u \in V_{\text{stat}}$, we divide the set of capacity positions $C^{\text{cap}}(u)$ into the set of substitutable capacity positions $C_s^{\text{cap}}(u)$ and the set $C_{\text{ns}}^{\text{cap}}(u)$ of nonsubstitutable capacity positions. Equation (14.12) is then replaced, for any point $u \in V_{\text{stat}}$, by

$$P_u = \sum_{c \in C_s^{\text{cap}}(u)} P_u^c + \sum_{c \in C_{\text{ns}}^{\text{cap}}(u)} P_u^c, \qquad (14.16)$$

i.e., we now distinguish between substitutable and nonsubstitutable capacity positions. Ideally, we would like to have the power nominated on $C_s^{\text{cap}}(u)$ to be given by s_u, i.e., $\sum_{c \in C_s^{\text{cap}}(u)} P_u^c = s_u$. To cover the potential need for adjustment, we introduce a slack variable $\Delta_u \in \mathbb{R}$ for any point $u \in V_{\text{stat}}$. The capacity constraints (14.13) together with the constraint

$$\sum_{c \in C_s^{\text{cap}}(u)} P_u^c = s_u + \Delta_u \quad \text{for all } u \in V_{\text{stat}}, \qquad (14.17)$$

ensure that we obtain booking-compliant values for the substitutable capacity positions $C_s^{\text{cap}}(u)$.

We determine values for the Δ_u variables whose sum of absolute values is as small as possible, i.e., that change the original sample vector s the least, using the following MILP:

$$\min \quad \sum_{u \in V_{\text{stat}}} \Delta_u' \qquad (14.18)$$

s.t. $(14.11), (14.16), (14.13), (14.14), (14.15), (14.17),$

$$\Delta_u' \geq \Delta_u \quad \text{for all } u \in V_{\pm},$$
$$\Delta_u' \geq -\Delta_u \quad \text{for all } u \in V_{\pm},$$
$$P_u, P_u^c, \Delta_u \geq 0 \quad \text{for all } u \in V_{\pm},$$
$$x^c \in \{0,1\} \quad \text{for all } c \in C^{\text{FAC}} \cup C^{\text{RAC}}.$$

For each sampled statistical load scenario $s = (s_u)_{u \in V_{\text{stat}}}$, we solve the (easy) MILP (14.18) and use its optimal solution to construct the adjusted statistical load scenario $s' = (s_u')_{u \in V_{\text{stat}}}$ defined by

$$s_u' := s_u + \Delta_u.$$

Note that we do not restrict the values of the variables at nonstatistical points—these are just computed to ensure that there is a feasible extension of the adjusted statistical load scenario.

14.5. An adversarial heuristic for generating booking-compliant nominations

Having obtained the adjusted statistical load scenario s', we now need to choose booking-compliant extending nominations. The goal is to construct such extending nominations that are likely to be a subset of the technically infeasible extending nominations of s'. Usually, more extreme nominations, i.e., high-load and low-load nominations, but also ones with high regional imbalances, are more critical w.r.t. technical feasibility than less extreme ones. To obtain such nominations, we construct nominations that are extreme points of the set of booking-compliant nominations that extend s', which we just formulated as a MILP. To this end, we choose a random direction in the space of all nominations and compute a solution of the extension MILP that maximizes the value in this direction.

Network planners often have some expert knowledge about the gas network (see Section 4.2.3), for instance about points which are "equivalent" from a gas network point of view in the following sense: Given a total load flow for a set of points V', it does not matter (much) how this total load flow is distributed among the single points in V' (e.g., since the points are rather close). Thus nominations with similar total load flow at a set of equivalent points should be considered similar as well, despite differently distributing the total load flow among the nodes. Moreover, when constructing a challenging nomination it does not make sense to use an entry and an exit that are close to each other at the same time. Hence, either the entry or the exit should be used. To model both aspects, we assume that the points V are partitioned into subsets $V^1, \ldots, V^r$ of "equivalent" points. This allows us to focus on the vector $\bar{P}$ defined by

$$\bar{P} := \left(\sum_{u \in V^i \cap V_+} P_u - \sum_{u \in V^i \cap V_-} P_u \right)_{1 \le i \le r},$$

capturing equivalence of the points by considering the net injections (which may be negative) of each point set V^i. Indeed, we will determine the final nomination P such that the vector $\bar{P}$ is extreme in the sense explained above, i.e., equivalent entries and exits will never be used at the same time if this is not enforced by some capacity positions.

In general we do not know which nominations are likely to be technically infeasible. We therefore try to select a set of nominations that are rather separate. To this end, we choose the direction θ for determining each extending nomination uniformly at random from the r-dimensional hypersphere. Since we generate just a few random directions, we use the same randomized QMC techniques to sample the direction that are also used for sampling statistical load scenarios. Again, the rationale is that QMC methods provide a better approximation to the uniform distribution than pseudorandom MC methods. We use the direction θ to determine a single extending booking-compliant nomination, i.e., an element of $\pi^{-1}(s') \cap \mathcal{B}_=$, that maximizes the value of that nomination in this direction. This is done by solving the following MILP that is similar to (14.18):

$$\max \sum_{1 \le i \le r} \theta_i \left(\sum_{u \in V^i \cap V_+} P_u - \sum_{u \in V^i \cap V_-} P_u \right) \tag{14.19a}$$

$$\text{s.t.} \quad (14.11), (14.16), (14.13), (14.14), (14.15), \tag{14.19b}$$

$$\sum_{c \in C_s^{\text{cap}}(u)} P_u^c = s'_u \quad \text{for all } u \in V_{\text{stat}}, \tag{14.19c}$$

$$P_u, P_u^c, \Delta_u \ge 0 \quad \text{for all } u \in V_\pm, \tag{14.19d}$$

$$x^c \in \{0, 1\} \quad \text{for all } c \in C^{\text{FAC}} \cup C^{\text{RAC}}. \tag{14.19e}$$

Note that we use the constraint (14.19c) instead of (14.17), i.e., we require that the feasible solutions are booking-compliant nominations extending s'. Observe further that this construction does *not* provide a uniform sampling of the extreme points of the feasible set of the MILP.

To obtain complete inputs for the nomination validation methods, we provide the pressure limits corresponding to each nomination as well. These pressure limits depend on the actual use of the network, i.e., the load flows. However, once a nomination is determined, it is clear which entries and exits are used and the corresponding pressure limits can be taken into account when checking technical feasibility.

14.6 • Methods to verify booked capacities

We now described all ingredients of the two methods outlined in Section 14.1 to check inequality (14.3), i.e., to verify that the validity probability β of a given booking B is at least α. In the following, we present our overall approach to this problem. As discussed in Chapter 13, we construct a stochastic model for the load flows at the statistical points for each temperature class in order to deal with the temperature dependency of the load flows.

The two methods we propose differ in how they incorporate the stochastic information:

1. The first method uses sampled statistical load scenarios to represent the stochastic nature of exit loads (see Section 14.2).
2. The second method uses multivariate quantiles to represent the stochastic nature of exit loads (see Section 14.3).

Formally, both methods require convexity of the set $\mathscr{T}$ of technically feasible nominations to be valid. In addition, the quantile-based method is formally only valid if the monotonicity assumption (14.10) is fulfilled. However, it may require validating significantly fewer nominations to certify that the validity probability β is at least α. In contrast, the sampling-based method does not rely on the monotonicity assumption, but requires many statistical load scenarios and extending nominations, resulting in considerable computational effort for validating those nominations.

So far we assumed that checking a nomination for technical feasibility either establishes feasibility or proves infeasibility. In practice, however, it may also happen that there is no conclusive answer, e.g., since the time limit for the computation has been reached before technical feasibility could be decided. In the description of our methods we assume that each check for technical feasibility yields one of the answers "feasible," "infeasible," or "unknown."

In both methods we need to check whether all extending booking-compliant nominations of s are technically feasible (i.e., evaluate $\Upsilon_B(s)$ in the notation of Section 14.1); we do this by considering a set of n booking-compliant nominations extending s generated by the adversarial heuristic explained in Section 14.5). We consider s to be technically feasible (i.e., $\Upsilon_B(s) = 1$), if all n extending nominations of s are "feasible." In the case that at least one extending nomination is "infeasible," s is infeasible as well (i.e., $\Upsilon_B(s) = 0$). The remaining case is that some extending nominations are "feasible" and the rest are "unknown," so the feasibility of s is unknown, too.

14.6.1 • Sampling-based verification of booked capacities

This method estimates the validity probability β of the given booking B using sampled statistical load scenarios. Since the feasibility of a statistical load scenario may be unknown, we actually compute two estimates $\hat{\beta}$ and $\hat{\beta}^*$ for the validity probability β: $\hat{\beta}$ is a pessimistic estimate, assuming that all of the statistical load scenarios whose feasibility is unknown are infeasible. In contrast, $\hat{\beta}^*$ is optimistic since it assumes that all of the statistical load scenarios with unknown feasibility are in fact feasible.

Using sampling to approximately check the probability requirement (14.3), the overall procedure to verify a booking is as follows. Note that we decided to base the final decision for or against validity of the booking on the pessimistic estimate $\hat{\beta}$.

A. For each temperature class perform the following steps:

1. Based on the stochastic model for this temperature class, sample a set of statistical load scenarios $\tilde{s}^1, \ldots, \tilde{s}^M$ using the randomized QMC method described in Section 14.2. For each statistical load scenario $\tilde{s}^i$, $1 \leq i \leq M$, compute an adjusted statistical load scenario s^i for which at least one booking-compliant nomination extending s^i exists (see Section 14.5).
2. Use the scenario reduction technique outlined in Section 14.2.2 to find a smaller representative subset $s^1, \ldots, s^N$ of the adjusted statistical load scenarios ($N \ll M$) together with updated probabilities $p_1, \ldots, p_N$.
3. Set $I = I^* = \emptyset$.
4. For each adjusted statistical load scenario $s = s^i$, $1 \leq i \leq N$:
   (a) Generate a set $P^1, \ldots, P^n$ of booking-compliant nominations extending s (see Section 14.5).
   (b) Check the technical feasibility of $P^1, \ldots, P^n$. If all of them are "feasible," add i to the index set I. If none of them is "infeasible," add i to the index set I^*.
5. Compute the estimates $\hat{\beta}$ and $\hat{\beta}^*$ for the validity probability β as

$$\hat{\beta} = \sum_{i \in I} p_i, \qquad \hat{\beta}^* = \sum_{i \in I^*} p_i.$$

B. The overall booking is considered feasible if we have $\hat{\beta} \geq \alpha$ for all temperature classes.

14.6.2 ▪ Quantile-based verification of booked capacities

This method uses multivariate quantiles to ensure that the validity probability β is at least α. To be correct, it requires the monotonicity assumption (14.10) to hold. Again, it may happen that the feasibility of a statistical load scenario (now a quantile) is unknown.

The overall procedure for quantile-based verification of the probability requirement (14.3) is as follows. To arrive at an overall conclusion in case the feasibility of all considered quantiles is unknown, we take the pessimistic view again and decide for invalidity then.

A. For each temperature class perform the following steps:

1. Based on the stochastic model for this temperature class, generate a set of multivariate quantiles $q^1, \ldots, q^N$.
2. Try each multivariate quantile $q = q^i$, $1 \leq i \leq N$, in turn, stopping if q^i has been successfully validated (see below):
   (a) Compute an *adjusted quantile* q' for which at least one booking-compliant nomination extending q' exists (see Section 14.5).
   (b) Generate a set $P^1, \ldots, P^n$ of booking-compliant nominations extending q' (see Section 14.5).

(c) Check whether *all* nominations $P^1, \ldots, P^n$ are technically feasible. If this is the case, we successfully validated q^i, i.e., q^i certifies (assuming monotonicity) that the desired probability level α is attained and stop. Otherwise, i.e., if at least one nomination is "infeasible" or "unknown," continue with the next quantile.

B. The overall booking is considered feasible if, for all temperature classes, we found a quantile for which all booking-compliant extensions are technically feasible. It is considered infeasible otherwise.

14.7 ▪ Computational results for verifications of booked capacities

In this section we present some preliminary computational results regarding the methods for the verification of booked capacities, which were developed in this chapter. We first describe the common computational setup before providing and discussing our computational results.

14.7.1 ▪ Computational setup

In Chapter 12 we have seen that the various approaches to the first stage part of validating a nomination, i.e., finding suitable discrete decisions, all have their strengths and weaknesses. Since there is no clear "winner," we combine these approaches and the high-detail ValNLP method to the following overall method to check a single nomination for technical feasibility:

1. Run the MILP, sMINLP, RedNLP, and MPEC approaches in parallel. If a solution is found, use the ValNLP to perform the nonlinear program (NLP) validation.
2. If the MILP or sMINLP approaches are able to prove infeasibility, we declare the nomination to be *infeasible*. We consider a nomination to be *feasible* if a solution with slack 0 has been found. Finally, if none of the four decision approaches found a solution with slack 0 (this includes the case that no decision approach solution was found within the time limit), the validity of the nomination is *unknown*.

The four approaches are used with the settings as in Section 12.7. All computations have been performed on the same cluster with the setup as described in Section 12.1.2. The only difference is that nodes of the cluster have been used nonexclusively, since we are not interested in precise performance comparisons, but aim to assess the practical feasibility of our methods.

As before, our computations are done for the northern high calorific gas network of Open Grid Europe GmbH (OGE). OGE also provided data about their booked capacity contracts, from which we used the FAC and RAC contracts to set up the model for booking-compliant nominations as described in Section 14.4. As mentioned there, handling interruptible capacities correctly requires extending the validation of nomination (NoVa) models: Interruptible capacities allow to reduce the load flow at the corresponding entry points or exit points if this helps to maintain technical feasibility. Hence, interrupting is an operational option like opening or closing a valve and needs to be considered during validation of nominations. However, OGE sold more firm exit capacity than firm entry capacity. Thus interruptible entry capacities need to be used in case of high load situations to balance the gas demand. For this study, we treated interruptible capacities as firm capacities, since the approaches for NoVa described in this book are not yet able to handle interruptible capacities. We also ignored further operationally relevant contract

14.7. Computational results for verifications of booked capacities

Table 14.1. *Overview of the distributions used by the stochastic model for the load flows for each temperature class. In OGE's H-gas network, for 295 exits there was enough data available to model them as statistical points. The load flow at a large part of those is modeled by three multivariate normal distributions (MND); the sizes of those exit point sets are given in column MND. The load flow at the remaining exits is modeled by various univariate distributions: normal (ND), shifted normal (shND), Dirac (DD), uniform (UD), shifted uniform (shUD). See Section 13.3 for details.*

Temp. class	Stat. exits	MND	ND	shND	DD	UD	shUD
(−15 °C, −4 °C]	295	43, 35, 30	121	29	25	4	8
(−4 °C, −2 °C]	295	43, 33, 30	117	35	25	3	9
(−2 °C, 0 °C]	295	41, 30, 30	117	39	25	2	11
(0 °C, 2 °C]	295	39, 27, 27	118	48	21	2	13
(2 °C, 4 °C]	295	37, 29, 24	101	69	19	2	14
(4 °C, 6 °C]	295	39, 36, 27	84	75	18	2	14
(6 °C, 8 °C]	295	35, 33, 26	86	82	17	2	14
(8 °C, 10 °C]	295	31, 23, 23	84	100	18	1	15
(10 °C, 12 °C]	295	29, 23, 22	82	104	19	1	15
(12 °C, 14 °C]	295	28, 21, 20	80	109	21	1	15
(14 °C, 16 °C]	295	27, 22, 21	75	119	15	1	15
(16 °C, 18 °C]	295	25, 22, 20	77	115	20	1	15
(18 °C, 20 °C]	295	25, 22, 20	74	111	27	1	15
(20 °C, 40 °C]	295	25, 20, 20	83	102	29	2	14

types like flow commitments for the same reason. Consequently, the presented results are too pessimistic in comparison to reality, since there may be other means to resolve infeasibilities detected by our methods.

To obtain the required distribution data, we used an earlier version of the methods described in Chapter 13 on measured gas flow data for the period 01/2004–05/2010 provided by OGE. Based on that data, 14 temperature classes were determined. Table 14.1 gives an overview on the stochastic model obtained for each temperature class. We used these stochastic models together with the booked capacities for OGE's entire H-gas network in the methods described in Section 14.6. Due to the structure of this network, a unique nomination for the northern part of this network can be extracted from a nomination for the whole network. These nominations for the northern part of the H-gas network were used to test our methods. Finally, we use a partitioning of the entry and exit points of the northern H-gas network into sets of equivalent points which was provided by OGE (see Section 14.5). The resulting power nominations are converted to flow nominations as described in Section 5.3.3.

Based on this distribution and capacity contract data, we ran both methods with the aim to verify that the validity probability β is at least 95%.

14.7.2 ▪ Verifying booked capacities using sampling

Recall from Section 14.6 that, for each temperature class, the sampling-based method samples M statistical load scenarios using a randomized quasi–Monte Carlo (QMC) method. These are in turn reduced to N statistical load scenarios via scenario reduction. For each remaining statistical load scenario, n extending booking-compliant nominations are generated by the adversarial heuristic based on the capacity contracts in the booking. We chose $M = 500$, $N = 100$, $n = 5$, resulting in 500 nominations per temperature class that are validated by the four approaches developed in this book. From these nominations,

Table 14.2. *Results of the sampling-based method for verifying booked capacities. Shown are, for each temperature class, the numbers of nominations for which a feasible solution has been found ("Feas."), infeasibility has been proved ("Infeas."), or for which no conclusive answer was found within the time limit ("Time limit"). Moreover, the table shows the estimated probability interval based on the pessimistic and the optimistic estimate as well as the final conclusion.*

Temperature class	Feas.	Infeas.	Time limit	Estimated prob. interval	Valid?
$(-15\,°C, -4\,°C]$	361	74	65	$[0.25, 0.48]$	no
$(-4\,°C, -2\,°C]$	376	67	57	$[0.2, 0.47]$	no
$(-2\,°C, 0\,°C]$	358	83	59	$[0.22, 0.47]$	no
$(0\,°C, 2\,°C]$	376	56	68	$[0.23, 0.51]$	no
$(2\,°C, 4\,°C]$	407	41	52	$[0.44, 0.67]$	no
$(4\,°C, 6\,°C]$	418	29	53	$[0.43, 0.75]$	no
$(6\,°C, 8\,°C]$	411	45	44	$[0.47, 0.65]$	no
$(8\,°C, 10\,°C]$	499	0	1	$[0.98, 1.0]$	yes
$(10\,°C, 12\,°C]$	499	0	1	$[0.98, 1.0]$	yes
$(12\,°C, 14\,°C]$	496	0	4	$[0.96, 1.0]$	yes
$(14\,°C, 16\,°C]$	491	0	9	$[0.92, 1.0]$	unknown
$(16\,°C, 18\,°C]$	496	0	4	$[0.96, 1.0]$	yes
$(18\,°C, 20\,°C]$	497	0	3	$[0.99, 1.0]$	yes
$(20\,°C, 40\,°C]$	496	0	4	$[0.95, 1.0]$	yes

the sampling-based method computes two estimates for the validity probability of the booking. The first estimate is pessimistic in the sense that all nominations for which no solution was found are considered technically infeasible. For the second estimate, these nominations are considered technically feasible and the estimate is hence optimistic.

The results of this experiment are shown in Table 14.2. The run time per temperature class is roughly 10 h wall clock time on average. Note that this is the total run time of all four approaches, since every approach ran until a solution was found or the time limit was reached. In a production system, however, all other approaches would be stopped as soon as one is successful, resulting in significant time savings.

The results show that nominations for lower temperatures are not technically feasible with a higher probability than those for higher temperatures, as indicated by the number of infeasible nominations. Moreover, for lower temperature nominations there are also more timeouts, i.e., the nominations are harder for the solvers to decide, too. There is a sharp increase of the estimates for the validity probability beyond 8 °C. Interestingly, there are certain special contracts that have not been considered here that are valid below 8 °C and offer OGE additional operational opportunities. The low probabilities at the lower temperature are to be expected, since in these cases there is a high gas demand, some of which is met from interruptible capacities, which are not taken into account so far.

14.7.3 ▪ Verifying booked capacities using quantiles

The quantile-based method for verifying booked capacities generates N multivariate quantiles. For each multivariate quantile, n extending booking-compliant nominations obtained from the adversarial heuristic are checked for technical feasibility. If all of these nominations are technically feasible, the booking is considered feasible and the process is

Table 14.3. *Results of the quantile-based method for verifying booked capacities. Except for the column "Iterations," the columns are the same as in Table* 14.2. *Column "Iterations" gives the maximum number of multivariate quantiles that have been considered. Since there is no multivariate quantile for which all extending booking-compliant nominations are technically feasible, this number is always* 10.

Temperature class	Feas.	Infeas.	Time limit	Iterations	Valid?
$(-15\,°C, -4\,°C]$	357	81	62	10	no
$(-4\,°C, -2\,°C]$	365	78	57	10	no
$(-2\,°C, 0\,°C]$	364	87	49	10	no
$(0\,°C, 2\,°C]$	358	83	59	10	no
$(2\,°C, 4\,°C]$	365	80	55	10	no
$(4\,°C, 6\,°C]$	364	81	55	10	no
$(6\,°C, 8\,°C]$	363	78	59	10	no
$(8\,°C, 10\,°C]$	377	73	50	10	no
$(10\,°C, 12\,°C]$	376	64	60	10	no
$(12\,°C, 14\,°C]$	384	61	55	10	no
$(14\,°C, 16\,°C]$	400	44	56	10	no
$(16\,°C, 18\,°C]$	389	43	68	10	no
$(18\,°C, 20\,°C]$	401	43	56	10	no
$(20\,°C, 40\,°C]$	401	38	61	10	no

stopped. In case that for each of the N multivariate quantiles at least one extending nomination is infeasible, the booking is considered infeasible. We chose $N = 10$ and $n = 50$, resulting in at most 500 nominations being verified per temperature class.

Table 14.3 summarizes the results of this experiment. Again, the run time is roughly 10 h wall clock time on average per temperature class. The rather constant number of instances that ran into the time limit indicates that the computational hardness of the NoVa problems does not depend much on the temperature and thus on the network load. Moreover, the obvious qualitative difference above 8 °C observed for the sampling-based method cannot be observed with the quantile-based method. The booking is considered invalid for all temperature classes. There may be two reasons for this difference to the sampling-based method. First, the quantile-based method is in a sense more pessimistic since more extending nominations are checked and it is thus more likely to find an infeasible one. Second, the multivariate quantiles themselves have to feature a higher load than the sampled statistical load scenarios in order to fulfill the quantile-defining property (14.8).

14.7.4 ▪ Discussion and further work

The presented methods may already be used to verify booked capacities in special cases where interruptible capacities and operational options like flow commitments need not be considered. For a real production system, however, both aspects need to be taken into account.

One way to model interruptible capacities is as follows. Recall from Section 14.7.1 that interruptible capacities are only taken into account since, in general, the amount of booked firm entry and exit capacities does not match. Thus, interruptible capacities need to be used on the deficit side to ensure overall balancing. If interruptible capacities are really necessary to match demand and supply, the TSO may, in principle, interrupt any allocation that may not be technically realized. To ensure technical feasibility on the surplus side, it is thus sufficient that there is at least one partial nomination for the interruptible capacities such that the overall nomination is technically feasible. In the NoVa models, we may thus relax the nominated power values corresponding to interruptible capacity to power intervals, adding a new balancing constraint that ensures that the (fixed) demand of the surplus side is met. However, the NoVa approaches then need to deal with nonfixed nominations, which is not possible in the sMINLP and RedNLP models.

Extending the NoVa methods presented in this book to deal, e.g., with flow commitments, is mathematically challenging. The reason is that flow commitments and similar measures grant a TSO the right to influence the load flow at certain entry or exit points, but the TSO has no control over how flows are redistributed among the remaining points in response to changing supply or demand at a few points. Thus, NoVa as discussed in this book needs to be interfaced with both, a model for how load flows may be adjusted based on contractual agreements and another model for the reaction of the transport customers. Straightforward modeling of flow commitments leads to a bilevel problem, which is likely to be intractable for the scale of problems considered here.

In addition to enhancing the scope of the mathematical models and methods, a thorough validation of the methods is needed, assessing their capability of providing reliable estimates for the validity probability. Finally, the run time has to be improved further to scale to large production networks.

14.8 • Conclusions

The presented approach to the problem of verifying booked capacities has been designed with applicability to real-world gas networks in mind. In order to obtain a tractable method, several aspects are done in a heuristic fashion, which we discuss in the following.

▷ The set $\mathcal{T}$ of technically feasible nominations is nonconvex, in general. However, many sources of nonconvexity of the stationary problem of validating nominations can be removed when operating the network in a transient regime (see Section 11.1). Thus, this assumption is justified to some extent from a practical point of view. In fact, it is one way to limit the complexity of the problem and is implicitly used in all industrial methods known to us, see Section 4.2.2. Alternatively, it might be possible to deal with the nonconvexity by some kind of space-partitioning method.

▷ In general, the adjustment of statistical load scenarios or multivariate quantiles destroys their stochastic properties. In principle it is possible to use some form of rejection sampling to ensure that the statistical load scenarios without adjustment allow booking-compliant extending nominations. However, this might be rather inefficient.

▷ The adversarial heuristic non-uniformly samples among the extreme points corresponding to the mixed-integer hull of the MILP (14.19). In contrast to explicitly given polytopes, it is currently not known how to efficiently sample uniformly from such a mixed-integer set.

Moreover, we see several places in which more mathematical research can lead to improvements of the presented methods:

▷ The effectiveness of the adversarial heuristic in finding technically infeasible nominations and the formal justification of ingredients, if only in special cases, should be investigated. For instance, does the proposed method eventually find a technically infeasible nomination if one exists?

▷ One goal is to obtain rigorous results establishing the assumed convergence properties of our randomized QMC method for our setting.
▷ Existing stability results for scenario reduction techniques establish that the optimal solution w.r.t. the reduced scenario set does not deviate much from that for the original scenario set. These should be adapted to the presented setting.

In any case, the proposed methods are a substantial improvement over the current state of the art in industry. Our results indicate that the methods may already be applied to gas networks of industrially relevant size.

Chapter 15
Perspectives

Christine Hayn, Jesco Humpola, Thorsten Koch,
Lars Schewe, Jonas Schweiger, Klaus Spreckelsen

Abstract *After we discussed approaches to validate nominations and to verify bookings, we consider possible future research paths. This includes determining technical capacities and planning of network extensions.*

The preceding chapters have shown how mathematical programming and statistical analysis can help in evaluating the capacities of natural gas networks. We can validate given nominations and decide the feasibility of bookings based on freely allocable capacities (FACs) and restrictively allocable capacities (RACs). Certain difficulties and possible improvements have been discussed (see e.g., Chapter 12 and Chapter 14). However, a number of more general mathematical and practical problems remain:

▷ Can we improve the quality of the physical model?
▷ Can we improve our contract model and our model of realistic, yet difficult, flow situations?
▷ Can we decide what the *maximal* technically feasible capacity is for the given network?
▷ Can we determine what the most cost-effective extensions of the network are?

These questions need to be answered in order to address the wider range of tasks a planning department of a TSO has to deal with. In this chapter we outline how the methods described in this book can be used or extended to answer them. Note that most of the following ideas are "work in progress." We first discuss the role of more accurate physical models in the planning process. We then show how to define sensible notions of "maximal" capacities and how to determine them. We furthermore discuss the different possibilities for network extensions and describe ideas to determine cost-effective extensions. We close the chapter with a short discussion of what we need to do to make all our models work in practice.

15.1 ▪ Physical models and transient effects

We have already discussed that the question "What is precise enough?" is very difficult to answer (see Chapter 11). There are, nevertheless, a number of obvious improvements in the physical models that need to be investigated. The main questions were discussed in Section 5.3:

▷ Are stationary models sufficient and if not, can we extend our models to incorporate transient effects?
▷ How do we deal with different gas types and with other deviations from (volumetric) flow balance?
▷ How do we incorporate gas temperature?

Transient effects become more relevant for the mid- and long-term planning process since the gas transmission network can be used as a cost-effective storage facility for renewable energy from intermittent renewable sources. Under the heading "power-to-gas" a variety of technologies to transform electricity into hydrogen or methane are proposed. They all have in common that they intend to use the natural gas transmission network either to smoothen regional imbalances in intermittent energy generation or to simply use it as a storage facility for a couple of days. On a technical level, the latter option means to increase the pressure in the pipes and operate the network at a higher overall pressure, which can lead to a decrease of the capacity of the network. This makes the operation of the network much more difficult, in general. It is, thus, of practical interest to investigate the impact of these new developments and to incorporate them into the planning process. Unfortunately, effects like the increased pressure level *cannot* be modeled in our stationary approach.

Our NoVa models make the simplifying assumption that the given nominations are balanced with respect to volumetric flow under normal conditions. Further assuming that we deal with only one gas type allows us an easy translation of the nomination in one that is balanced with respect to mass flow. These assumptions are, however, too strong for the purposes of a practical planning tool. Capacity contracts are given in terms of power, so for the purposes of deciding the feasibility of bookings, power, and neither volumetric nor mass flow is the relevant quantity. Furthermore, biogas and hydrogen input will make it necessary to consider a detailed model of mixture effects at the planning stage. The straightforward way of dealing with these questions is, of course, to use mixture models as discussed in Section 10.1.2. Incorporating these models into the first stage approaches seems to be far too costly at this point in time. We think, however, that we can incorporate the mixture effects for the calorific value in the first stage by iteratively refining the correct flow values at the exits, instead of using the crude averaging procedure of Section 5.3.3. In the second stage we then solve a sequence of more detailed nonlinear program (NLP) models instead of just the one ValNLP model, which does not include the mixture effects. Similar ideas can be used to deal with the fuel gas needed for the operation of gas-driven compressors. For the sMINLP and RedNLP models a technical difficulty arises: An iterative refinement of the flow values of a nomination works most easily when the underlying models can deal not only with fixed nominations, but still work when only given bounds at the entries and exits. Both the sMINLP and the RedNLP models assume, however, fixed nominations and therefore need to be modified.

In summary, to model the transient effects one needs to directly consider the PDE model described in Section 2.3.1 for the flow in pipes and to adapt the other component models accordingly. Dealing with mixture models leads to very difficult MINLPs; we think, however, that further developing iterative refining techniques will allow us to circumvent this problem.

15.2 ▪ Modeling flow situations

To make the booking validation approach outlined in Chapter 14 practical, we need to balance the requirements of being able to check all realistic, yet difficult load flow scenarios and keeping the runtime low enough to be of practical value. Additionally, the contract model needs to be expanded to deal with the contracts that are actually in force at this point in time.

As discussed in Section 14.5 we use a heuristic to generate the load flow scenarios that we check using our NoVa approaches. We first use our stochastic models to generate realistic loads for those nodes where we have sufficient data. The heuristic then needs to generate a set that "covers" the set of all contractually valid load flow scenarios in such a way that all difficult scenarios are checked. We would, of course, like to have an exact approach for this problem. Currently, this seems to be out of reach.

Our approach of verifying bookings has another shortcoming: We have not incorporated interruptible capacities and flow commitments so far. Both are indispensable measures for transmission system operators (TSOs) to manage their network. In both cases we need to modify the underlying NoVa models to incorporate the additional flexibility the TSO is granted through these measures. As mentioned in Section 14.7.4, interruptible capacities can be incorporated in the NoVa models, if we allow nonfixed nominations.

The difficult problem, however, is to deal with flow commitments. The TSO may demand a certain load flow from a customer at a certain entry or exit point. The customer, however, can then choose the point at which the additional demand resp. supply is realized. The straightforward way to incorporate these possiblities leads to a bilevel problem.

As we have seen, both to improve our main approach to verify bookings and to incorporate flow commitments in our model of contracts, there is one major mathematical obstacle: We need to deal with bi- or even trilevel optimization problems. These can be viewed in two different ways: One point of view is to see our problem as a game that the TSO and its customers are playing. Another point of view is the view of robust optimization: In this setting we are dealing with *adjustable* robustness. The TSO can act to counterbalance the effects of the nomination. Regardless of the point of view: The problems we have to deal with are extremely challenging, since these types of problems are pretty difficult even when the basic problem is an easy optimization problem and not a complex mixed-integer nonlinear program (MINLP).

15.3 • Determining maximal capacities

As described in Section 3.2.1, TSOs are required to offer the maximum capacity that can be provided on a firm basis, the so-called *technical capacity* to the market (see [GasNZV 2010, §2 (13) and §9]). We call the problem of determining the technical capacity the *capacity problem*.

From a mathematical point of view the concept of technical capacity is still ill-defined. The capacities at different points in the network are, of course, physically interdependent. The least restrictive interpretation is to allow any Pareto-optimal allocation of capacities to the different points. Another interpretation is to require that the sum over all allocated capacities is maximal. This is, however, too restrictive in general, since there are a number of points for which market participants are not interested in buying more capacity. In the following we only consider a basic variant and try to allocate FACs. We do not consider extensions such as also allocating RACs or allowing the buying of further flow commitments to allow higher capacities.

The *space of FAC* has a complicated structure, even in the case of trivial networks consisting of, e.g., one pipe (Willert (2014)). Specifically, the monotonicity assumption of (14.10) does not hold in general. Whether weakened assumptions hold is currently under investigation.

From a mathematical point of view, the capacity problem is an optimization variant of the problem of validating a booking, which we described in Chapter 14. Most of the discussion on how to incorporate knowledge about typical loads on the gas network from Chapter 14 still applies: We can either use a number of scenarios as described in Section 14.2 or quantiles as described in Section 14.3.

A natural approach to the capacity problem would be to extend the method presented in Chapter 14 and use it as an oracle for an optimization algorithm. The verification of a single booking is, however, very time consuming. Moreover, it is likely that a number of very similar nominations need to be verified, when evaluating the oracle at points with small distance in the FAC space.

One approach to solve the capacity problem starts with picking a number of points in the network for which we want to compute the capacity. For the remaining nodes, we either use the statistical information on their typical usage or we assume that they are used in a worst-case sense within their booked capacities. To incorporate statistical information, we either use scenarios or quantiles as described above. We then compute an approximation of the set of firm FAC that can be allocated to the points of interest.

In order to be able to compute this approximation to the set of firm FAC vectors, we make a crucial assumption: If a "neighborhood" around a feasible nomination is "small enough," actually *every* nomination in this neighborhood is technically feasible. This assumption can be justified by the fact that small alternations of the gas flow to avoid infeasibilities can be achieved using additional dispatching measures that are typically not considered for planning purposes (see Section 11.1). We furthermore assume that we are given an upper bound on the firm FAC that is requested or can be used. These assumptions allow us to reduce the capacity problem to a finite number of validations of interval nominations.

The validation of nomination (NoVa) problems are solved using one of our decision approaches as a black-box-algorithm. The algorithm itself proceeds as follows: We subdivide the space spanned by the given upper bounds, the FAC space, in small boxes. For each of these boxes, we try to establish feasibility for at least one nomination. To achieve this, the input to the NoVa tool is not a fixed nomination but intervals at each FAC point corresponding to the current box of the FAC space. The NoVa tool then effectively tries to find a nomination within these FAC intervals that extends the given scenario or quantile and is technically feasible. If all nominations in a box are infeasible, bounds on the available capacity can be derived.

15.4 • Extending the network

In the case that capacity of the network does not suffice, the network can be extended. To this end, we assume to be given an infeasible nomination and have to determine network extensions such that the nomination can be realized in the extended network.

Even though the task seems clear at first sight, several questions arise: What types of network extensions do we allow? Can originally feasible nominations turn infeasible in the extended network? What is the trade-off between building costs and future operation cost, e.g., for compressors? How high will the future operation cost be? Is it, for instance, worthwhile to remove a compressor group and build a new pipe that does not cause any operation cost?

15.4.1 • Available network extensions

Let us first review the measures that are available to increase the capacity of the network.

New pipe connections In order to increase the capacity, a pipe can be built connecting two nodes which were previously unconnected or only connected by a long path in the network. Since we are not restricted to the actual nodes in our current network model, but can place additional nodes on any arc, we already face a continuum of possibilities. Furthermore, the specification of the new pipe, most importantly the diameter, needs to be determined and might vary over the entire pipe. In practice, the diameter is chosen from a discrete set of standard sizes.

The course of the pipe is chosen with regard to the cost of different soil conditions that influence the construction process and costs for the purchase of land-use rights. Some areas, such as highly populated areas or nature reserves, may not be traversed for pipe construction. The construction cost of a new pipe can be estimated from the length of the pipe, its diameter, and its exact course in the landscape.

New pipes can even connect two unconnected transmission networks. In this case the question is where to connect both networks in order to not decrease the overall capacity (if possible) as described in Section 3.2.2.

Looping A new pipe in parallel to an existing one is called a loop. In practice, loops are preferred over new connections, since the existence of a gas pipe reduces the bureaucratic and administrative work and greatly facilitates the planning process. This is reflected by reduced building costs for this type of network extension.

Loops can be dimensioned flexibly. Not only the diameter of the loop has to be specified, but also the exact length of the loop can be chosen from a continuous range. This leads to a model with variable pipe capacity.

The same pipe can also be looped multiple times; even with all loops having different diameters. Fujiwara and Dey (1987) show that not all combinations of diameters have to be considered, which reduces the complexity of the resulting models to some extent.

New active elements All types of active elements (valves, compressors, and control valves) can be added almost arbitrarily in the network, not only at existing nodes, but on any arc and also in combination with new pipes. One also can add complex stations with multiple, possibly distinct elements and subnetwork operation modes.

Dimensioning new active elements is a challenging task. For compressors, the type, i.e., turbo compressor or piston compressor, and the exact specification needs to be decided and influences the building costs.

Also operation costs have to be taken into account. Typically one would add the discounted operation costs to the building costs, but the future operation costs are hard to predict. How do we know how much a compressor will be used in the future, especially in a changing market environment?

Retrofitting existing stations Retrofitting existing stations is often a cost-effective measure to increase capacity. This can range from minor changes in the piping to allow additional subnetwork operation modes to adding active elements.

Deconstruction Even though not increasing the capacity of the network, deconstruction of existing infrastructure can be a reasonable measure to ensure an effective network operation. With changes in the market environment, the characteristics of the transport requests might permanently render parts of the network infrastructure obsolete. Moreover, after connecting previously unconnected networks, deconstructions might save operation and maintenance costs.

15.4.2 ▪ Planning for multiple nominations

In practice, network operators face considerable uncertainty in the future use of the network and thus also in future flow nominations. Network planning methods should take care of this uncertainty by using robust methods which allow the operators to hedge against unforeseen nominations and developments.

As infrastructure investments are very cost intensive and long-living, network extensions should not only focus on one bottleneck nomination, but should increase the flexibility to fulfill different demand situations. Handling multiple nominations simultaneously increases the robustness of the solution and avoids overfitting of the extensions to one particular nomination.

Mathematically, we deal with a two-stage stochastic program (see for example Birge and Louveaux (2011) for an introduction to Stochastic Programming). In the first stage, the extension decisions have to be taken. These decisions have to coincide for all nominations. With these decisions fixed, we have to make sure that each nomination can be actually operated. This is the second stage, where operational decisions are to be taken on a per-nomination basis. Cost, i.e., for compressor operation, can also be part of the second stage. The cost of the solution is then computed as the sum of the first stage investment costs and the expected second stage operation costs.

Decomposition is a common tool to solve such problems (see Birge and Louveaux (2011) and the references therein). With fixed first stage decisions, the second stage problems can be solved independently. Decomposition methods use this to formulate a master problem which focuses on the first stage decisions. The nominations are treated as independent subproblems which are solved for a fixed set of first stage decisions.

15.5 ▪ Making it work in practice

Having read to this point one may ask the following questions:

What has to be done to produce a tool that works in practice at a TSO?
Bringing mathematical/algorithmic developments like the ones presented in this book into practice requires additional efforts. First, as already mentioned in Chapter 14, several requirements on the generation of nominations have not yet been incorporated into our model. Starting from flow commitments, through flow relocations, to interruptions, there are a large number and variety of flow-adjustment measures that are applied in practice and have to be modeled and incorporated. Second, coping with gas qualities is necessary, since gas is traded by energy content and not by volume.

Third, once the necessary functionality is present, a significant amount of software development effort is necessary to produce a stable, usable tool integrated into the IT infrastructure of a TSO. Stability in this case does not only refer to the software itself, but also to the results. A small change in the input data should not lead to large changes in the outcome. This of course depends on the number of nominations generated and computed. On the other hand, the time for a complete verification of bookings is limited, as are the computational resources. There are a large number of parameters involved and these have to be tuned as the sequence of computations has to be optimized to reduce computation time. To get anywhere near a usable running time, a distributed computing setup is necessary.

Finally, the biggest challenge is to obtain correct input data. The amount of interconnected data, from historic measurements to produce the distributions needed for the generation of nominations, to the network data and finally contract data is daunting. Of course, all of it is present somewhere at a TSO. But converting it into a coherent data set is a challenge and finding the errors in the data requires significant additional efforts.

What are the future plans for putting the developed methods into practice?
As of this writing, the ForNe project is still running and has the ultimate goal of getting a booking verification tool into daily use at a major TSO. The functional requirements are mostly met, but getting everything verified, the computational engine set up, and the data ready still needs considerable time and effort.

What will be the expected benefit?
Firstly, having a fully automated booking verification will save much tedious manual work currently done by network planners. Secondly, in contrast to the approach described in Chapter 4, the method presented in the rest of the book is capable of going beyond qualitative investigations and able to provide quantitative answers. Finally, TSOs are required to offer ever more complicated capacity products, and we believe this tool and its paradigm will only be the start of a whole chain of tools capable of coping with these demands.

15.6 ▪ Outlook

As we have briefly sketched in this chapter, there are many open mathematical questions that need to be answered in order to handle the discussed problems in planning of gas transport. Some of these are even more challenging than the problems to validate nominations and verifying bookings. They are, however, of great theoretical and practical interest.

In a sense, this book is only a milestone. Due to the importance of natural gas for the European energy supply, there are many initiatives to investigate the topic. Starting from the status described herein, there are currently two roads into the future:

▷ Many theoretical questions posed by gas transport will be investigated in the new collaborative research center/transregio 154 *Mathematical modelling, simulation and optimization using the example of gas networks*,[17] supported by the German Research Foundation (DFG).

▷ As mentioned above, the ForNe project is working on building the first real world usable system based on the methods presented in this book. We envision this system to become productive at a TSO in 2015.

As can be seen, there is a lot of work to do—possibly for several generations of future mathematicians and engineers. Stay tuned.

---

[17]See http://www.trr154.fau.de

Appendix
Background on gas market regulation

Jessica Rövekamp

Abstract *This appendix provides further information about the organizations and stakeholders involved in the European gas market and its regulations. Moreover, the historical development of the regulation in Europe is described.*

A.1 ▪ Legislative power, authorities, and organizations

The European Union (EU) was established in the Maastricht Treaty in 1992. Since then, the EU legislative bodies are the European Parliament and the Council of the European Union with their responsibility for regulations, directives, and guidelines. The European Commission is involved in legislation insofar as it can propose legal acts by means of its right to initiate legislation. In addition, supervision of compliance with European antitrust law and sanctions in the event of infringements are the responsibility of the European Commission [EU Treaty, Article 105].

In 2000, the Council of European Energy Regulators (CEER) was set up to pool the national regulatory authorities at European level and to improve cooperation between the regulatory authorities themselves. In 2003, the European Regulators Group for Electricity and Gas (ERGEG) was set up by the European Commission as its own *advisory group*. Established by the so-called third EU Internal Energy Market Package (EIMP), which was passed by the European Parliament and the Council of the European Union, the Agency for the Cooperation of Energy Regulators (ACER) started its work on March 3, 2011, largely replacing the work of CEER.

Since the beginning of the 1990s, organizations like GEODE[18] and CEDEC[19] representing local distribution or energy companies, respectively, took part in the development of the market. Additionally, different organizations of traders and asset operators formed up in the energy market. European trading companies founded the European Federation of Energy Traders (EFET) in 1999 to improve the conditions for the energy trade in Europe and promote an open, liquid, and transparent internal market in the EU. In 1999, the European Gas Regulatory Forum, or Madrid Forum, was set up to promote the exchange between regulatory authorities and market players [EEForumMadrid]. EASEE-gas with

---

[18] Groupement Européen des entreprises et Organismes de Distribution d'Énergie.
[19] European Federation of Local Energy Companies.

members along the whole gas value chain and regulatory authorities was "*set up in 2002 to develop and promote the simplification and streamlining of both the physical transfer and the trading of gas across Europe*" [EASEEbg]. In 2002 as well, the organization Gas Infrastructure Europe (GIE) was established by asset operators in the transmission, storage, and liquified natural gas (LNG) business to represent their interests in European institutions such as the commission, parliament, and council as well as the European regulatory organizations such as ACER and CEER.

On the basis of the provisions of the Gas Directive 715/2009, Articles 4 and 5, as part of the third EIMP, the European Network of Transmission System Operators for Gas (ENTSO-G) was set up in 2009. As the European representative of the transmission system operators (TSOs), its work includes advising politicians, drafting network codes and network development plans as well as providing market information. It was founded as a sister organization to the European Network of Transmission System Operators for Electricity (ENTSO-E).

At national level, the most important legislative body in Germany is the Bundestag. In view of the federal system of government, the Bundesrat is also involved in legislation; it receives all laws for approval and, depending on the type of the law, can also reject a draft bill. The Bundesrat, the federal government as well as members of parliament and parliamentary parties of the Bundestag can introduce new or revised draft bills to the Bundestag.

The national regulatory authority is the German Network Agency (BNetzA),[20] which was initially formed from the Federal Ministry of Post and Telecommunications and the Federal Office for Post and Telecommunications. In 2005, it also took over the regulation of the electricity and gas market, in addition to post and telecommunications. The railway infrastructure market was added to its mandate in 2006. In contrast to other European countries, Germany has a Federal Antitrust Office (BKartA)[21] and, at federal state level, also state antitrust offices to protect competition. The BNetzA and BKartA, as independent superior federal authorities, are assigned to the Federal Ministry for Economic Affairs and Energy,[22] the state antitrust offices to their respective state ministries of economy.

Germany also has different associations and interest groups of customers, traders, and network operators which influence the general conditions in the gas sector. More than 60 years ago, the Federation of German Industry (BDI)[23] and the Association of the Energy and Power Industry (VIK)[24] were established to represent the interests of German industry and industry-related services as well as industrial and commercial energy customers in Germany vis-à-vis various political bodies. EFET Germany and GEODE Germany represent German energy traders and local distribution companies as subsidiary associations of the European organizations. The energy and water industries, including national, regional, and municipal network operators, are represented by the Federal Association of the Energy and Water Industry (BDEW)[25] and the Association of Municipal Utilities (VKU).[26]

---

[20] Bundesnetzagentur.
[21] Bundeskartellamt.
[22] Bundesministerium für Wirtschaft und Energie.
[23] Bundesverband der Deutschen Industrie.
[24] Verband der Industriellen Energie- und Kraftwirtschaft.
[25] Bundesverband der Energie- und Wasserwirtschaft.
[26] Verband kommunaler Unternehmen.

A.2 • Chronology of European and German gas market regulation

In the history of European and German legislation, several laws, regulations, and directives concerning the energy sector had strong impact on network planning and capacity calculation in gas transport. They cover the creation of an internal market, the regulation of asset-based transmission, as well as today's unbundling regulations and allocation mechanisms.

The regulatory environment of the gas business in Germany is closely connected to the European history. The European Economic Community (EEC), founded in 1957, became one of three pillars of the European Community (EC) in 1967. The EC consisted of the European Atomic Energy Community (EAEC or Euratom), the European Coal and Steel Community (ECSC), and the European Economic Community (EEC) and existed from 1967 to 1992. Articles 85 and 86 of the Treaty establishing the European Economic Community (TEEC) already addressed the formation of a competitive internal European market with the prohibition of abuse of dominant positions and discriminatory behavior. In 1992 the EU was founded as an umbrella organization in Maastricht where the EC became one pillar, but stayed its own legal entity rather than being incorporated in the EU. This changed with the signing of the consolidated versions of the Treaty on European Union (TEU) and the Treaty on the Functioning of the European Union (TFEU) in Lisbon in 2007. They entered into force in 2009 and the EU became the legal successor of the EC in 2009.

With the idea of establishing an internal European market, also the European energy sector came into regulatory focus. Thus, the Council of the European Communities and later the European Parliament and the Council of the European Union passed several council directives and decisions regarding power and gas since 1990.

Especially known is gas directive 98/30/EC as part of the first EU Internal Energy Market Package (EIMP) which laid *"down the rules relating to the organization and functioning of the natural gas sector, including LNG, access to the market, the operation of systems, and the criteria and procedures applicable to the granting of authorizations for transmission, distribution, supply and storage of natural gas"* [Directive 98/30/EC, Article 1]. It gave member states the choice between negotiated or regulated access to gas transport systems. It made nondiscriminatory behavior, the publication of main commercial conditions, and strict confidentiality obligatory for all TSOs, especially for undertakings of gas traders and suppliers. Additionally, the directive determined the legal compulsion for necessary enhancements of the network if investments were economically reasonable and paid by transport customers [Directive 98/30/EC].

With upcoming regulatory influence, German associations representing different national stakeholders (see Section A.1) passed the first association agreement (VV)[27] between energy producers and industrial consumers for gas in 2000. The agreement was on a voluntary basis since gas directive 98/30/EC was not implemented into German law. It determined the commercial and technical conditions for third party access (TPA), but was criticized for blocking the market instead of advancing it. Due to political pressure from Brussels on negotiated network access, the second advanced agreement was signed in 2002. In the same year and again legally not binding, the Madrid-Forum passed the *Recommendations on Guidelines for Good Practice in relation to TPA services, tariffication, balancing, etc.* (RGGPTPA) [EESC].

---

[27] Verbändevereinbarung.

Disregarding these approaches, the European Parliament and the Council of the European Union passed gas directive 2003/55/EC as part of the second EIMP to accelerate the unsatisfactory market development and to repeal directive 98/30/EC in 2003. Major changes to 1998 were the obligation of regulated TPA for transport systems (see Article 18 ff.) and the legal and operational unbundling of transmission system operators (see Article 9). Additionally, the directive emphasized the attention to be paid to final costumer and consumer protection, environmental protection, and security of supply. One year later, gas directive 2004/67/EC concerning concrete measures to safeguard security of natural gas supply was passed, which was seen as a crucial element of the liberalized internal market.

These regulations and directives have had a strong impact on national legislation in Germany. The German Energy Industry Act (EnWG)[28] is the basis of legislation concerning the gas sector. It was first passed in 1935. Especially due to military reasons, its goal was to organize the energy supply in a safe, inexpensive, and decental manner by codifying the prevalent conditions of concession and demarcation contracts. In this context, vertical concession contracts settled the exclusive energy supply of municipalities by one supplier and horizontal demarcation contracts divided regions between these monopoly suppliers. Its amendment in 1998 focused on the liberalization of the electricity, not on the gas market. With the amendment of 2005, however, the obligatory national implementations of gas acceleration directive 2003/55/EC and security of supply directive 2004/67/EC were realized. The revised EnWG constituted legal and operational unbundling, accompanying data confidentiality, and unbundled accounting, as well as measures for the security and reliability of supply. Also, EnWG Sec. 20 1(b) provided the definition of freely allocable capacity (FAC) in an entry-exit model and additional obligations for the cooperation of network operators and for market area reductions (for detailed information see Section 3.2.2). The EnWG of 2005 was thereby path-breaking for the further development of network planning and TSO cooperation in Germany.

On its basis the BMWi passed the Gas Network Access Ordinance (Gasnetzzugangsverordnung – GasNZV) and the Gas Network Fee Ordinance (Gasnetzentgeltverordnung – GasNEV) in the same year. GasNZV provided amongst other things further details for the different national capacity products and mechanisms. GasNEV introduced a *revenue-cap regulation* with incentive elements to advance efficiency. With this form of regulation the allowed revenues, i.e., the income of network operators, are essentially determined by the BNetzA.

Because of the new cooperation requirements of the revised German regulatory framework, several network operators and energy associations signed the first cooperation agreement (KoV)[29] in July 2006. It provided two possibilities of organizing gas transmission. In the so-called *single capacity booking model*, available gas transport capacity from exactly one entry point to exactly one exit point was offered by so-called *point-to-point (P2P)* contracts. In this model, the gas transport customer is responsible for organizing the gas transport by obtaining a set of P2P contracts that realize a transportation path between an entry and an exit point; see Figure A.1 for an illustration. The alternative model is the *two-contract model* (2CM), in which a gas transport customer obtains two independent capacity contracts, namely one *entry contract* and one *exit contract*. In this model, the gas transport customer is no longer responsible for organizing a transport path, which is now the sole responsibility of the TSOs. This is illustrated in Figure 3.4. In November 2006, the regulatory authority decided that only the two-contract model was a valid form of

---

[28] Short title *Energiewirtschaftsgesetz*, official title *Gesetz über die Elektrizitäts- und Gasversorgung*.
[29] Kooperationsvereinbarung.

A.2. Chronology of European and German gas market regulation

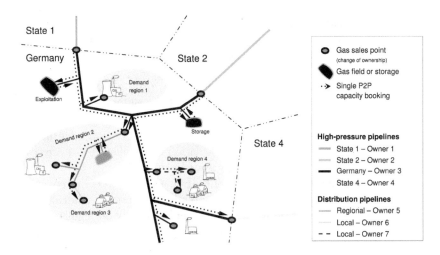

Figure A.1. *Example of the single capacity booking model with P2P contracts. Each P2P contract is represented by a dotted arrow. For supplying, e.g., end-consumers in demand region 3 at least a set of three single P2P contracts has to be booked to follow the path the gas takes through the different connected high- and low-pressure pipeline systems.*

capacity marketing and banned the use of the single capacity booking model [BNAGNM]. Since the two-contract model is the model used today, we discuss it in more detail in Section 3.2.2.

Due to this decision, the second revised cooperation agreement had already to be signed in early 2007. With the new Incentive Regulation Ordinance (ARegV)[30] and the new decisions on Change of Supplier Rules (GeLiGas)[31] and Balancing Energy Rules (GABi Gas)[32] by Ruling Chamber 7 of the BNetzA, again the cooperation agreement had to be modified and the third version was passed in 2008.

In 2009, the European Parliament and the Council of the European Union passed the third EIMP. It consists of several decisions and regulations for the electricity and gas market. Moreover, it focuses on facilitating cross-border trading and market integration between member states. Regulation (EC) No. 713/2009 establishes the Agency for the Cooperation of Energy Regulators (ACER) for power and gas. Directive 2009/73/EC concerns common rules for the internal market in natural gas and repeals directive 2003/55/EC. Regulation No. 715/2009 repeals regulation (EC) No 1775/2005, establishes ENTSO-G, and sets conditions for the access to the natural gas transmission networks.

Again directive 2009/73/EC had to be transferred into national law, thus the EnWG and the GasNEV were amended, and the GasNZV was revised in the years 2010 and 2011. The idea of having capacity auctions and a national auctioning capacity platform was introduced in the new GasNZV in addition to the European regulation. On its basis a new BNetzA decision regarding Capacity and Auctioning Rules, so-called KARLA Gas,[33] was passed in February 2011. The new set of regulation contains different obligations for network operators beginning on August 1, 2011, e.g., to constrain renomination rights and to establish a platform for primary and secondary capacities. Moreover, it allows offering bundled capacities on cross-border and market area entry/exit points, firm day-

---

[30] Anreizregulierungsverordnung.
[31] Geschäftsprozesse Lieferantenwechsel Gas.
[32] Grundmodell für Ausgleichsleistungen und Bilanzierungsregel im Gassektor.
[33] Kapazitätsregelung und Auktionsverfahren im Gassektor.

ahead capacities, and products for defined time periods (e.g., daily, monthly, quarterly, and yearly products). This was again followed by a basic revision of the cooperation agreement between the network operators. The fourth version entered into force on October 1, 2011.

On European level, the new regulation (EU) No. 994/2010 concerning measures to safeguard security of gas supply repealed Council Directive 2004/67/EC in 2010. It introduced new elements like common Preventive Action Plans and Emergency Plans, the enabling of permanent bidirectional capacity on all cross-border interconnections between member states, as well as the calculation of so-called $N-1$ scenarios in the event of a disruption of the single largest gas infrastructure.

Since 2009, ACER, CEER, ERGEG, and ENTSO-G in consultation with all stakeholders have been working on nonbinding framework guidelines and binding network codes on the basis of the third EIMP and especially on regulation No. 715/2009. *"These network codes [...] are not intended to replace the necessary national network codes for non cross-border issues"* [Regulation (EC) No. 715/2009, point (16)], but have to establish a functioning European wholesale market with a harmonized set of rules. Parts of an overall gas target model (GTM) are network codes on capacity allocation mechanisms, congestion management procedures, balancing, tariffication, as well as interoperability and data exchange.

A.3 ▪ Ongoing and future activities

The network code on capacity allocation mechanisms was already passed by Commission Regulation (EU) No. 984/2013 in October 2013 and will apply starting November 1, 2015. Its objective is to ensure more efficient capacity allocation on crossborder interconnection points in order to facilitate trade in gas and thus support the creation of an efficient EU-wide gas market. Inter alia, instruments are harmonized auctions, online-based booking platforms, and bundled products at cross-border points.

In August 2012, the European Commission adopted rules to reduce congestion by amending the existing annex to regulation (EC) No. 715/2009. Most of the new mechanisms already have to be applied since October 2013. Instruments to incentivize operators to sell extra capacity to the market above the technical capacity are risk-based overbooking and capacity buy-back, as well as long-term and firm day-ahead use-it-or-lose-it mechanisms. Additionally, TSOs have to facilitate the surrender of contracted capacity by traders.

While the framework guidelines for balancing were already passed in October 2011 by ACER, the binding network codes are still under consultation. The goals are to defragment the several balancing zones across the EU and different balancing arrangements applying in neighboring markets as well as the provision of improved informational display on whether portfolios are balanced or not.

For tariffication, an analysis of the national practices was carried out in order to evolve concrete policy options for different issues. In December 2013, ENTSO-G was invited by the European Commission to draft a network code on rules regarding harmonized transmission tariff structures for gas and to submit it to ACER by the 31st of December 2014.

On the basis of ACER framework guidelines, ENTSO-G submitted a draft for network codes on interoperability and data exchange rules in December 2013, which aims at improving the technical and operational compatibility of the European networks.

Thus, the European GTM is finalized in important areas like capacity allocation and congestion management, while the regulation of balancing, tariffication, interoperability, and data exchange is still under consultation and can be expected over the next few years.

Acronyms

DSO distribution system operator

ENTSO-G European Network of Transmission System Operators for Gas

EnWG German Energy Industry Act

FAC freely allocable capacity

GNDP German network development plan

GTM gas target model

IIS irreducible infeasible subsystem

LNG liquified natural gas

LP linear programming

MILP mixed-integer linear program

MINLP mixed-integer nonlinear program

MPEC mathematical program with equilibrium constraints

NLP nonlinear program

OGE Open Grid Europe GmbH

P2P point-to-point

QMC quasi–Monte Carlo

RAC restrictively allocable capacity

toe ton of oil equivalent

TSO transmission system operator

VTP virtual trading point

Glossary

active element An element of a gas network is called active if there is a possibility to control it. Active elements are valves, control valves and compressors.

adiabatic An *adiabatic* process is a thermodynamical process that transforms one state into another state without exchange of thermal energy.

booking Transaction of a transport customer of buying capacity.

bypass Compressor groups or control valve stations can be bypassed, meaning that the gas is routed around the element. The compressor group or control valve station contained is then assumed to be passed without pressure loss.

calorific value The *calorific value* of a gas mixture is the amount of heat generated by combustion of a mass unit amount of it. Its value depends on the particular mixture. It is measured in J/kg.

capacity Amount of gas that can be transported by a network operator.

capacity contract A contract defining the right to transport certain amounts of gas subject to certain conditions.

compressor A compressor is synonymous with a compressor machine.

compressor group A set of compressors, which can be operated in different configurations or can be bypassed by a parallel valve. Compressor groups are used to increase the gas pressure. They contain several compressors and drives. The operation modes are closed, bypass, and active. In the active mode, the network operators can choose from a set of internal modes, called configurations. Compressor groups are used to model compressor stations.

compressor machine A compressor machine (or compressor, sometimes compressor unit) is used to increase the gas pressure. One mainly distinguishes turbo compressors and piston compressors. A set of compressors can be combined in a compressor group. Every compressor has an associated drive that delivers the power required for the compression process.

compressor station A compressor station is an often complex facility at which the gas pressure is increased to allow long-distance transportation. Typically, compressor stations are located at connection points of several pipeline systems where they are used to route the gas between the systems. In simulation and optimization models, a compressor station is represented by a subnetwork that contains a number of compressor groups, which are interconnected by pipes. There may be additional valves to route the gas through the station/subnetwork.

configuration An internal operation mode of a compressor group. Typically a certain series-parallel interconnection of compressors.

congestion A situation where the transportation capacity of a gas network is insufficient. Sometimes also refers to the particular part of the gas network that limits the transportation capacity. A *contractual congestion* occurs if there is more demand for firm capacity at a single point than the network operator can offer. The reason for a contractual congestion is usually a *technical congestion*, i.e., too much gas flow for a part of the gas network.

control setting In a simulation, control settings (indirectly) describe the behavior of active elements in a gas network. One main control setting is to preset a certain outlet pressure or flow at a compressor machine or control valve. The simulation software automatically determines the physical settings in order to meet this preset value, if this is possible.

control valve A control valve (also known as pressure regulator) is an active network element capable of reducing the gas pressure. It can be in the operation modes active, bypass, or closed.

control valve station A control valve station is a subnetwork that consists of a control valve, a number of additional valves, and, optionally, other network facilities like gas preheaters. Pressure loss due to station piping, gas preheaters, etc. is modeled by resistors. All these elements are interconnected by (short) pipes.

controllable network element see active element

cotree arc Given a graph and an associated spanning tree, all arcs that are not within the spanning tree are called cotree arcs.

distribution Distribution refers to local or regional gas transportation from upstream transmission networks to customers [Directive 2009/73/EC].

distribution system operator A distribution system operator is a company that is responsible for operating a local or regional network downstream to high-pressure transmission networks of transmission system operators.

drive Drives deliver the power that is required for the compression process at the compressor machine. Thus, a drive is always associated with one or more compressor machines. One mainly distinguishes between gas turbines, gas driven motors, electric motors, and steam turbines.

electric motor A certain type of drive.

entry [point] Point at which gas is fed into a network.

entry-exit model Entry-exit model refers to a model for accounting the cost of gas transport. In the entry-exit model transport customers pay for injecting gas into the network and for withdrawing gas from the network. Thus the cost for transporting gas does not depend on the transportation path within the network.

entry zone Internal upstream interconnection points within one market area for which there is a single capacity contract.

exit [point] Point at which gas is withdrawn from a network.

exit zone Internal downstream interconnection points within one market area for which there is a single capacity contract.

firm capacity Gas flows nominated on firm capacity have to be transported by the network operator.

flow commitment Contractual agreements with third parties (i.e., transport customers) that guarantee certain load flows which are adequate and necessary to enhance the offer of firm FAC.

gas driven motor A certain type of drive.

gas storage Facility to store gas, often underground.

gas turbine A certain type of drive.

H-gas High calorific gas with a calorific value of about 41 MJ/kg.

hybrid point Point which can be used both as an entry and an exit.

interconnection point Point where gas can be exchanged between two different gas networks or market areas.

interruptible capacity A network operator has the right to reduce the transported amount of gas that is nominated on interruptible capacity if this helps to ensure proper network operation.

Joule–Thomson effect Cooling effect of gas when its pressure is reduced.

Kirchhoff's laws Kirchhoff's laws were originally valid in electrical networks. The first law is also called nodal rule and requires that at any node the sum of gas flows out of that node is equal to the sum of gas flows into that node. The second law is also called loop rule and says that the pressure drop around any loop within the given network is zero.

L-gas Low calorific gas with a calorific value of about 36 MJ/kg.

linepack The amount of gas stored in a pipeline. Linepack may be used as a buffer to deal with peaks in the gas flow.

loop A loop is a set of two or more parallel pipes which are connected by a set of valves. Loops are often built to increase the transmission capacity, but they provide additional control possibilities, since the two pipes may be operated independently. For instance, the pipes may be used (partly) in different directions or at different pressure levels.

market area A region of the gas network that realizes the entry-exit model, i.e., a transport customer can transport gas independently of the transportation path by only holding two capacity contracts, namely one entry and one exit contract. Typically, parts of the gas network in a market area belong to different network operators.

network operator A network operator is a company that is responsible for running a gas network. A network operator may either be a transmission system operator or a distribution system operator.

node-arc-incidence matrix Given a directed graph $G = (V, A)$ with node set V and arc set A, the node-arc-incidence matrix $\mathscr{A} \in \{+1, -1, 0\}^{V \times A}$ is a full description of that graph. For each arc $a = (u, v) \in A$ the entries in column a are given as follows: $\mathscr{A}_{ua} = 1$, $\mathscr{A}_{va} = -1$, and 0 otherwise.

nomination There are two meanings:
- ▷ In the gas business, nomination refers to the notification of the network operator about the usage of capacity contracts by transport customers, which is typically specified per hour. The desired load flows are specified in terms of thermal power.
- ▷ In the technical chapters of this book, nomination refers to a vector of flow values at all entries and exits which is balanced. Note that here the gas flows are usually given in terms of mass flow instead of thermal power.

normal condition Certain environmental conditions as specified in DIN 1343: 1.013 25 bar, 0 °C, and 0% humidity.

operating range The set of all possible combinations of throughput and specific change in adiabatic enthalpy of a compressor machine.

passive element An element of a gas network is called passive if there is no possibility to control it, i.e., its behavior is completely determined by certain physical laws. The most frequent passive element is a pipe; other passive elements are resistors and short cuts.

pipe/pipeline Basic network element that connects two endpoints. Pipes are usually assumed to be of cylindrical shape.

piston compressor A certain type of compressor machine.

resistor A resistor is a fictitious network element, representing sources of pressure loss not covered by the models of the other elements.

scenario A scenario is a possible outcome of some future event, i.e., usually a possible gas flow situation. Apart from this general meaning, *scenario* is used to describe slightly different things in this book:
- ▷ In Chapter 4, *scenario* refers to a balanced load flow situation for all entry and exit points specified in terms of thermal power. See nomination.
- ▷ In Chapter 14, a statistical load scenario is a load flow situation specifying the thermal power for those points that are modeled statistically.

shipper see transport customer

short cut A short cut is a fictitious network element introduced for modeling reasons. There is no pressure drop for gas flowing along a short cut and there are no restrictions on the amount of flow.

spanning tree A subgraph consisting of all nodes and a subset of the edges of the original graph such that all nodes are reached by the edges and there is no cycle within the subgraph is called a spanning tree.

station A station is a facility where the gas flow is controlled, i.e., either a compressor station or a control valve station. At a station, there are usually a number of active elements connected by more or less complex piping. These active elements are typically located in the geographical vicinity.

station subnetwork A station subnetwork is a subnetwork representing a station. It is defined via its boundary nodes and includes active elements. The possible ways for the gas to flow in a station subnetwork are described by a set of subnetwork operation modes.

stationary A state of a system is called stationary if it does not change over time. See steady state.

steady state [model] A steady state (or stationary) model or analysis assumes that the state of the network does not change over time. In particular, all injections and withdrawals occur at a fixed rate and the network operates with constant control settings. The steady state is the state in which the system would be after an arbitrarily long time without changes to the system from the outside.

steam turbine A certain type of drive.

subnetwork A subnetwork consists of a subgraph of the directed graph modeling the overall gas network together with the data associated to all elements of this subgraph, e.g., the operating ranges for the compressor machines and the relevant subnetwork operation modes. Subnetworks usually represent compressor stations or control valve stations.

subnetwork operation mode A description of joint settings for several active elements (compressors, valves, control valves) that are used together to realize a certain gas flow pattern.

technical capacity Maximal freely allocable firm capacity that can be offered at an entry or exit.

technically feasible Mode of operation where no technical limits are violated in the entire network.

temperature class A (small) temperature interval for which a stochastic model of the load flow is derived. Temperature classes are used to take the temperature dependency of load flows into account. A temperature class is small enough to allow temperature dependency to be neglected within the temperature class, yet big enough to contain sufficient measurement data for estimating a stochastic model reliably.

transient A transient or dynamic system changes its state over time—in contrast to a stationary or steady state system.

transmission Transmission refers to interregional gas transportation from import, export, or production points to exit points belonging to industrial consumers, public utilities, or distribution system operators [Directive 2009/73/EC].

transmission system operator A transmission system operator is a company that runs a (typically high-pressure) inter-regional gas network connecting import, export, or production points with exit points belonging to industrial consumers, public utilities, or distribution system operators [Directive 2009/73/EC].

transport customer A natural or legal person who transports gas through the network. A transport customer is sometimes also called shipper.

tree arc Given a graph and a spanning tree, all edges of the spanning tree are called tree arcs.

turbo compressor A certain type of compressor machine.

unbundling Separation of different support and management functions or business sectors along the value chain.

ValNLP The concrete nonlinear programming (NLP) model that is used to validate the solution candidates produced by the solution approaches.

valve A valve is an active network element that can be in the operation modes open or closed. Valves allow switching on or off certain parts of the network in order to route the gas through the network.

virtual trading point The virtual point in a market area where entry and exit gas flows are interchanged.

Regulation and gas business literature

AFT: Applied Flow Technology. *AFT - Arrow Titan*. URL: http://www.aft.com/products. Last visited: Sep. 26, 2013 (cit. on p. 78).

BGR2010: Bundesanstalt für Geowissenschaften und Rohstoffe. *Short term study: Reserven, Ressourcen und Verfügbarkeit von Energierohstoffen 2010*. URL: http://www.bgr.bund.de/DE/Themen/Energie/Downloads/Energiestudie-Kurzstudie2010.pdf. Downloaded on April 8, 2013 (cit. on p. 7).

BGR2011: Bundesanstalt für Geowissenschaften und Rohstoffe. *Short term study: Reserven, Ressourcen und Verfügbarkeit von Energierohstoffen 2011*. URL: http://www.bgr.bund.de/DE/Themen/Energie/Downloads/Energiestudie-Kurzf-2011.pdf. Downloaded on April 8, 2013 (cit. on pp. 7 sq.).

BNAGNM: M. Kurth. *Gas network access model - Decision on gas network access model. Single capacity booking model in cooperation agreement is not permissible*. Bundesnetzagentur. URL: http://www.bundesnetzagentur.de/cln_1932/SharedDocs/Pressemitteilungen/EN/2006/061117GasNetworkAccessModel.html?nn=149010. Downloaded on April 3, 2013 (cit. on p. 329).

BNetzA Capacity Management: Bundesnetzagentur. *Neugestaltung des Kapazitätsmanagements im deutschen Gasmarkt*. 2009. URL: http://www.bundesnetzagentur.de/DE/DieBundesnetzagentur/Beschlusskammern/BK7/Kapazitaetsmanagements_Gasmarkt/KonsultationEckpunkteKapazitaetsmanagementsGasmarkt.pdf?__blob=publicationFile. Downloaded on July 28, 2011 (cit. on p. 57).

BNetzA Monitoring 2009: Bundesnetzagentur. *Monitoringbericht 2009*. URL: http://www.bundesnetzagentur.de/SharedDocs/Downloads/DE/BNetzA/Presse/Berichte/2009/Monitoringbericht2009EnergieId17368pdf.pdf?__blob=publicationFile. Downloaded on April 8, 2013 (cit. on p. 11).

BP Review 2012: BP. *BP Statistical Review of World Energy June 2012*. URL: http://www.bp.com/liveassets/bp_internet/globalbp/globalbp_uk_english/reports_and_publications/statistical_energy_review_2011/STAGING/local_assets/pdf/statistical_review_of_world_energy_full_report_2012.pdf. Downloaded on April 8, 2013 (cit. on p. 10).

CEER Gas Target Model: Council of European Energy Regulators. *CEER Vision for a European Gas Target Model*. Conclusions Paper C11-GWG-82-03. 2011. URL: http://www.energy-regulators.eu/portal/page/portal/EER_HOME/EER_CONSULT/CLOSED%20PUBLIC%20CONSULTATIONS/GAS/Gas_Target_Model/CD. Downloaded on October 19, 2013 (cit. on p. 64).

DVGW-G 260: Deutscher Verein des Gas- und Wasserfaches. *Arbeitsblatt DVGW-G 260 Gasbeschaffenheit*. 2013. URL: http://www.stadtwerke-coesfeld.de/fileadmin/Medienablage/Stadtwerke/Netze/Gas/arbeitsblatt_g_260.pdf (cit. on p. 75).

Directive 2009/73/EC: Council European Parliament. *Directive 2009/73/EC*. URL: http://eur-lex.europa.eu/LexUriServ/LexUriServ.do?uri=OJ:L:2009:211:0094:0136:EN:PDF. Downloaded on August 24, 2013 (cit. on pp. 60, 334, 338).

Directive 98/30/EC: European Parliament Council. *Directive 98/30/EC*. 1998. URL: http://eur-lex.europa.eu/LexUriServ/LexUriServ.do?uri=OJ:L:1998:204:0001:0012:EN:PDF (cit. on p. 327).

EESC: European Economic and Social Committee. *Recommendations on Guidelines for Good Practice in relation to TPA Services, Tarification, Balancing etc.* 2007. URL: http://www.eesc.europa.eu/self-and-coregulation/documents/codes/private/101-private-act.pdf. Downloaded on April 8, 2013 (cit. on p. 327).

EASEEbg: EASEE-gas – The European Association for the Streamlining of Energy Exchange-gas. *Background on EASEE-gas*. URL: http://easee-gas.eu/background. Downloaded on August 24, 2013 (cit. on p. 326).

EASEE CBP: European Association for the Streamlining of Energy exchanges (EASEE). *Gas Common Business Practices*. 2013. URL: http://easee-gas.eu/cbps/implementation. Downloaded on July 20, 2013 (cit. on p. 75).

EEG2012: *Renewable Energy Act (EEG)*. URL: http://www.gesetze-im-internet.de/bundesrecht/eeg_2009/gesamt.pdf. Downloaded on November 15, 2012 (cit. on p. 8).

EEForumMadrid: European Commission. *Single market for gas & electricity*. 2013. URL: http://ec.europa.eu/energy/gas_electricity/gas/forum_gas_madrid_en.htm. Downloaded on April 3, 2013 (cit. on p. 325).

Energinet. *Gas in Denmark 2010 – Security of supply and development*. URL: http://www.energinet.dk/SiteCollectionDocuments/Engelske%20dokumenter/Gas/Gas%20i%20Danmark%20UK.pdf. Downloaded on July 6, 2013 (cit. on p. 68).

ENTSOGMap2012: ENTSOG. *ENTSOG Transmission Capacity Map*. URL: http://www.entsog.eu/public/uploads/files/maps/transmissioncapacity/2012/ENTSOG_CAP_MAY2012_UPDATED.pdf. Downloaded on August 24, 2013 (cit. on p. 12).

EnWG 2005: *Energiewirtschaftsgesetz – Gesetz über die Elektrizitäts- und Gasversorgung*. URL: http://www.gesetze-im-internet.de/enwg_2005/index.html. Version released July 7, 2005 (cit. on p. 51).

EnWG 2013: *Energiewirtschaftsgesetz – Gesetz über die Elektrizitäts- und Gasversorgung*. Version released July 7, 2005 with changes of October 4, 2013 (cit. on pp. 52 sq., 60 sq.).

EU Treaty: *Consolidated versions of the Treaty on the European Union and the Treaty on the Functioning of the European Union*. URL: http://eur-lex.europa.eu/en/treaties/index.htm. Downloaded on August 24, 2013 (cit. on p. 325).

Eurogas2010: Eurogas. *Global LNG markets*. URL: http://www.eurogas.org/up loaded/Eurogas_LNG\%20markets_101110.pdf. Downloaded on April 8, 2013 (cit. on p. 9).

Eurostat GIC: Eurostat. *Gross inland consumption of primary energy*. 2013. URL: http ://epp.eurostat.ec.europa.eu/tgm/table.do?tab=table&init=1 &language=en&pcode=ten00086&plugin=1. Downloaded on April 3, 2013 (cit. on p. 45).

Eurostat GIEC: Eurostat. *Gross inland energy consumption, by fuel*. 2013. URL: http:// epp.eurostat.ec.europa.eu/tgm/refreshTableAction.do?tab=tab le&plugin=1&pcode=tsdcc320&language=en. Downloaded on April 3, 2013 (cit. on p. 8).

Exxon2009: ExxonMobil. *Energieprognose 2009*. URL: http://www.exxonmobil.c om/Germany-German/PA/Files/Energieprognose_09.pdf. Downloaded on April 8, 2013 (cit. on p. 9).

Flowmaster: Flowmaster Group. *Flowmaster Simulation Software*. URL: http://www. flowmaster.com. Last visited: Sep. 26, 2013 (cit. on p. 78).

FNBGas: Vereinigung der Fernleitungsnetzbetreiber Gas e. V. *Netzentwicklungsplan Gas 2013. Konsultationsdokument der deutschen Fernleitungsnetzbetreiber*. URL: http:// www.netzentwicklungsplan-gas.de/. Downloaded on August 26, 2013 (cit. on pp. 13 sq.).

Gascade. *Berechnung der technischen Kapazität*. 2013. URL: http://www.gascade.d e/fileadmin/downloads/kapazitaetsplanung/GASCADE_Berechnung _technische_Kapazitaet_120302.pdf. Downloaded on July 6, 2013 (cit. on p. 68).

GasNZV 2005: *Verordnung über den Zugang von Gasversorgungnetzen (Gasnetzzugangsverordnung – GasNZV)*. URL: http://www.bmwi.de/BMWi/Redaktion/PDF/ Gesetz/energiewirtschaftsgesetz-verordnung-gasnetz-gasnzv -april-2005,property=pdf,bereich=bmwi2012,sprache=de,rwb=t rue.pdf. Version released Apr. 14, 2005 (cit. on pp. xiv, 51).

GasNZV 2010: *Verordnung über den Zugang von Gasversorgungnetzen (Gasnetzzugangsverordnung – GasNZV)*. URL: http://bundesrecht.juris.de/bundesrecht /gasnzv_2010/gesamt.pdf. Version released Sep. 03, 2010 (cit. on pp. 49, 51, 53, 55 sqq., 59 sqq., 291, 319).

Gaspool: G. B. Services. *GasPool Homepage*. URL: http://www.gaspool.de/fusi on_aequamus.html. Downloaded on April 4, 2013 (cit. on p. 52).

KARLA Gas: Bundesnetzagentur. *Beschluss der Bundesnetzagentur BK7-10-001 (KARLA Gas): Neugestaltung des Kapazitätsmanagements im deutschen Gasmarkt*. 2011. URL: http://www.bundesnetzagentur.de/cln_1912/DE/Service-Funktio nen/Beschlusskammern/Beschlusskammer7/BK7_71_Kapazitaetsman agement/BK7_Kapazitaetsmanagement_basepage.html. Version of Feb. 24, 2011 (cit. on p. 56).

KEMA 2013/Corvinus University of Budapest. *Study on Methodologies for Gas Transmission Network Tariffs and Gas Balancing Fees in Europe*. 2013. URL: http://ec.eur opa.eu/energy/gas_electricity/studies/doc/gas/2009_12_gas_t ransmission_and_balancing_annex_fact_sheets.pdf. REKK Regional Centre for Energy Policy and Research, Downloaded on July 6, 2013 (cit. on p. 68).

KoV: BDEW – Bundesverband der Energie- und Wasserwirtschaft e.V. *Kooperationsvereinbarung zwischen den Betreibern von in Deutschland gelegenen Gasversorgungsnetzen.* 2014. URL: https://www.bdew.de/internet.nsf/id/33EEC2362FA39C3AC1257D04004ED1C2/$file/14-06-30_KoV\%20VII_Gesamtdokument_final2_clean.pdf. Downloaded on November 18, 2014 (cit. on pp. 57, 287).

MYNTS: Fraunhofer SCAI. *Multiphysical Network Simulator.* URL: http://www.scai.fraunhofer.de/geschaeftsfelder/high-performance-analytics/produkte/mynts.html. Last visited: Sep. 18, 2013 (cit. on p. 78).

National Grid. *Gas Demand Forecasting Methodology.* 2013. URL: http://www.nationalgrid.com/NR/rdonlyres/71CFD0F6-3607-474B-9F37-0952404976FB/52071/GasDemandForecastingMethodologyFeb12.pdf. Downloaded on July 6, 2013 (cit. on p. 68).

Net4Gas. *Calculations of technical capacities of the transmission system of Net4Gas.* 2013. URL: http://www.net4gas.cz/en/media/Calculation_technical_capacities.pdf. Downloaded on July 6, 2013 (cit. on p. 68).

NetConnect: NetConnect Germany. *Portrait und Gesellschafter.* URL: http://www.net-connect-germany.de/cps/rde/xchg/SID-0AB92080-B81006C6/ncg/hs.xsl/804.htm. Downloaded on April 3, 2013 (cit. on p. 52).

PipelineStudio: Energy Solutions International Inc. *PipelineStudio.* URL: http://www.energy-solutions.com/products/pipelinestudio. Last visited: Sep. 26, 2013 (cit. on p. 78).

PIPESIM: Schlumberger Software. *PIPESIM – Steady-State Multiphase Flow Simulator.* URL: http://www.software.slb.com/products/foundation/Pages/pipesim.aspx. Last visited: Sep. 26, 2013 (cit. on p. 78).

PSIG: Pipeline Simulation Interest Group (PSIG). *Home page of the PSIG.* URL: http://www.psig.org/ (cit. on p. 17).

PSIGanesi: PSI. *PSIGanesi – Grid Analysis.* URL: http://www.psioilandgas.com/en/gas-management/energy-trading-and-sales-system/#c27583. Last visited: Sep. 18, 2013 (cit. on p. 78).

PricewaterhouseCoopers AG WPG, ed. *Entflechtung und Regulierung in der deutschen Energiewirtschaft – Praxishandbuch zum Energiewirtschaftsgesetz.* 3rd ed. Haufe-Lexware GmbH & Co. KG, 2012 (cit. on p. 45).

Regulation (EC) No. 715/2009: Council European Parliament. *Regulation (EC) No 715/2009 of the European Parliament and of the Council of 13 July 2009 on conditions for access to the natural gas transmission networks and repealing Regulation (EC) No 1775/2005 (Text with EEA relevance).* URL: http://eur-lex.europa.eu/LexUriServ/LexUriServ.do?uri=OJ:L:2009:211:0036:0054:EN:PDF. Downloaded on August 24, 2013 (cit. on pp. 56, 59 sqq., 330).

Regulation (EU) No. 994/2010: Council European Parliament. *Regulation (EU) No 994/2010 of the European Parliament and of the Council of 20 October 2010 concerning measures to safeguard security of gas supply and repealing Council Directive 2004/67/EC Text with EEA relevance.* 2010. URL: http://eur-lex.europa.eu/LexUriServ/LexUriServ.do?uri=OJ:L:2010:295:0001:0022:EN:PDF. Downloaded on August 24, 2013 (cit. on pp. 48, 60).

J. P. Schneider and C. Theobald, eds. *Recht der Energiewirtschaft – Praxishandbuch.* 4th ed. C. H. Beck, 2013 (cit. on p. 45).

SIMONE: LIWACOM Informationstechnik GmbH. *SIMONE Software*. URL: http://www.liwacom.de/index.php?id=236. Last visited: Sep. 18, 2013 (cit. on pp. 78, 216).

SPS: GL Noble Denton. *SPS – Stoner Pipeline Simulator*. URL: http://www.gl-nobledenton.com/en/software/sps.php. Last visited: Sep. 26, 2013 (cit. on p. 78).

STANET: Ingenieurbüro Fischer-Uhrig. *STANET Netzberechnung*. URL: http://www.stafu.de/. Last visited: Sep. 18, 2013 (cit. on p. 78).

Terranets. *Verfahren zur Ermittlung der technischen Ein- und Ausspeisekapazitäten*. 2013. URL: http://www.terranets-bw.de/fileadmin/public/redakteure/pdf/Verfahren_Berechnung_der_technischen_Kapazit%C3%A4t.pdf. Downloaded on July 6, 2013 (cit. on p. 68).

Thyssengas: Thyssengas. *Beschreibung und Darstellung der Prozesse und Methoden zur Ermittlung der technischen Kapazität*. 2013. URL: http://www.thyssengas.com/web/cms/de/582522/thyssengas/netzauskunft/transparenzinformation/kapazitaetsermittlung/. Downloaded on July 6, 2013 (cit. on p. 68).

UBA2011: Umweltbundesamt. *Einschätzung der Schiefergasförderung in Deutschland*. URL: http://www.umweltbundesamt.de/wasser-und-gewaesserschutz/publikationen/stellungnahme_fracking.pdf. Downloaded on April 3, 2013 (cit. on p. 7).

WinTran: Gregg Engineering. *WinTran*. URL: http://www.greggengineering.com/index.php/products/winflow-wintran-product-suite/wintran. Last visited: Sep. 26, 2013 (cit. on p. 78).

Bibliography

T. Achterberg. "SCIP: Solving constraint integer programs." In: *Math. Programming Comput.* 1, No. 1, 2009, pp. 1–41 (cit. on pp. 101, 123, 235).

J. D. Anderson, Jr., J. Degroote, G. Degrez, E. Dick, R. Grundmann, and J. Vierendeels. *Computational Fluid Dynamics: An Introduction.* 3rd ed. Springer-Verlag, 2009 (cit. on p. 24).

J. André, F. Bonnans, and L. Cornibert. "Optimization of capacity expansion planning for gas transportation networks." In: *Eur. J. Oper. Res.* 197, No. 3, 2009, pp. 1019–1027 (cit. on p. 101).

P. Avella, A. Sassano, and I. Vasil'ev. "Computational study of large-scale p-median problems." In: *Math. Programming* 109, 2007, pp. 89–114 (cit. on p. 301).

D. Avis, D. Bremner, and R. Seidel. "How good are convex hull algorithms?" In: *Comput. Geom.* 7, No. 5–6, 1997, pp. 265–301 (cit. on p. 41).

P. Aymanns, A. Stolte, D. Wilms, and W. Dörner. "Online simulation of gas distribution networks." In: *Proceedings of 9th SIMONE Congress.* Dubrownik, Croatia, 2008 (cit. on p. 78).

F. Babonneau, Y. Nesterov, and J.-P. Vial. "Design and operations of gas transmission networks." In: *Oper. Res.* 60, No. 1, 2012, pp. 34–47 (cit. on p. 101).

R. Bagnara, P. M. Hill, and E. Zaffanella. "The Parma Polyhedra Library: Toward a complete set of numerical abstractions for the analysis and verification of hardware and software systems." In: *Sci. Comput. Programming* 72, No. 1–2, 2008, pp. 3–21 (cit. on p. 140).

P. Bales. "Hierarchische Modellierung der Eulerschen Flussgleichungen in der Gasdynamik." Diplomarbeit. Technische Universität Darmstadt, 2005 (cit. on p. 26).

M. K. Banda and M. Herty. "Multiscale modeling for gas flow in pipe networks." In: *Math. Methods Appl. Sci.* 31, 2008, pp. 915–936 (cit. on p. 101).

M. K. Banda, M. Herty, and A. Klar. "Gas flow in pipeline networks." In: *Netw. Heterog. Media* 1, No. 1, 2006, pp. 41–56 (cit. on p. 101).

G. K. Batchelor. *An Introduction to Fluid Dynamics.* Cambridge Mathematical Library series. Cambridge University Press, 2000 (cit. on p. 24).

B. T. Baumrucker and L. T. Biegler. "MPEC strategies for cost optimization of pipeline operations." In: *Comput. Chem. Eng.* 34, No. 6, 2010, pp. 900–913 (cit. on pp. 100, 165).

B. T. Baumrucker, J. G. Renfro, and L. T. Biegler. "MPEC problem formulations and solution strategies with chemical engineering applications." In: *Comput. Chem. Eng.* 32, No. 12, 2008, pp. 2903–2913 (cit. on p. 100).

P. Belotti, J. Lee, L. Liberti, F. Margot, and A. Wächter. "Branching and bounds tightening techniques for non-convex MINLP." In: *Optim. Methods Softw.* 24, No. 4–5, 2009, pp. 597–634 (cit. on p. 125).

M. Benedict, G. B. Webb, and L. C Rubin. "An empirical equation for thermodynamic properties of light hydrocarbons and their mixtures: I. Methane, ethane, propane, and n-butane." In: *J. Chem. Phys.* 8, No. 4, 1940, pp. 334–345 (cit. on p. 21).

H. Y. Benson, D. F. Shanno, and R. J. Vanderbei. *Interior-Point Methods for Nonconvex Nonlinear Programming: Complementarity Constraints.* Technical Report ORFE 02-02. Princeton University, 2002 (cit. on p. 253).

J. R. Birge and F. Louveaux. *Introduction to Stochastic Programming.* Springer-Verlag, 2011 (cit. on p. 322).

K. Bollaerts, H. Eilers, and I. van Mechen. "Simple and multiple P-splines regression with shape constraints." In: *British J. Math. Stat. Psychol.* 59, 2006, pp. 451–469 (cit. on pp. 286, 288 sq.).

P. Bonami, L. T. Biegler, A. R. Conn, G. Cornuéjols, I. E. Grossmann, C. D. Laird, J. Lee, A. Lodi, F. Margot, N. Sawaya, and A. Wächter. "An algorithmic framework for convex mixed integer nonlinear programs." In: *Discrete Optim.* 5, No. 2, 2008, pp. 186–204 (cit. on p. 234).

J. Bonnans, G. Spiers, and J.-L. Vie. *Global Optimization of Pipe Networks by the Interval Analysis Approach: The Belgium Network Case.* Rapport de Recherche RR 7796. INRIA, 2011 (cit. on p. 100).

C. Borraz-Sánchez and R. Z. Ríos-Mercado. "A non-sequential dynamic programming approach for natural gas network optimization." In: *WSEAS Trans. Syst.* 3, No. 4, 2004, pp. 1384–1389 (cit. on p. 100).

C. Borraz-Sánchez and R. Z. Ríos-Mercado. "A tabu search approach for minimizing fuel consumption on cyclic natural gas pipeline systems." In: *Proceedings of 12th CLAIO Conference.* La Habana, Cuba, 2004 (cit. on p. 100).

P. J. Brockwell and R. A. Davis. *Introduction to Time Series and Forecasting.* Springer Verlag, 2002 (cit. on p. 274).

J. Brouwer, I. Gasser, and M. Herty. "Gas pipeline models revisited: Model hierarchies, nonisothermal models, and simulations of networks." In: *Multiscale Model. Sim.* 9, No. 2, 2011, pp. 601–623 (cit. on p. 101).

J. Burgschweiger, B. Gnädig, and M. C. Steinbach. "Nonlinear programming techniques for operative planning in large drinking water networks." In: *Open Appl. Math. J.* 3, 2009, pp. 14–28 (cit. on p. 188).

J. Burgschweiger, B. Gnädig, and M. C. Steinbach. "Optimization models for operative planning in drinking water networks." In: *Optim. Eng.* 10, No. 1, 2009, pp. 43–73 (cit. on pp. 167, 188).

R. Byrd, J. Nocedal, and R. Waltz. "KNITRO: An integrated package for nonlinear optimization." In: *Large-Scale Nonlinear Optimization.* Ed. by G. Pillo, M. Roma, and P. Pardalos. Vol. 83. Nonconvex Optimization and Its Applications. Springer-Verlag, 2006, pp. 35–59 (cit. on p. 234).

C. Cadorso-Suárez, L. Meira-Machado, T. Kneib, and F. Gude. "Flexible hazard ratio curves for continuous predictors in multi-state models: an application to breast cancer data." In: *Stat. Model.* 10, No. 3, 2010, pp. 291–314 (cit. on p. 288).

R. E. Caflisch, W. Morokoff, and A. Owen. "Valuation of mortgage-backed securities using Brownian bridges to reduce effective dimension." In: *J. Comput. Finance* 1, No. 1, 1997, pp. 27–46 (cit. on p. 299).

R. G. Carter. "Pipeline optimization: Dynamic programming after 30 years." In: *Proceedings of the 30th PSIG Annual Meeting. Pipeline Simulation Interest Group*. 1998 (cit. on p. 100).

R.G. Carter. *Compressor Station Optimization: Computational Accuracy and Speed.* Tech. rep. PSIG 9605. Pipeline Simulation Interest Group, 1996 (cit. on pp. 101, 138).

R.G. Carter, D. Schroeder, and T. Harbick. *Some Causes and Effects of Discontinuities in Modeling and Optimizing Gas Transmission Networks.* Tech. rep. PSIG 9308. Pipeline Simulation Interest Group, 1993 (cit. on pp. 101, 138).

G. Cerbe. *Grundlagen der Gastechnik: Gasbeschaffung – Gasverteilung – Gasverwendung.* Technik. Hanser Verlag, Leipzig, 2008 (cit. on pp. 6, 17, 30, 35, 205, 274, 286, 288).

A. Chebouba, F. Yalaoui, A. Smati, L. Amodeo, K. Younsi, and A. Tairi. "Optimization of natural gas pipeline transportation using ant colony optimization." In: *Comput. Oper. Res.* 36, No. 6, 2009, pp. 1916–1923 (cit. on p. 100).

J. W. Chinneck. *Feasibility and Infeasibility in Optimization: Algorithms and Computational Methods.* Vol. 118. International Series in Operations Research and Management Science. Springer-Verlag, 2008 (cit. on p. 231).

W. Cleveland. "Robust locally weighted regression and smoothing scatterplots." In: *J. Am. Stat. Assoc.* 74, No. 368, 1979, pp. 829–836 (cit. on p. 288).

M. Collins, L. Cooper, R. Helgason, J. Kennington, and L. LeBlanc. "Solving the pipe network analysis problem using optimization techniques." In: *Management Sci.* 24, No. 7, 1978, pp. 747–760 (cit. on pp. 126 sq.).

R. M. Colombo, G. Guerra, M. Herty, and V. Schleper. "Optimal control in networks of pipes and canals." In: *SIAM J. Control Optim.* 48, No. 3, 2009, pp. 2032–2050 (cit. on p. 215).

R. M. Colombo, M. Herty, and V. Sachers. "On 2×2 conservation laws at a junction." In: *SIAM J. Math. Anal.* 40, No. 2, 2008, pp. 605–622 (cit. on p. 215).

G. Cornuéjols, M. Fisher, and G. Nemhauser. "Location of bank accounts to optimize float: An analytic study of exact and approximate algorithms." In: *Management Sci.* 23, No. 8, 1977, pp. 789–810 (cit. on p. 301).

CPLEX. *User's Manual for CPLEX.* 12.5. IBM Corporation. Armonk, USA, 2011 (cit. on p. 235).

R. B. D'Agostino and M. A. Stephens. *Goodness-of-Fit Techniques.* Marcel Dekker, 1986 (cit. on p. 277).

N. de Nevers. *Fluid Mechanics.* Addison-Wesley Series in Chemical Engineering. Addison-Wesley, 1970 (cit. on p. 166).

D. de Wolf and Y. Smeers. "Optimal dimensioning of pipe networks with application to gas transmission networks." In: *Oper. Res.* 44, No. 4, 1996, pp. 596–608 (cit. on p. 101).

D. de Wolf and Y. Smeers. "The gas transmission problem solved by an extension of the simplex algorithm." In: *Management Sci.* 46, No. 11, 2000, pp. 1454–1465 (cit. on p. 100).

V. DeMiguel, M. P. Friedlander, F. J. Nogales, and S. Scholtes. "A two-sided relaxation scheme for mathematical programs with equilibrium constraints." In: *SIAM J. Optim.* 16, No. 2, 2005, pp. 587–609 (cit. on pp. 164, 175).

J. Dick and F. Pillichshammer. *Digital Nets and Sequences: Discrepancy Theory and Quasi-Monte Carlo Integration.* Cambridge University Press, 2010 (cit. on pp. 296 sq.).

E. Doering, H. Schedwill, and M. Dehli. *Grundlagen der Technischen Thermodynamik.* Springer Verlag, 2012 (cit. on pp. 21, 200).

P. Domschke, B. Geißler, O. Kolb, J. Lang, A. Martin, and A. Morsi. "Combination of nonlinear and linear optimization of transient gas networks." In: *INFORMS J. Comput.* 23, 2011, pp. 605–617 (cit. on pp. xiv, 100, 212).

A. S. Drud. "CONOPT – A large-scale GRG code." In: *ORSA J. Comput.* 6, 1992, pp. 207–216 (cit. on pp. 235, 252, 254).

A. S. Drud. "CONOPT – A large-scale GRG code." In: *INFORMS J. Comput.* 6, No. 2, 1994, pp. 207–216 (cit. on pp. 235, 252, 254).

A. S. Drud. *CONOPT: A System for Large Scale Nonlinear Optimization, Reference Manual for CONOPT Subroutine Library.* Tech. rep. ARKI Consulting and Development A/S, Bagsvaerd, Denmark, 1996 (cit. on pp. 235, 252, 254).

A. S. Drud. *CONOPT: A System for Large Scale Nonlinear Optimization, Tutorial for CONOPT Subroutine Library.* Tech. rep. ARKI Consulting and Development A/S, Bagsvaerd, Denmark, 1995 (cit. on pp. 235, 252, 254).

K. Ehrhardt and M. C. Steinbach. "Nonlinear optimization in gas networks." In: *Modeling, Simulation and Optimization of Complex Processes.* Ed. by H. G. Bock, E. Kostina, H. X. Phu, and R. Rannacher. Springer-Verlag, 2005, pp. 139–148 (cit. on pp. 100, 212).

K. Ehrhardt and M. C. Steinbach. "KKT systems in operative planning for gas distribution networks." In: *Proc. in App. Math. and Mech.* 4, No. 1, 2004, pp. 606–607.

H. Eilers and B. Marx. "Flexible smoothing with B-splines and penalties." In: *Stat. Sci.* 11, No. 2, 1996, pp. 89–121 (cit. on pp. 286, 288).

H. Eilers and B. Marx. "Splines, knots, and penalties." In: *Wiley Interdisc. Rev. Comput. Stat.* 2, No. 6, 2010, pp. 637–653 (cit. on pp. 286, 288).

M. Elad. *Sparse and Redundant Representations: From Theory to Applications in Signal and Image Processing.* Springer-Verlag, 2010 (cit. on p. 226).

L. Fahrmeir, R. Künstler, I. Pigeot, and G. Tutz. *Statistik.* Springer-Verlag, 2007 (cit. on pp. 277 sq.).

H.-G. Fasold. "Langfristige Gasbeschaffung für Europa – Pipelineprojekte und LNG-Ketten." In: *GWF-Gas/Erdgas* 151, No. 9, 2010 (cit. on p. 11).

H.-G. Fasold and H.-N. Wahle. "Physikalische Grundlagen des Transports von Fluiden in Rohrleitungen mit Folgerungen für die Leitungsplanung (aufgezeigt an den Medien Erdgas, Wasser und Mineralöl), Teil I." In: *3R International Pipeline Technology Conference* 31, 1992, pp. 637–644 (cit. on p. 17).

H.-G. Fasold and H.-N. Wahle. "Ein Modell zur planerischen Berechnung von Offshore-Pipelines unter Verwendung eines Personal-Computers." In: *GWF-Gas/Erdgas* 139, No. 2, 1998 (cit. on p. 11).

M. Feistauer. *Mathematical Methods in Fluid Dynamics.* Vol. 67. Pitman Monographs and Surveys in Pure and Applied Mathematics Series. Longman Scientific & Technical, 1993 (cit. on pp. 24, 166, 187).

E. J. Finnemore and J. E. Franzini. *Fluid Mechanics with Engineering Applications*. 10th ed. McGraw-Hill, 2002 (cit. on pp. 6, 25, 28, 168).

A. Fischer. "A special Newton-type optimization method." In: *Optimization* 24, No. 3-4, 1992, pp. 269–284 (cit. on p. 253).

R. Fletcher and S. Leyffer. "Nonlinear programming without a penalty function." In: *Math. Programming* 91, 2000, pp. 239–269 (cit. on p. 253).

R. Fletcher and S. Leyffer. "Solving mathematical programs with complementary constraints as nonlinear programs." In: *Optim. Methods Softw.* 19, No. 1, 2004, pp. 15–40 (cit. on p. 253).

L. Ford and D. Fulkerson. "Maximal flow through a network." In: *Can. J. Math.* 8, 1956, pp. 399–404 (cit. on p. 3).

H. Friedl, R. Mirkov, and A. Steinkamp. "Modeling and forecasting gas flow on exits of gas transmission networks." In: *Int. Statist. Rev.* 80, No. 1, 2012, pp. 24–39 (cit. on p. 288).

A. Fügenschuh, B. Geißler, A. Martin, and A. Morsi. "The transport PDE and mixed-integer linear programming." In: *Models and Algorithms for Optimization in Logistics*. Ed. by C. Barnhart, U. Clausen, U. Lauther, and R. H. Möhring. Dagstuhl Seminar Proceedings 09261. Dagstuhl, Germany, 2009 (cit. on p. 100).

A. Fügenschuh, B. Geißler, R. Gollmer, C. Hayn, R. Henrion, B. Hiller, J. Humpola, T. Koch, T. Lehmann, A. Martin, R. Mirkov, A. Morsi, J. Rövekamp, L. Schewe, M. Schmidt, R. Schultz, R. Schwarz, J. Schweiger, C. Stangl, M. C. Steinbach, and B. M. Willert. "Mathematical optimization for challenging network planning problems in unbundled liberalized gas markets." In: *Energy Syst.* 5, No. 3, 2013, pp. 449–473 (cit. on p. 101).

O. Fujiwara and D. Dey. "Two adjacent pipe diameters at the optimal solution in the water distribution network models." In: *Water Resources Res.* 23, No. 8, 1987, pp. 1457–1460 (cit. on p. 321).

B. Furey. "A sequential quadratic programming-based algorithm for optimization of gas networks." In: *Automatica* 29, No. 6, 1993, pp. 1439–1450 (cit. on p. 100).

M. R. Garey and D. S. Johnson. *Computers and Intractability. A Guide to the Theory of NP-Completeness*. W. H. Freeman and Company, 1979 (cit. on p. 226).

Gaslib. *A Library of Gas Network Instances*. http://gaslib.zib.de. 2013 (cit. on p. 234).

B. Geiger and M. Hellwig. *Enwicklung von Lastprofilen für die Gaswirtschaft Gewerbe, Handel und Dienstleistungen*. Tech. rep. Technische Universität München, Department of Electrical Engineering and Information Technology, Institute for Power Engineering, 2002 (cit. on pp. 274, 286, 288).

B. Geißler. "Towards Globally Optimal Solutions for MINLPs by Discretization Techniques with Applications in Gas Network Optimization." PhD thesis. Friedrich-Alexander-Universität Erlangen-Nürnberg, 2011 (cit. on pp. 115 sqq., 239).

B. Geißler, O. Kolb, J. Lang, G. Leugering, A. Martin, and A. Morsi. "Mixed integer linear models for the optimization of dynamical transport networks." In: *Math. Methods Oper. Res.* 73, 2011, pp. 339–362 (cit. on p. 100).

B. Geißler, A. Martin, A. Morsi, and L. Schewe. "Using piecewise linear functions for solving MINLPs." In: *Mixed Integer Nonlinear Programming*. Ed. by J. Lee and S. Leyffer. Vol. 154. The IMA Volumes in Mathematics and its Applications. Springer-Verlag, 2012, pp. 287–314 (cit. on pp. 100 sq., 103, 115 sq.).

General Algebraic Modeling System (GAMS). http://www.gams.com/ (cit. on p. 235).

A. Genz and F. Bretz. *Computation of Multivariate Normal and t-Probabilities*. Springer-Verlag, 2009 (cit. on p. 303).

P. E. Gill, W. Murray, and M. S. Saunders. "SNOPT: An SQP algorithm for large-scale constrained optimization." In: *SIAM J. Optim.* 12, No. 4, 2002, pp. 979–1006 (cit. on pp. 235, 252, 254).

B. Gilmour, C. Luongo, and D. Schroeder. *Optimization in Natural Gas Transmission Networks: A Tool to Improve Operational Efficiency*. Tech. rep. Stoner Associates Inc., 1989 (cit. on p. 100).

P. Glasserman. *Monte Carlo Methods in Financial Engineering*. Vol. 53. Applications of Mathematics. Stochastic Modelling and Applied Probability. Springer-Verlag, 2004 (cit. on p. 299).

H. J. Greenberg and F. H. Murphy. "Approaches to diagnosing infeasible linear programs." In: *ORSA J. Comput.* 3, No. 3, 1991, pp. 253–261 (cit. on p. 224).

M. Griebel, F. Y. Kuo, and I. H. Sloan. "The smoothing effect of integration in R^d and the ANOVA decomposition." In: *Math. Comput.* 82, 2013, pp. 383–400 (cit. on p. 299).

M. Griebel, F. Y. Kuo, and I. H. Sloan. "The smoothing effect of the ANOVA decomposition." In: *J. Complexity* 26, No. 5, 2010, pp. 523–551 (cit. on p. 299).

A. Griewank, L. Lehmann, H. Leovey, and M. Zilberman. "Automatic Evaluations of Cross-Derivatives." In: *Math. Comput.* 2013 (cit. on p. 297).

Z. Gu, E. Rothberg, and R. Bixby. *Gurobi Optimizer Reference Manual, Version 5.1*. Gurobi Optimization Inc. Houston, USA, 2013 (cit. on pp. 235, 237).

M. Gugat, M. Dick, and G. Leugering. "Gas flow in fan-shaped networks: Classical solutions and feedback stabilization." In: *SIAM J. Control Optim.* 49, No. 5, 2011, pp. 2101–2117 (cit. on p. 148).

M. Gugat. "Boundary controllability between sub- and supercritical flow." In: *SIAM J. Control Optim.* 42, No. 3, 2003, pp. 1056–1070 (cit. on p. 101).

M. Gugat, M. Herty, A. Klar, G. Leugering, and V. Schleper. "Well-posedness of networked hyperbolic systems of balance laws." In: *Constrained Optimization and Optimal Control for Partial Differential Equations*. Ed. by G. Leugering, S. Engell, A. Griewank, M. Hinze, R. Rannacher, V. Schulz, M. Ulbrich, and S. Ulbrich. Vol. 160. International Series of Numerical Mathematics. Springer-Verlag, 2012, pp. 123–146 (cit. on p. 215).

M. Gugat and G. Leugering. "Global boundary controllability of the de St. Venant equations between steady states." In: *Ann. de l'Inst. Henri Poincare (C) Non Linear Anal.* 20, No. 1, 2003, pp. 1–11 (cit. on p. 101).

M. Gugat, G. Leugering, K. Schittkowski, and E. J. P. G. Schmidt. "Modelling, stabilization, and control of flow in networks of open channels." In: *Online Optimization of Large Scale Systems*. Ed. by M. Grötschel, S. O. Krumke, and J. Rambau. Springer-Verlag, 2001, pp. 251–270 (cit. on p. 101).

O. Guieu and J. W. Chinneck. "Analyzing infeasible mixed-integer and integer linear programs." In: *INFORMS J. Comput.* 11, No. 1, 1999, pp. 63–77 (cit. on p. 231).

Y. Hamam and A. Brameller. "Hybrid method for the solution of piping networks." In: *Proc. Inst. Electr. Eng.* 118, No. 11, 1971, pp. 1607–1612 (cit. on pp. 100, 145 sq.).

C. T. Hansen, K. Madsen, and H. B. Nielsen. "Optimization of pipe networks." In: *Math. Programming* 52, No. 1–3, 1991, pp. 45–58 (cit. on p. 101).

J. Hartung. *Statistik*. Oldenbourg, München, 2005 (cit. on pp. 277 sqq.).

W. Hauenherm. "Konstruktion von überregionalen Gasverteilungsleitungen." In: *Handbuch der Gas-Rohrleitungs-Technik*. Ed. by K. Homann and R. Hüning. 2nd ed. R. Oldenbourg, 1977 (cit. on p. 11).

H. Heitsch, H. Leövey, and W. Römisch. *Are Quasi-Monte Carlo algorithms efficient for two-stage stochastic programs?* Preprint. Stochastic Programming E-Print Series 5. 2012. URL: www.speps.org (cit. on p. 299).

H. Heitsch and W. Römisch. "A note on scenario reduction for two-stage stochastic programs." In: *Oper. Res. Lett.* 35, 2007, pp. 731–738 (cit. on p. 301).

H. Heitsch and W. Römisch. "Scenario reduction algorithms in stochastic programming." In: *Comput. Optim. Appl.* 24, 2003, pp. 187–206 (cit. on p. 301).

M. Hellwig. "Enwicklung und Anwendung parametrisierter Standard-Lastprofile." PhD thesis. Technische Universität München, Department of Electrical Engineering and Information Technology, Institute for Power Engineering, 2003 (cit. on p. 286).

R. Henrion, C. Küchler, and W. Römisch. "Discrepancy distances and scenario reduction in two-stage stochastic mixed-integer programming." In: *J. Indust. Management Optim.* 4, 2008, pp. 363–384 (cit. on p. 301).

R. Henrion, C. Küchler, and W. Römisch. "Scenario reduction in stochastic programming with respect to discrepancy distances." In: *Comput. Optim. Appl.* 43, 2009, pp. 67–93 (cit. on p. 300).

T. L. Hill. *An Introduction to Statistical Thermodynamics*. Courier Dover Publications, 1960 (cit. on p. 19).

G. Hölzel. "Gewinnung und Aufbereitung der Brenngase." In: *Grundlagen der Gastechnik – Gasbeschaffung, Gasverteilung, Gasverwendung*. Ed. by G. Cerbe. 3rd ed. Carl Hanser Verlag, 1988 (cit. on p. 9).

P. Hofer. "Beurteilung von Fehlern in Rohrnetzberechnungen (Error evaluation in calculation of pipelines)." In: *GWF-Gas/Erdgas* 11, 1973, pp. 113–119 (cit. on p. 25).

K. Homann and R. Hüning, eds. *Handbuch der Gas-Rohrleitungstechnik*. 4th ed. Oldenbourg, 1997 (cit. on p. 6).

HSL. *HSL. A Collection of Fortran Codes for Large Scale Scientific Computation*. 2013. URL: http://www.hsl.rl.ac.uk (cit. on p. 235).

HSL. *HSL MA27 Package Specification*. The HSL Mathematical Software Library. 2003 (cit. on p. 235).

X. M. Hu and D. Ralph. "Convergence of a penalty method for mathematical programming with complementarity constraints." In: *J. Optim. Theory Appl.* 123, 2004, pp. 365–390 (cit. on pp. 164, 175, 253).

R. Hüning and R. Eberhard. *Handbuch der Gasversorgungs-Technik — Gastransport und Gasverteilung*. 2nd ed. Oldenbourg, 1990 (cit. on p. 6).

M. Jach, D. Michaels, and R. Weismantel. "The convex envelope of (n-1)-convex functions." In: *SIAM J. Optim.* 19, No. 3, 2008, pp. 1451–1466 (cit. on p. 109).

R. Jarrow, D. Ruppert, and Y. Yu. "Estimating the interest rate term structure of corporate debt with a semiparametric penalized spline model." In: *J. Am. Stat. Assoc.* 99, No. 465, 2004, pp. 57–66 (cit. on pp. 286, 288 sq.).

T. Jeníček. "Steady-state optimization of gas transport." In: *Proceedings of 2nd International Workshop SIMONE on Innovative Approaches to Modeling and Optimal Control of Large Scale Pipeline Networks.* Prague, 1993, pp. 26–38 (cit. on pp. 78, 100).

G. Jones, Y. Leung, and H. Robertson. "A mixed model for investigating a population of asymptotic growth curves using restricted B-splines." In: *J. Agr. Biol. Envir. Stat.* 14, No. 1, 2009, pp. 66–78 (cit. on pp. 286, 288).

I. Joormann. "Infeasibilities in gas networks." Unpublished manuscript. 2013 (cit. on p. 231).

V. Kamnev. "Integral evaluation of surface roughness." In: *Measurement Techniques* 9, No. 2, 1966, pp. 261–263 (cit. on p. 24).

O. Kariv and S. L. Hakimi. "An algorithmic approach to network location problems. II: The p-medians." In: *SIAM J. Appl. Math.* 37, No. 3, 1979, pp. 539–560 (cit. on p. 300).

D. L. V. Katz. *Handbook of Natural Gas Engineering.* McGraw-Hill Series in Chemical Engineering. McGraw-Hill, 1959 (cit. on pp. 6, 166).

J. L. Kennedy. *Oil and Gas Pipeline Fundamentals.* Penn Well Publishing Company, 1993 (cit. on p. 28).

S. Kim, R. Z. Ríos-Mercado, and E. A. Boyd. "Heuristics for minimum cost steady-state gas transmission networks." In: *Computing Tools for Modeling, Optimization and Simulation.* Ed. by M. Laguna and J. L. González-Velarde. Vol. 12. Interfaces in Computer Science and Operations Research. Kluwer, 2000. Chap. 11, pp. 203–213 (cit. on p. 100).

B. Korte and J. Vygen. *Combinatorial Optimization: Theory and Algorithms.* Springer Verlag, 2007 (cit. on pp. 3, 127).

J. Králik. "Compressor stations in SIMONE." In: *Proceedings of 2nd International Workshop SIMONE on Innovative Approaches to Modeling and Optimal Control of Large Scale Pipeline Networks.* Prague, 1993, pp. 93–117 (cit. on p. 78).

J. Králik, P. Stiegler, Z. Vostrý, and J. Záworka. "A universal dynamic simulation model of gas pipeline networks." In: *IEEE Trans. Systems Man Cybernet.* 14, No. 4, 1984, pp. 597–606 (cit. on p. 78).

J. Králik, P. Stiegler, Z. Vostrý, and J. Záworka. *Dynamic Modeling of Large-Scale Networks with Application to Gas Distribution.* Vol. 6. Studies in Automation and Control. Elsevier, 1988 (cit. on pp. 20, 78, 215, 218).

J. Králik, P. Stiegler, Z. Vostrý, and J. Záworka. "Modeling the dynamics of flow in gas pipelines." In: *IEEE Trans. Systems Man Cybernet.* SMC-14, No. 4, 1984, pp. 586–596 (cit. on p. 78).

O. Kunz, R. Klimeck, W. Wagner, and M. Jaeschke. "The GERG-2004 wide-range equation of state for natural gases and other mixtures." In: *Fortschritt-Berichte VDI* 6, No. 557, 2007 (cit. on p. 20).

O. Kunz and W. Wagner. "The GERG-2008 wide-range equation of state for natural gases and other mixtures: An expansion of GERG-2004." In: *J. Chem. Eng. Data* 57, No. 11, 2012, pp. 3032–3091 (cit. on p. 20).

F. Kuo, C. Schwab, and I. Sloan. "Quasi-Monte Carlo methods for high-dimensional integration: the standard (weighted Hilbert space) setting and beyond." In: *ANZIAM J.* 53, 2011, pp. 1–37 (cit. on pp. 296 sq., 299).

H. Lall and P. Percell. "A dynamic programming based gas pipeline optimizer." In: *Analysis and Optimization of Systems*. Ed. by A. Bensoussan and J. Lions. Vol. 144. Lecture Notes in Control and Information Sciences. Springer-Verlag, 1990, pp. 123–132 (cit. on p. 100).

LaMaTTO++. *LaMaTTO++: A Framework for Modeling and Solving Mixed-Integer Nonlinear Programming Problems on Networks.* http://www.mso.math.fau.de/edom/projects/lamatto.html. 2014 (cit. on p. 235).

L. Landau and E. Lifshitz. *Fluid Mechanics*. 2nd ed. Vol. 6. Course of Theoretical Physics. Butterworth-Heinemann, 1987 (cit. on p. 24).

G. Lappus. "Analyse und Synthese eines Zustandsbeobachtersystems für große Gasverteilnetze." PhD thesis. Technische Universität München, 1984 (cit. on p. 78).

P. L'Ecuyer. "Good parameters and implementations for combined multiple recursive random number generators." In: *Oper. Res.* 47, No. 1, 1999, pp. 159–164 (cit. on p. 296).

P. L'Ecuyer and C. Lemieux. "Recent advances in randomized quasi-Monte Carlo methods." In: *Modeling Uncertainty*. Ed. by M. Dror, P. L'Ecuyer, and F. Szidarovszky. Vol. 46. International Series in Operations Research & Management Science. Springer-Verlag, 2005, pp. 419–474 (cit. on pp. 296 sq.).

J. S. Lee and D. D. Cox. "Robust smoothing: Smoothing parameter selection and applications to fluorescence spectroscopy." In: *Comput. Stat. Data Anal.* 54, No. 12, 2010, pp. 3131–3143 (cit. on p. 290).

E. Lehmann and J. Romano. *Testing Statistical Hypotheses*. Springer Texts in Statistics, 3rd ed. Springer-Verlag, 2005 (cit. on p. 278).

H. Leövey, W. Römisch, A. Steinkamp, and I. Wegner-Specht. "Modellierung der Gasabnahme als Funktion der Temperatur: Optimierung der Temperaturgewichte." In: *GWF-Gas/Erdgas* 11, 2011, pp. 778–785 (cit. on pp. 274, 288).

G. Leugering and E. J. P. G. Schmidt. "On the modelling and stabilization of flows in networks of open canals." In: *SIAM J. Control Optim.* 41, No. 1, 2002, pp. 164–180 (cit. on p. 101).

S. Leyffer, G. López-Calva, and J. Nocedal. "Interior methods for mathematical programs with complementarity constraints." In: *SIAM J. Optim.* 17, No. 1, 2006, pp. 52–77 (cit. on p. 253).

C. Li, W. Jia, Y. Yang, and X. Wu. "Adaptive genetic algorithm for steady-state operation optimization in natural gas networks." In: *J. Software* 6, No. 3, 2011, pp. 452–459 (cit. on p. 100).

X. Li, E. Armagan, A. Tomasgard, and P. I. Barton. "Stochastic pooling problem for natural gas production network design and operation under uncertainty." In: *AIChE J.* 57, No. 8, 2011, pp. 2120–2135 (cit. on p. 100).

X. Liu and J. Sun. "A robust primal-dual interior-point algorithm for nonlinear programs." In: *SIAM J. Optim.* 14, No. 4, 2004, pp. 1163–1186 (cit. on p. 253).

LIWACOM. *Gleichungen und Methoden*. Benutzerhandbuch. LIWACOM Informations GmbH and SIMONE Research Group s.r.o. 2004 (cit. on pp. 21, 37, 201).

M. Lloyd, J. VanZelfden, A. Brodsky, and M. Tsai. *Tennessee Gas Pipeline's Experience with Optimization*. Tech. rep. PSIG 0603. Pipeline Simulation Interest Group, 2006 (cit. on p. 100).

Z.-Q. Luo, J.-S. Pang, and D. Ralph. *Mathematical Programs with Equilibrium Constraints*. Cambridge University Press, 1996 (cit. on p. 163).

M. V. Lurie. *Modeling of Oil Product and Gas Pipeline Transportation*. Wiley-VCH, 2008 (cit. on pp. 6, 17, 24 sqq., 28, 166, 168, 187).

M. Mackenzie, C. Donovan, and B. McArdle. "Regression spline mixed model: A forestry example." In: *J. Agr. Biol. Envir. Stat.* 10, No. 4, 2005, pp. 394–410 (cit. on p. 288).

D. Mahlke, A. Martin, and S. Moritz. "A mixed integer approach for time-dependent gas network optimization." In: *Optim. Methods Softw.* 25, No. 4, 2010, pp. 625–644 (cit. on p. 100).

D. Mahlke, A. Martin, and S. Moritz. "A simulated annealing algorithm for transient optimization in gas networks." In: *Math. Methods Oper. Res.* 66, No. 1, 2007, pp. 99–116 (cit. on p. 100).

J. Mallinson, A. Fincham, S. Bull, J. Rollet, and M. Wong. "Methods for optimizing gas transmission networks." In: *Ann. Oper. Res.* 43, 1993, pp. 443–454 (cit. on pp. 100, 145 sq.).

H. Markowitz and A. Manne. "On the solution of discrete programming problems." In: *Econometrica* 25, 1957, pp. 84–110 (cit. on pp. 106, 116).

A. Martin, B. Geißler, C. Hayn, J. Humpola, T. Koch, T. Lehman, A. Morsi, M. Pfetsch, L. Schewe, M. Schmidt, R. Schultz, R. Schwarz, J. Schweiger, M.C. Steinbach, and B.M. Willert. "Optimierung Technischer Kapazitäten in Gasnetzen." In: *VDI-Berichte: Optimierung i. d. Energiewirtschaft.* 2157, 2011, pp. 105–115 (cit. on p. 101).

A. Martin, M. Möller, and S. Moritz. "Mixed integer models for the stationary case of gas network optimization." In: *Math. Programming* 105, No. 2, 2006, pp. 563–582 (cit. on p. 100).

P. Massart. "The tight constant in the Dvoretzky-Kiefer-Wolfowitz inequality." In: *Ann. Probability* 18, No. 3, 1990, pp. 1269–1283 (cit. on p. 278).

M. Matsumoto and T. Nishimura. "Mersenne twister: A 623-dimensionally equidistributed uniform pseudo-random number generator." In: *ACM Trans. Model. Comput. Sim.* 8, No. 1, 1998, pp. 3–30 (cit. on p. 296).

J. J. Maugis. "Etude de réseaux de transport et de distribution de fluide." In: *RAIRO Oper. Res.* 11, No. 2, 1977, pp. 243–248 (cit. on p. 126).

E. Menon. *Gas Pipeline Hydraulics*. Taylor & Francis, 2005 (cit. on pp. 6, 19, 23, 192).

J. Mischner. "Notices about hydraulic calculations of gas pipelines." In: *GWF-Gas/Erdgas* 4, 2012, pp. 158–273 (cit. on p. 25).

J. Mischner. "Zur analytischen Berechnung des Temperaturverlaufs in Gastransportleitungen – Teil 1." In: *GWF-Gas/Erdgas* 150, No. 10, 2009 (cit. on p. 11).

J. Mischner, L. Huke, and G. Möhlen. "Zur Bewertung von Einflussparametern bei der Kapazitätsermittlung von Gastransportleitungen – Teil 1." In: *GWF-Gas/Erdgas* 151, No. 4, 2010 (cit. on p. 11).

J. Mischner and S. Schewe. "Zur Ermittlung von Stoffdaten für die hydraulische Berechnung von Gasrohrleitungen." In: *GWF-Gas/Erdgas* 150, No. 4, 2009 (cit. on p. 9).

M. Möller. "Mixed Integer Models for the Optimisation of Gas Networks in the Stationary Case." PhD thesis. Technische Universität Darmstadt, Fachbereich Mathematik, 2004 (cit. on p. 100).

S. Moritz. "A Mixed Integer Approach for the Transient Case of Gas Network Optimization." PhD thesis. Technische Universität Darmstadt, 2007 (cit. on p. 100).

A. Morsi. "Solving MINLPs on Loosely-Coupled Networks with Applications in Water and Gas Network Optimization." PhD thesis. Friedrich-Alexander-Universität Erlangen-Nürnberg, 2013 (cit. on pp. 235, 239).

G. L. Nemhauser and L. A. Wolsey. *Integer and Combinatorial Optimization*. John Wiley & Sons, New York, 1988 (cit. on p. 41).

H. Niederreiter. *Random Number Generation and Quasi-Monte Carlo Methods*. Vol. 63. CBMS-NSF Regional Conference Series in Applied Mathematics. SIAM, 1992 (cit. on pp. 295 sqq.).

J. Nikuradse. *Strömungsgesetze in Rauhen Rohren*. Forschungsheft auf dem Gebiete des Ingenieurwesens. VDI-Verlag, 1933 (cit. on p. 25).

J. Nikuradse. *Laws of Flow in Rough Pipes*. Technical Memorandum. Vol. 1292. National Advisory Committee for Aeronautics, Washington, 1950 (cit. on p. 25).

J. Nocedal and S. J. Wright. *Numerical Optimization*. 2nd ed. Springer-Verlag, 2006 (cit. on p. 252).

V. S. Nørstebø, F. Rømo, and L. Hellemo. "Using operations research to optimise operation of the Norwegian natural gas system." In: *J. Nat. Gas Science Eng.* 2, 2010, pp. 153–162 (cit. on p. 100).

E. Novak and H. Woźniakowski. *Tractability of Multivariate Problems. Volume II: Standard Information for Functionals*. Vol. 12. EMS Tracts in Mathematics. European Mathematical Society (EMS) Publishing House, 2010 (cit. on p. 296).

F. M. Odom and G. L. Muster. *Tutorial on Modeling of Gas Turbine Driven Centrifugal Compressors*. Tech. rep. PSIG #09A4. Pipeline Simulation Interest Group, 2009 (cit. on pp. 17, 34 sq.).

M. J. Oliveira. *Equilibrium Thermodynamics*. Graduate Texts in Physics. Springer-Verlag, 2013 (cit. on pp. 21 sq.).

A.J. Osiadacz. "Nonlinear programming applied to the optimum control of a gas compressor station." In: *Int. J. Numer. Methods Eng.* 15, No. 9, 1980, pp. 1287–1301 (cit. on p. 101).

A. J. Osiadacz. *Simulation and Analysis of Gas Networks*. E. & F.N. Spon, London, 1987 (cit. on pp. 6, 17, 100).

A.J. Osiadacz and K. Pienkosz. "Methods of steady-state simulation for gas networks." In: *Int. J. Syst. Sci.* 19, No. 7, 1988, pp. 1311–1321 (cit. on pp. 146, 148).

M. Padberg. "Approximating separable nonlinear functions via mixed zero-one programs." In: *Oper. Res. Lett.* 27, No. 1, 2000, pp. 1–5 (cit. on p. 116).

I. Papay. *A termeléstechnológiai paraméterek változása a gáztelepek muvelése során*. In: OGIL Musz. Tud. Kozl., 1968 (cit. on p. 20).

D.-Y. Peng and D. B. Robinson. "A new two-constant equation of state." In: *Ind. Eng. Chem. Fundam.* 15, No. 1, 1976, pp. 59–64 (cit. on p. 21).

J. Peng and Y. Wei. "Approximating K-means-type clustering via semidefinite programming." In: *SIAM J. Optim.* 18, No. 1, 2007, pp. 186–205 (cit. on p. 301).

P. B. Percell and M. J. Ryan. "Steady-state optimization of gas pipeline network operation." In: *Proceedings of the 19th PSIG Annual Meeting*. 1987 (cit. on p. 34).

P. B. Percell and M. J. Ryan. *Steady-State Optimization of Gas Pipeline Network Operation.* Tech. rep. PSIG 8703. Pipeline Simulation Interest Group, 1987 (cit. on p. 100).

T. A. Perdicoúlis and L. Fletcher. "Decentralised dynamic optimisation of gas." In: *31st Annual Meeting of Pipeline Simulation Interest Group*. St. Louis, MO, USA, 1999 (cit. on p. 100).

M. E. Pfetsch, A. Fügenschuh, B. Geißler, N. Geißler, R. Gollmer, B. Hiller, J. Humpola, T. Koch, T. Lehmann, A. Martin, A. Morsi, J. Rövekamp, L. Schewe, M. Schmidt, R. Schultz, R. Schwarz, J. Schweiger, C. Stangl, M. C. Steinbach, S. Vigerske, and B. M. Willert. "Validation of nominations in gas network optimization: Models, methods, and solutions." In: *Optim. Methods Softw.* 30, No. 1, 2015, pp. 15–53 (cit. on pp. 101, 126, 139 sq., 233 sq.).

M. Plischke and B. Bergersen. *Equilibrium Statistical Physics*. 3rd ed. World Scientific Publishing, 2006 (cit. on p. 19).

A. Prékopa. "Multivariate value at risk and related topics." In: *Ann. Oper. Res.* 193, 2012, pp. 49–69 (cit. on p. 302).

S. Rachev. *Probability Metrics and the Stability of Stochastic Models*. John Wiley & Sons, 1991 (cit. on p. 299).

A. U. Raghunathan and L. T. Biegler. "An interior point method for mathematical programs with complementarity constraints (MPCCs)." In: *SIAM J. Optim.* 15, No. 3, 2005, pp. 720–750 (cit. on p. 253).

N. Ramchandani. "Optimisation of Gas Networks Using Nash Equlibria Derived from Dynamic Non-Cooperative Game Theory." PhD thesis. Stanford University, 1993 (cit. on p. 100).

Y. V. C. Rao. *An Introduction to Thermodynamics*. 2nd ed. Universities Press, 2003 (cit. on p. 21).

O. Redlich and J. N. S. Kwong. "On the thermodynamics of solutions. V. An equation of State. Fugacities of gaseous solutions." In: *Chemical Rev.* 44, No. 1, 1949, pp. 233–244 (cit. on p. 21).

M. G. C. Resende and R. F. Werneck. "A hybrid heuristic for the p-median problem." In: *J. Heuristics* 10, 2004, pp. 59–88 (cit. on p. 301).

K. Riedel and K. Imre. "Smoothing spline growth curves with covariates." In: *Commun. Stat. Theory Methods* 22, No. 7, 1993, pp. 1795–1818 (cit. on p. 288).

R. Z. Ríos-Mercado and C. Borraz-Sánchez. "Optimization problems in natural gas transmission systems: A state-of-the-art survey." Submitted. 2012 (cit. on p. 100).

R. Z. Ríos-Mercado, S. Kim, and E. A. Boyd. "Efficient operation of natural gas transmission systems: A network-based heuristic for cyclic structures." In: *Comput. Oper. Res.* 33, No. 8, 2006, pp. 2323–2351 (cit. on p. 100).

R. Z. Ríos-Mercado, S. Wu, L. R. Scott, and E. A. Boyd. "A reduction technique for natural gas transmission network optimization problems." In: *Ann. Oper. Res.* 117, No. 1, 2002, pp. 217–234 (cit. on pp. 100, 145 sq., 157).

C. Ritz and J. Streibig. *Nonlinear Regression with R*. Use R! Springer-Verlag, 2008 (cit. on p. 288).

R. Rockafellar. *Convex Analysis*. Princeton Mathematical Series. Princeton University Press, 1970 (cit. on p. 109).

H. S. Rodrigues and M. T. T. Monteiro. "Solving mathematical programs with complementarity constraints with nonlinear solvers." In: *Recent Advances in Optimization*. Ed. by A. Seeger. Vol. 563. Lecture Notes in Economics and Mathematical Systems. Springer-Verlag, 2006, pp. 415–424 (cit. on p. 253).

W. Römisch. "Stability of stochastic programming problems." In: *Stochastic Programming*. Ed. by A. Ruszczyński and A. Shapiro. Elsevier, 2003. Chap. 8, pp. 483–554 (cit. on pp. 281, 299).

J. Rövekamp. "*Transportnetzberechnung zur Feststellung der Erdgasversorgungssicherheit in Deutschland unter regulatorischem Einfluss.*" Dissertation. Technische Universität Clausthal, 2015 (cit. on p. 13).

J. Saleh, ed. *Fluid Flow Handbook*. McGraw-Hill Handbooks. McGraw-Hill, New York, 2002 (cit. on pp. 20, 25, 27, 166).

H. Scheel and S. Scholtes. "Mathematical programs with complementarity constraints: Stationarity, optimality and sensitivity." In: *Math. Oper. Res.* 25, 2000, pp. 1–22 (cit. on p. 252).

D. Scheibe and A. Weimann. "Dynamische Gasnetzsimulation mit GANESI." In: *GWF Gas/Erdgas* 9, 1999, pp. 610–616 (cit. on p. 78).

M. Schmidt. "A Generic Interior-Point Framework for Nonsmooth and Complementarity Constrained Nonlinear Optimization." PhD thesis. Gottfried Wilhelm Leibniz Universität Hannover, 2013 (cit. on pp. 102, 165, 254).

M. Schmidt, M. C. Steinbach, and B. M. Willert. "A primal heuristic for nonsmooth mixed integer nonlinear optimization." In: *Facets of Combinatorial Optimization*. Ed. by M. Jünger and G. Reinelt. Springer-Verlag, 2013, pp. 295–320 (cit. on pp. 102, 165, 265).

M. Schmidt, M. C. Steinbach, and B. M. Willert. "High detail stationary optimization models for gas networks." In: *Optim. Eng. Published online March 18*, 2014 (cit. on pp. 22, 32, 34, 37, 166 sq., 171, 188 sq., 191 sq., 196, 198, 201).

M. Schmidt, M. C. Steinbach, and B. M. Willert. "High detail stationary optimization models for gas networks: Validation and results." Submitted. Preprint available at Optimization Online. 2014 (cit. on pp. 217, 221).

S. Scholtes. "Convergence properties of a regularization scheme for mathematical programs with complementarity constraints." In: *SIAM J. Optim.* 11, No. 4, 2001, pp. 918–936 (cit. on pp. 164, 175, 253, 259).

SCIP: Solving Constraint Integer Programs. http://scip.zib.de/. 2013 (cit. on pp. 101, 123, 234 sq.).

G. Seber and C. Wild. *Nonlinear Regression*. Wiley's Series in Probability and Statistics. John Wiley & Sons, 2003 (cit. on pp. 274, 286).

E. Sekirnjak. *Practical Experiences with Various Optimization Techniques for Gas Transmission and Distribution Systems*. Tech. rep. PSIG 9603. Pipeline Simulation Interest Group, 1996 (cit. on p. 100).

D. Shaw. *Pipeline System Optimization: A Tutorial*. Tech. rep. PSIG 9405. Pipeline Simulation Interest Group, 1994 (cit. on p. 100).

E. Smith and C. Pantelides. "A symbolic reformulation/spatial branch-and-bound algorithm for the global optimization of nonconvex MINLPs." In: *Comput. Chem. Eng.* 23, 1999, pp. 457–478 (cit. on p. 125).

G. Soave. "Equilibrium constants from a modified Redlich-Kwong equation of state." In: *Chem. Eng. Science* 27, No. 6, 1972, pp. 1197–1203 (cit. on p. 21).

I. M. Sobol'. "Global sensitivity indices for nonlinear mathematical models and their Monte Carlo estimates." In: *Math. Comput. Sim.* 55, 2001, pp. 271–280 (cit. on p. 298).

K. Starling and J. Savidge. *Compressibility Factors of Natural Gas and Other Related Hydrocarbon Gases*. Transmission Measurement Committee report. AGA, American Gas Association, 1992 (cit. on pp. 20, 90).

M. C. Steinbach. "On PDE solution in transient optimization of gas networks." In: *J. Comput. Appl. Math.* 203, No. 2, 2007, pp. 345–361 (cit. on pp. 100, 212).

O. Streicher and A. Feßmann. "Konstruktion von Gasverteilungsleitungen." In: *Handbuch der Gas-Rohrleitungs-Technik*. Ed. by K. Homann and R. Hüning. 2nd ed. R. Oldenbourg, 1977 (cit. on p. 11).

O. Strusberg and S. Engell. "Optimal control of switched continuous systems using mixed-integer programming." In: *15th IFAC World Congress of Automatic Control*. Barcelona, Spain, 2002 (cit. on p. 100).

R. Stryjek and J. H. Vera. "PRSV: An improved Peng–Robinson equation of state for pure compounds and mixtures." In: *Can. J. Chem. Eng.* 64, No. 2, 1986, pp. 323–333 (cit. on p. 21).

R. Stryjek and J. H. Vera. "PRSV2: A cubic equation of state for accurate vapor-liquid equilibria calculations." In: *Can. J. Chem. Eng.* 64, No. 5, 1986, pp. 820–826 (cit. on p. 21).

C. K. Sun, V. Uraikul, C. W. Chan, and P. Tontiwachwuthikul. "An integrated expert system/operations research approach for the optimization of natural gas pipeline operations." In: *Eng. Appl. Artif. Intell.* 13, No. 4, 2000, pp. 465–475 (cit. on p. 100).

D. Sun and L. Qi. "On NCP-functions." In: *Comput. Optim. Appl.* 13, No. 1-3, 1999. Computational Optimization – a Tribute to Olvi Mangasarian, Part II, pp. 201–220 (cit. on p. 253).

J. Szabó. *The Set of Solutions to Nomination Validation in Passive Gas Transportation Networks with a Generalized Flow Formula*. ZIB-Report 11-44. Zuse Institute Berlin, 2012 (cit. on p. 292).

R. Tahir-Kheli. *General and Statistical Thermodynamics*. Graduate Texts in Physics. Springer-Verlag, 2012 (cit. on pp. 32 sq.).

M. Tawarmalani and N. Sahinidis. "A polyhedral branch-and-cut approach to global optimization." In: *Math. Programming* 103, No. 2, 2005, pp. 225–249 (cit. on pp. 125, 234).

M. Tawarmalani and N. Sahinidis. *Convexification and Global Optimization in Continuous and Mixed-Integer Nonlinear Programming: Theory, Algorithms, Software, and Applications*. Kluwer Academic Publishers, 2002 (cit. on pp. 125, 234).

M. Tawarmalani and N. Sahinidis. "Global optimization of mixed-integer nonlinear programs: A theoretical and computational study." In: *Math. Programming* 99, No. 3, 2004, pp. 563–591 (cit. on pp. 125, 234).

H. C. Thode. *Testing for Normality*. M. Dekker, 2002 (cit. on pp. 277 sq.).

A. Tomasgard, F. Rømo, M. Fodstad, and K. Midthun. "Optimization models for the natural gas value chain." In: *Geometric Modelling, Numerical Simulation, and Optimization*. Ed. by G. Hasle, K.-A. Lie, and E. Quak. Springer Verlag, 2007, pp. 521–558 (cit. on p. 100).

T. van der Hoeven. "Math in Gas and the Art of Linearization." PhD thesis. Rijksuniversiteit Groningen, 2004 (cit. on p. 78).

J. D. Van der Waals. "De Continuiteit van den Gas- en Vloeistoftoestand." PhD thesis. Universiteit Leiden, 1873 (cit. on p. 21).

R. Vershynin. "How close is the sample covariance matrix to the actual covariance matrix?" In: *J. Theor. Prob.* 25, No. 3, 2012, pp. 655–686 (cit. on p. 285).

J. Vielma and G. Nemhauser. "Modeling disjunctive constraints with a logarithmic number of binary variables and constraints." In: *Math. Programming* 128, No. 1, 2011, pp. 49–72 (cit. on p. 239).

S. Vigerske. "Decomposition in Multistage Stochastic Programming and a Constraint Integer Programming Approach to Mixed-Integer Nonlinear Programming." PhD thesis. Humboldt-Universität zu Berlin, 2012 (cit. on pp. 125, 234 sq.).

Z. Vitezica, C. Marie-Etancelin, M. D. Bernadet, X. Fernandez, and C. Robert-Granie. "Comparison of nonlinear and spline regression models for describing mule duck growth curves." In: *Poultry Sci.* 89, 2010, pp. 1778–1784 (cit. on pp. 286, 288).

Z. Vostrý. "Transient optimization of gas transport and distribution." In: *Proceedings of 2nd International Workshop SIMONE on Innovative Approaches to Modeling and Optimal Control of Large Scale Pipeline Networks*. Prague, 1993, pp. 53–62 (cit. on pp. 78, 100).

Z. Vostrý and J. Záworka. "Heat dynamics in gas transport." In: *Proceedings of the SIMONE Congress '95*. 1995, pp. 64–75 (cit. on p. 78).

A. Wächter and L. T. Biegler. "On the implementation of a primal-dual interior point filter line search algorithm for large-scale nonlinear programming." In: *Math. Programming* 106, No. 1, 2006, pp. 25–57 (cit. on pp. 235, 242, 252, 254).

U. Wagner and B. Geiger. *Gutachten zur Festlegung von Standardlastprofilen Haushalte und Gewerbe für BGW und VKU*. Tech. rep. Technische Universität München, Department of Electrical Engineering and Information Technology, Institute for Power Engineering, 2005 (cit. on p. 286).

X. Wang and K.-T. Fang. "The effective dimension and quasi-Monte Carlo integration." In: *J. Complexity* 19, No. 2, 2003, pp. 101–124 (cit. on p. 299).

X. Wang and I. H. Sloan. "Why are high-dimensional finance problems often of low effective dimension." In: *SIAM J. Sci. Comput.* 27, No. 1, 2005, pp. 159–183 (cit. on p. 299).

E. Wegman and I. Wright. "Splines in statistics." In: *J. Am. Stat. Assoc.* 78, No. 382, 1983, pp. 351–365 (cit. on pp. 286, 288).

A. Weimann. "Modellierung und Simulation der Dynamik von Gasnetzen im Hinblick auf Gasnetzführung und Gasnetzüberwachung." PhD Thesis. Technische Universität München, 1978 (cit. on p. 78).

T. Westerlund and R. Pörn. "Solving pseudo-convex mixed integer optimization problems by cutting plane techniques." In: *Optim. Eng.* 3, No. 3, 2002, pp. 253–280 (cit. on p. 234).

T. R. Weymouth. "Problems in natural gas engineering." In: *Trans. Am. Soc. Mech. Eng.* 34, No. 1349, 1912, pp. 185–231 (cit. on p. 166).

J. F. Wilkinson, D. V. Holliday, E. H. Batey, and K. W. Hannah. *Transient Flow in Natural Gas Transmission Systems.* American Gas Association, 1964 (cit. on p. 26).

B. M. Willert. "Validation of Nominations in Gas Networks and Properties of Technical Capacities." PhD thesis. Gottfried Wilhelm Leibniz Universität Hannover, 2014 (cit. on pp. 221, 259, 319).

P. J. Wong and R. E. Larson. "Optimization of natural-gas pipeline systems via dynamic programming." In: *IEEE Trans. Automatic Control* 15, No. 5, 1968, pp. 475–481 (cit. on p. 100).

P. J. Wong and R. E. Larson. "Optimization of tree-structured natural-gas transmission networks." In: *J. Math. Anal. Appl.* 24, 1968, pp. 613–626 (cit. on p. 100).

S. Wright, M. Somani, and C. Ditzel. *Compressor Station Optimization.* Tech. rep. PSIG 9805. Pipeline Simulation Interest Group, 1998 (cit. on pp. 100 sq.).

S. Wu, R. Z. Ríos-Mercado, E. A. Boyd, and L. R. Scott. "Model relaxations for the fuel cost minimization of steady-state gas pipeline networks." In: *Math. Comput. Model.* 31, No. 2, 2000, pp. 197–220 (cit. on pp. 100, 138).

J. J. Ye and D. L. Zhu. "Optimality conditions for bilevel programming problems." In: *Optimization* 33, 1995, pp. 9–27 (cit. on pp. 163, 252).

J. Záworka. *Project SIMONE – Achievements and running development.* Tech. rep. Institute of Information Theory and Automation, 2004 (cit. on p. 78).

J. Záworka. "Project SIMONE—Achievements and running development." In: *Proceedings of 2nd International Workshop SIMONE on Innovative Approaches to Modeling and Optimal Control of Large Scale Pipeline Networks.* Prague, 1993, pp. 1–24 (cit. on pp. 78, 215, 218).

J. Zhang and D. Zhu. "A bilevel programming method for pipe network optimization." In: *SIAM J. Optim.* 6, No. 3, 1996, pp. 838–857 (cit. on p. 101).

Q. Zheng, S. Rebennack, N. Iliadis, and P. Pardalos. "Optimization models in the natural gas industry." In: *Handbook of Power Systems I.* Springer-Verlag, 2010, pp. 121–148 (cit. on p. 100).

G. M. Ziegler. *Lectures on Polytopes.* Springer Verlag, 1994 (cit. on p. 41).

H. Zimmer. *Calculating Optimum Pipeline Operations.* Tech. rep. SAND2009-5066C. El Paso Natural Gas Company, 1975 (cit. on p. 100).

R. Zucker and O. Biblarz. *Fundamentals of Gas Dynamics.* John Wiley & Sons, 2002 (cit. on p. 17).

Index

An italicized page number indicates the location of a term's definition.

active element, *18*, 39, 68, 87, 88, 91, 94, 95, 113, 123, 124, 130–133, 140, 143, 146, 149, 150, 177, 321, 334
adiabatic, 32, 108, 336
adiabatic efficiency, *34, 35*
AGA (American Gas Association), 90, 135
analysis of variance (ANOVA), 277, 298, *299*, 299
asset operator, 10
auctioning, 49, 55, 59

balancing group, *52*, 52
biogas, *8*, 318
booking, 5, *49*, 49, 74, 77, 291, 292, 293, 294, 304, 308, 309, 311
  validity probability, *293*, 294, 295, 308, 309, 311, 312, 314
booking-compliant, 310
boundary values, *68*
bundling, *53*, 53, 55, 63
bypass, 151, 333

calorific value, 7, 8, *22*, 22, 35, 36, 183, 205, 335
  lower, *35*, 205
capacity, *5*, 62, 65, 74, 83, 333
  buy-back, *61*
  conditional, *57*, 59, 67
  contract, *49*, 74, *304*, 318, 336
  day-ahead, *49*, 49, 56, 59, 64
  firm, *49*, 49, 57, 59, 68, 74, 305, 337
  freely allocable, *51*, 53, 54, 56–59, 64, 66, 69, 84, 291, *304*, 304, 305, 310, 317, 319, 320, 328, 335
  space of, *319*, 320
hoarding, *56*
interconnection, 53, 303
interruptible, *49*, 53, 59, 67, 74, 305, 310, 312–314, 319
nonsubstitutable, *77*, *306*, 306
position, *304*, 304, 305
product, 73
restrictively allocable, 75, *304*, 304, 305, 310, 317, 319
substitutable, *77*, *306*, 306
technical, *319*, 319
validity period, 73
capacity problem, *319*
chokeline, 93, 109, 111, 203
common capacity calculation model, 75
compressibility factor, *19*
compression ratio, 35, 40, 108, 112, 202, 204
compressor, 4, 17, 18, 31, 34–37, 39–41, 89, 91–94, 96–101, 103, 108, 109, 112–114, 118, 119, 121, 122, 124, 138, 139, 150, 204, 209, 212, 213, 223, 229, 320, 321, 333, 334, 337
characteristic diagram, *34*, 108, 112, 120, 138–140, 203
chokeline, *36*
isoline, *36*, 203
maximum power, *37*
maximum torque, *36*, 94
operating range, *32*, 203, 204
power, *34*, 108, 200
shaft torque, *36*, 93, 112, 204
surgeline, *36*
throughput, *32*, 92, 139, 202, 204
compressor group, 18, 19, 35, 38–41, 79, 89, 91, 92, 94, 97, 99, 101, 103, 112, 113, 138–141, 143, 150, 151, 154, 155, 162, 164, 171, 172, 176–178, 182, 198, 223, 229, 249, 262, 320, 333, 334
  postprocessing, 99
compressor machine, 13, 31, 32, 36, 39, 40, 68, 79–81, 92, 103, 130, 138–140, 164, 171, 176, 198–205, 207, 223, 333, 334, 336–338
compressor station, 11, 13, 18, 28, 38, 39, *40*, 40–42, 79, 90, 91, 94, 226, 333, 337
concave envelope, 110, 111, 116–119
concave overestimator, 109, 110
configuration, *40*, 79, 92, 93, 112, 113, 138–140, 171–173, 176–178, 198, 333
congestion, 63, *70*
  contractual, 48, *49*, 49, 59
  technical, *49*, 53, 59, 60
congestion management procedures, 49, 330
constraint qualification, 163, 175
  LICQ (linear independence constraint qualification), 175

361

MFCQ (Mangasarian–Fromowitz constraint qualification), 175
control setting, *68*, 79–83
control valve, 11, 18, *30*, 30, 31, 38–41, 68, 69, 79, 89, 94, 97, 106–108, 112, 118, 123, 136, 137, 141, 152, 154, 170, 171, 176, 177, 196–198, 213, 217, 220, 226, 229, 262, 321, 334, 337
   active, *30*, 31, 107, 136, 137, 170, 171, 196
   bypass, *30*, 107, 108, 136, 153, 170, 196
   closed, *30*, 31, 107, 136, 170, 196
   open, 107, 108, 136
   without remote access, 156
control valve station, 29, 30, *38*, 38, 89, 94, 108, 112, 164, 182, 195–198, 262, 333, 337
controllable network element, *18*, 41
convex envelope, 109, 116–118
convex hull, 138–140
convex underestimator, 109
cotree arc, 149, 155
cutting plane algorithm, 114, 118, 121

day classes, *280*
deposit
   conventional, *6*
   unconventional, *7*
design temperature, *57*, 57, 71, *286*, 288
digital b-ary shifting, *297*
discrepancy, *296*
discretization, 131, 134, 137, 140–142
distribution, 46
distribution system operator, 46, 48, 53, 55, 66, 74, 77, 336, 338
drive, 31, 35–37, 39, 92, 93, 103, 122, 138, 150, 171, 204, 209, 333–335, 337
dynamic viscosity, 25, 167, 188

effective dimension, *295*, *299*, 299
electric motor, 32, 35, 93, 206, 334

empirical distribution function, *277*, 281, 282
energy consumption rate, *35*, *37*
enthalpy, *32*
   adiabatic enthalpy, 200, 203, 207
entry [point], 5, *18*, 18, 19, 47, 51, 52, 54, 68–72, 74, 76, 80, 82–84, 88, 91, 95, 104, 146, 182, 291, 292, 310, 311, 314, 318, 319, 335–337
entry contract, *328*
entry-exit model, *49*, 51, 328, 336
entry zone, *53*
equation
   AGA, 20, 166, 183, 194, 207, 217
   Colebrook–White equation, 25
   continuity equation, *24*, 25, 166
   energy conservation, *24*
   energy equation, 166
   equation of state, *19*, 105, 110, 166, 169, 185, 189–191, 193
   equation of state for ideal gases, *19*
   Euler equations, *24*, 89, 90, 148, 166, 187, 194, 213, 217
   formula of Nikuradse, 25
   formula of Papay, 20, 183, 194, 207, 217
   GERG-2004 equation, 20
   GERG-2008 equation, 20
   Hagen–Poisseuille formula, 25
   of Hofer, 25
   momentum equation, *24*, 25, 26, 122, 166
   of state, 25
   thermodynamical standard equation of state for real gases, *20*
European Network of Transmission System Operators for Gas, *60*, 326, 329, 330
exit contract, *328*
exit [point], 5, *18*, 18, 19, 48, 51–54, 66, 68–77, 79, 80, 82–84, 88, 91, 95, 104, 146, 158, 182, 262, 291, 292, 310, 311, 314, 318, 319, 335–337
exit zone, *53*, 53, 55

flow commitment, *60*, 60, 61, 73–76, 319
   distribution, *75*
   minimum, *74*
   reduction, *74*
   storage, *74*
Fourier–Motzkin elimination, 140
fracking, *7*
friction factor, *25*
fundamental cycle, 145

gas
   density, *25*
   ideal gas, *19*
   mass flow, *22*, 104, 106, 108, 124, 139
   mixture, *22*
   normal condition, *28*
   real gas, *19*
   thermal power, *22*
gas driven motor, 31, 35, 334
gas storage, 11, 18
gas target model, *330*
gas turbine, 31, 35, 37, 38, 93, 334
German Energy Industry Act, *49*, 328, 329

H-gas, *9*, 13, 22, 52, 274
heat capacity coefficients, *21*
heat transfer coefficient, *24*

ideal gas law, *19*
incremental method, *106*, 116
   extended, *115*, 116, 121
indicator constraints, 134
interconnection agreements, *74*
interconnection point, 18, 47, 48, 53, 55, 75, 335
internal order, *53*, 53, 57
inversion temperature, *21*
irreducible infeasible subsystem, 225, 231, 232
isentropic exponent, *33*, *200*, 200, 201, 209
isothermal, 90, *96*, 101, 148

Joule–Thomson effect, *21*, 21, 29, 31, 80, 184, 195, 196, 198, 201

Kantorovich distance, *300*
Kirchhoff's laws, 145
Kolmogorov distance, *277*, 281, 282, *300*
Kolmogorov–Smirnov test, *277*, 281

L-gas, *9*, 13, 22, 52, 274
laminar, *25*, 166, 188
linear programming, 224, 235, 242, 247, 303
linepack, 5, *10*, 69, 73, 212
liquified natural gas, *9*, 9, 10, 60, 326, 327
load condition, *68*
load flow vector, *292*
    booking-compliant, *292*
load scenario
    adjusted statistical, *306*, 306, 307
    statistical, *293*, 294, 295, 297, 299, 303, 305–309, 311, 313, 314, 336
loop, *38*, 321

market area, *51*, 52, 53, 303, 304, 335
    coordinator, *52*, 52
mathematical program with equilibrium constraints, 164, 165, 171, 173–176, 179, 252–260, 264–267, 270
median problem, 300
mixed-integer linear program, 88, 100, 101, 103, 114–123, 224–226, 231, 235–241, 245, 265–267, 270, 303, 306–308, 314
mixed-integer nonlinear program, 88, 99–103, 112, 114, 118, 121–123, 125, 134, 143, 178, 181, 215, 224, 234, 236, 242, 319
molar isobaric heat capacity, *21*
monotonicity assumption, *303*, 308, 309, 319
Monte Carlo method (MC), 295

network, 138
network interconnection model, 75
network interconnection point, 75, 77

network operator, 5, 11, 18, 45, 46, 48, 51–53, 59, 77, 87, 162, 216, 287, 292, 326, 328–330, 333–336
node-arc-incidence matrix, 146
nominate, *49*, 292
nomination, viii, 5, 18, 68, 94, 103, 114, 123–126, 128, 130, 132, 143, 145, 146, 157, 185, 273, *292*, 304, 311, 320, 322, 336
    balanced, *18*, 292
    booking compliant, 304
nomination validation, 123
nonlinear program, 88, 90, 91, 93–103, 108, 121–123, 125, 130, 131, 157, 160, 161, 163–165, 175, 176, 179, 181–183, 189, 190, 192–196, 199, 203, 207–209, 211, 213, 215–221, 223, 233, 235, 237, 239–242, 244, 246–261, 264–269, 310, 318
normal condition, 28, 124
normal distribution, 281

Open Grid Europe GmbH, xv, 13, 15, 16, 23, 29, 38, 40, 65, 68, 98, 216, 217, 234, 274, 310–312
operating range, 34, 36, 108, 109, 112, 114, 120–122, 131, 133, 137–141, 143, 164, 203, 212, 242, 243
operating volume flow, 112, 114, 204
operation mode, *40*
outer approximation, 138
overestimator, *109*
oversubscription, *61*

passive element, *18*, 123, 135, 146
passive network, 146
passive pipe networks, *124*
path integral, *75*
peak load scenarios, *71*
penalized splines, 286, 288
piecewise polyhedral envelope, 115, 118, 121
pipe, 104
    cross-sectional area, *24*
    diameter, *23*
    inclination, *23*

    integral roughness, *24*
    length, *23*
pipe/pipeline, 23, 29, 103, 104, 114, 118, 119, 123–126, 129–131, 143, 209, 222, 320, 321, 363
piston compressor, 31, 34–36, 93, 112, 200, 204, 321, 333
point
    hybrid, *18*, 76, 105, 305
    statistical, 77, *293*, 294, 311
point-to-point, *328*, 328, 329
polytope, 137, 138, 140
preheater, *31*, 38
primal heuristic, 143
principal components analysis (PCA), 299
pseudocritical point, 20
pseudocritical pressure, 20
pseudocritical temperature, 20
pump prevention, *35*, 69

Q-Q (quantile-quantile) plots, *278*
quantile, 279, 302
    adjusted, *309*
    multivariate, *302*
    univariate, *302*
quantile-quantile (Q-Q) plots, *278*
quasi–Monte Carlo, *294*, 311
quasi–Monte Carlo method (QMC), 295, *296*

random load vector, *293*, 293
reduced pressure, 20
reduced temperature, 20
reference temperature, *279*
relieving, *71*
renomination, *49*, 49, 50, 56, 59
resistor, 17, 28, 29, 31, 38–40, 89, 91, 94, 105, 106, 108, 112, 114, 118, 134, 154, 164, 167, 168, 176–178, 182, 194, 196–198, 226, 229
revenue-cap regulation, *328*
Reynolds number, *25*, 167

scenario, 65, 66, 68, *69*
security of supply, *48*, 48, 57, 60, 328
shifted distribution function, 281
shifted normal distribution, 280, 281
shifted uniform distribution, 281
shipper, *46*, 49, 54, 338

short cut, *43*, 43, 89, *91*, 91, 105, 118, 121, 229, 262, 336
sigmoidal function, 276, 287, *288*
simulation software, *65*, 162
single capacity booking model, *328*, 328, 329
slack reformulation, *128*
Sobol' sequences, 297
soil temperature, *24*
spanning tree, 147, 334, 338
spatial branching, *123*, 123, 125, 126, 130, 134
specific change in adiabatic enthalpy, *32*
specific isobaric heat capacity, *21*
station, 321
station subnetwork, 157, 160
stationary, 4, 19, 25, 87, 96, 100, 103, 114, 123, 124, 212, 314, 337
steady state [model], 4, 19, 25, 337
steam turbine, 32, 93, 334
stressing, *71*
stressing and relieving, *71*
subnetwork, 18, 38–42, 91, 103, 106, 112, 113, 138–140, 142, 158, 196, 198, 333, 334, 337
subnetwork operation mode, 18, 41, 42, 91, 94, 95, 108, 112, 142, 157, 160, 179, 254, 260, 321, 337
surgeline, 93, 109, 111, 112, 203
switching state, *146*, 156

tangential cut, *118*
technical capacity, *49*, 83, 84
technically feasible, 49, 59, *69*, 69, 292
temperature class, 308, 311
temperature classes, *280*
ten-year network development plan, 60
ton of oil equivalent, 8
transient, 19, 96, 100, 101, 212, 314
transmission, 46, 334
transmission system, *8*
transmission system operator, 13, 46, 48–53, 55–64, 68, 70, 73–75, 96, 291, 292, 303, 305, 313, 314, 319, 322, 323, 326–328, 330, 334, 336
transport customer, *10*, 22, 46, 48, 49, 51, 52, 57, 59, 64, 291, 303, 314, 333, 334, 336
tree arc, 149
turbo compressor, 31, 34, 35, 37, 93, 109, 114, 120, 121, 200, 202, 203, 212, 236, 237, 245, 321, 333
turbulent, *25*, 166, 188
two-contract model, *328*, 328, 329

unbundling, *48*, 67, 328
underestimator, *109*
universal gas constant, *19*

use-it-or-lose-it, *56*, 63
use-it-or-sell-it, *56*

validation of nomination, *5*
validation of nominations, *94*
ValNLP (validation nonlinear program), 101, 102, 162, 179, 209, 211, 213, 215, 216, 218, 219, 221–224, 233, 235, 242, 248, 249, 251, 254–265, 310, 318
valve, 4, 11, 17, 18, 28, *29*, 29–31, 38–42, 68, 79, 80, 89, 91–94, 106–108, 112, 113, 118, 123, 130, 131, 136, 147, 167, 169–171, 176, 177, 182, 195, 196, 213, 226, 229, 262, 321, 337
   closed, *29*, 30, 106, 136, 169
   open, *29*, 30, 106, 136
verification of booked capacities, *6*
   deterministic version, *292*
   stochastic version, *293*
virial coefficient, *20*
virial expansion, *19*
virtual trading point, 5, *51*, 51, 53, 55, 57, 58, 67, 292

Weymouth constant, 125
Weymouth equation, 135

zoning, *53*, 53, 55, 63, 304